Cathedrals of Science

CATHEDRALS OF SCIENCE

The Personalities and Rivalries That Made Modern Chemistry

Patrick Coffey

2008

Oxford University Press, Inc., publishes works that further
Oxford University's objective of excellence
in research, scholarship, and education.

Oxford New York
Auckland Cape Town Dar es Salaam Hong Kong Karachi
Kuala Lumpur Madrid Melbourne Mexico City Nairobi
New Delhi Shanghai Taipei Toronto

With offices in
Argentina Austria Brazil Chile Czech Republic France Greece
Guatemala Hungary Italy Japan Poland Portugal Singapore
South Korea Switzerland Thailand Turkey Ukraine Vietnam

Published by Oxford University Press, Inc.
198 Madison Avenue, New York, New York 10016

www.oup.com

Library of Congress Cataloging-in-Publication Data
Coffey, Patrick.
Cathedrals of science / the personalities and rivalries that made modern chemistry /
Patrick Coffey.
p. cm.
Includes bibliographical references and index.
ISBN 978-0-19-532134-0
1. Discoveries in science. 2. Chemical weapons. 3. Chemists—Psychology.
4. Langmuir, Irving, 1881–1957. 5. Lewis, Gilbert Newton, 1875–1946.
6. Science—Moral and ethical aspects. 7. Chemistry—History—20th century. I. Title.
Q180.55.D57C64 2008
540—dc22 2007048304

9 8 7 6
Printed in the United States of America
on acid-free paper

To Ellen

Contents

There are ancient cathedrals which, apart from their consecrated purpose, inspire solemnity and awe. Even the curious visitor speaks of serious things, with hushed voice, and as each whisper reverberates through the vaulted nave, the returning echo seems to bear a message of mystery. The labor of generations of architects and artisans has been forgotten, the scaffolding erected for the toil has long since been removed, their mistakes have been erased, or have become hidden by the dust of centuries. Seeing only the perfection of the completed whole, we are impressed as by some superhuman agency. But sometimes we enter such an edifice that is still partly under construction; then the sound of hammers, the reek of tobacco, the trivial jests bandied from workman to workman, enable us to realize that these great structures are but the result of giving to ordinary human effort a direction and a purpose.

Science has its cathedrals, built by the efforts of a few architects and many workers. . . .

—Gilbert Newton Lewis, from the preface to *Thermodynamics and the Free Energy of Chemical Substances*

Prologue

It was Saturday afternoon, 23 March 1946. A group of students huddled together in a Berkeley chemistry laboratory, still sniffing nervously for the almond scent of hydrogen cyanide that had filled the laboratory earlier. A seventy-year-old man with a gash across his forehead lay dead on the laboratory floor. The fireman who had finally come, an hour after the students had gone seeking help, rolled the body over, looked at the students, and said, "Does anybody know who this guy is?"[1]

"This guy" was Gilbert Lewis, one of the scientists who had built modern chemical theory and a principal character in this book. He had been born in 1875, and his life span is roughly coincident with the transformation of chemical theory that is this book's subject. At the time of his birth, chemists synthesized compounds and classified them by type and characteristics, but basic questions—Why do atoms combine to form certain compounds? What forces hold atoms together? Would a particular reaction occur at all, and would it go to completion?—had different answers than they do today, if they had answers at all. By the time Lewis died in 1946, chemistry was firmly grounded in the same principles as physics, although the approach and practice of the two sciences remain distinct. Not only could chemists answer these questions by 1946—they could write (but not solve) the quantum mechanical equations that described any chemical system. And by the 1950s, when the structures of proteins and of DNA were determined, chemistry and biology were coming together as had chemistry and physics.

The branch of chemistry that ties chemistry to physics is called "physical chemistry." It was a hot science in 1920—much as molecular biology is today. I have examined five subspecialties within physical chemistry: dissociation theory, thermodynamics, surface chemistry, the chemical bond, and protein structure. This list of specialties is far from exhaustive—I have not treated kinetics or spectroscopy in any depth, for example. I have used physical chemistry to examine individual scientists, the making of science, and the interactions between science and the larger world. Scientists try to study an objective reality that exists distinct from humanity. But science is like any other human endeavor—not everyone is allowed to participate, the rules of the game are in dispute, the participants have different motivations, credit is not always assigned "fairly," and so on. Much of this book is concerned with the interactions between the personal and the scientific, especially with the idea of scientific discovery—what constitutes a discovery and who deserves credit. Arthur Koestler wrote a book about the 17th-century scientific revolution, *The Sleepwalkers*, in which he talked about "the psychological process of discovery as the most concise manifestation of man's creative faculty—and...that converse process that blinds him towards truths, which once perceived by a seer, become so heartbreakingly obvious."[2] The individual's moment of discovery is of interest, but so is the scientific community's perception of discovery—how quickly a discovery can progress from being seen as impossible, to accepted, to obvious, to obsolete. Many discoveries are later rediscovered, sometimes more than once, and the rediscoverers are often unwilling to cede credit to the original discoverers. And at the moment of discovery, there is often no way to determine whether a discovery will pan out. The brilliance of an insight is not a reliable predictor of its validity, and this book includes several theories that were as ingenious as any but that eventually collapsed. Their proponents, remembering the moment of discovery, were usually extremely unwilling to let their theories go.

During the seventy-year span reviewed in this book, 1883 to 1953, science changed the Western world, and chemistry had profound commercial and societal effects. I have tried to place chemistry within a larger context. Two world wars, the rise of Communism and Nazism, and the start of the cold war with its nuclear arms race changed the political landscape completely. Most scientists prior to the 20th century had not had to consider ethical questions regarding the effects of their work on society. World War I, beginning in 1914, was the first war in which scientists had a fundamental role. The Haber-Bosch process for synthesis of ammonia from atmospheric nitrogen, developed just before the war's start, enabled Germany to produce munitions for four years despite the British naval blockade. And this was the war that saw the introduction of chemical weapons. Fritz Haber, Walther Nernst, and Gilbert Lewis, all subjects of this book, were involved to some degree in chemical warfare, with Haber the most enthusiastic in its pursuit. A different sort of ethical issue arose when the Nazis came to power in Germany in 1933 and immediately dismissed most Jewish scientists. Non-Jewish German scientists were forced to decide how to respond. Some resigned their own positions and left the country, some stayed to help, some tried to ignore the situation, some were enthusiastic supporters of the Nazis, and some were happy to profit from the professorships that opened up when the Jews were pushed out. Scientists also responded differently to the development of atomic weapons during World War II. Some were enthusiastic, some were opposed on moral grounds, and some were interested primarily in the effects on their careers.

Most scientists are not much given to introspection about their own scientific methodology, but there is something to be learned by comparing the ways in which scientists actually work. There is an entire discipline, the philosophy of science, concerned with the meaning and the validation of scientific hypotheses. One formulation of the scientific method uses a hypothetico-deductive model, in which a scientist proposes a new hypothesis—for example, Copernicus's hypothesis that the planets revolve in circular orbits around the sun. If the hypothesis accounts for the same facts that are explained by the existing theory and also explains anomalous results, it is eventually promoted from a hypothesis to a theory—Copernicus's model could explain the same data as Ptolemy's earth-centered model and could better explain the occasional reversal of the motions of some planets in their paths. As observations are made that are consistent with the new theory, it becomes more trusted. New data that are not consistent with the theory—Kepler's observations that the planets' orbits could be better described as elliptical than circular, for example—cause the theory to be either modified or rejected. Theories that cannot be tested by observation or experiment are often said to be "not even wrong"—neither scientifically true nor false but meaningless. But it is not clear that all scientists agree on this as *the* scientific method. Theoretical physicists, for example, often seem to behave more like mathematicians, looking for internal symmetry in their theories rather than for experimental confirmation—the predictions of some modern string theories may not be testable with anything smaller than a particle accelerator the size of the galaxy, for example, which may put them in the realm of metaphysics rather than physics. The hypothetico-deductive model relies on the idea of reproducibility—another scientist who performs the same experiment or makes the same observation should observe the same result. But this is a physicist's idea of the scientific method. Roald Hoffmann has asked what a scientific method based on chemistry rather than physics would look like—what if the model for the scientific method were chemical synthesis rather than mathematical laws of motion?[3] Using a synthetic model, the primary point of view would be the changes that occur as constituents join and separate—using economics as an example, one might start from the changes in the states of market participants rather than from a mathematical analysis of the economy. And for medical and social sciences, the hypothetico-deductive model is often inapplicable. A five-year drug trial is not reproducible in the same way as a physical or chemical experiment, and it is often not possible to conclusively demonstrate a causal relationship among observations correlated by statistical data. Yet few would say that the statement "smoking causes cancer" is either scientifically meaningless or unproven. The scientists surveyed in this book did not all share a common philosophy of science or definition of the scientific method.

As scientists mature, they move through stages—student, young researcher making a mark, mature scientist with a reputation. Most history of science has focused on the middle stage, where important discoveries are usually made, but the first and last stages are also of interest. Students have problems breaking with their mentors and establishing themselves, and their career success, if it is to come, usually happens sometime before age forty, when as successful scientists they find that they have achieved what they thought they had wanted—the acclaim of their colleagues and of society, institutes of their own, political influence within the scientific community and perhaps even beyond. Those who have gotten that far usually look around for new challenges, and they do not always choose them wisely.

Even with a field as limited in subject, time, and people as physical chemistry, the problem of handling the mass of information is formidable. This book is neither a history of physical chemistry nor a biography of its scientists, although it includes elements of both. My models are Lytton Strachey's portrait of Victorian England, *Eminent Victorians*; Joseph Ellis's stories of the men who began the American government, *Founding Brothers*; and Phyllis Rose's narratives of the relationships of Victorian couples, *Parallel Lives*. These books all use selected vignettes to describe a much larger topic. By emulating these authors, I hope to catch the essence of both the science and the people involved in its development.

I have selected six scientists as principal actors—Svante Arrhenius, Walther Nernst, Gilbert Lewis, Irving Langmuir, Fritz Haber, and Linus Pauling. Eight other scientists—Jacobus van't Hoff, Wilhelm Ostwald, Paul Ehrlich, Theodore Richards, William Harkins, Harold Urey, Glenn Seaborg, and Dorothy Wrinch—have supporting roles, and many others walk on for a scene or two. This list of physical chemists is far from exhaustive; I have largely passed over such important contributors as Henri le Chatelier, Peter Debye, James Franck, Erich Hückel, and Louis Hammett, to name only a few. Like the workers who constructed the medieval cathedrals, the scientists who built physical chemistry were not saints, despite the beauty of what they built. They sometimes cooperated but often clashed.

Svante Arrhenius, a Swede, developed a revolutionary chemical theory—that many substances, when dissolved in water, dissociate into electrically charged ions. He collaborated with Wilhelm Ostwald and Jacobus van't Hoff, and the three of them, known as the "Ionists," started the modern physical chemical revolution. In his late thirties, Arrhenius developed a quantitative model for what is now called the greenhouse effect—global warming due to an increase in atmospheric carbon dioxide. In his forties, he applied the principles of his dissociation theory to immunochemistry and became embroiled in a dispute with Paul Ehrlich, who eventually received the Nobel Prize in Medicine or Physiology over Arrhenius's efforts to block it.

Walther Nernst, a German, was responsible for many of the early advances in the theory and practice of physical chemistry, and his greatest achievement was what came to be known as the "third law of thermodynamics." On the practical side, he developed an electric lamp that was a serious competitor in the lighting market. He and Arrhenius were friends as young men, but they became bitter enemies as they moved into middle age, and Arrhenius's unforgiving opposition forced Nernst to wait fifteen years for his Nobel Prize. Nernst became a political adviser to the Kaiser and the German General Staff during World War I and to the Weimar Republic thereafter. Despite his reputation for what Einstein would call "childlike vanity and self-complacency," he was in the end a moral man, refusing to have anything to do with the Nazis and their expulsion of Jews from German science.

Gilbert Lewis, an American chemist, proposed the idea of the covalent bond and turned thermodynamics into a tool for working chemists. He had been a student at Harvard of Theodore Richards, whom Lewis believed had unfairly claimed credit for his ideas. Lewis spent a semester in Nernst's laboratory, where he and Nernst developed an enmity that would last through both their careers. Lewis was uncomfortable in the larger scientific world and built his chemistry department at Berkeley as a support system for himself, staffing it almost exclusively with scientists who had trained there. Despite thirty-five nominations, more than for any other chemist who had been passed over, he did not receive a Nobel Prize, possibly due in part to the machinations of one of Nernst's friends on the Nobel Committee for Chemistry. At age forty-eight, Lewis stopped work in physical chemistry to make an unsuccessful attempt to solve one of the great problems in theoretical physics, the dual wave–particle nature of light. He returned to physical chemistry ten years later and worked productively in the field until age seventy. When he died in a laboratory filled with cyanide, many of his colleagues, who had thought him depressed, suspected suicide.

Irving Langmuir, an American chemist, spent his career at General Electric, where he developed the gas-filled incandescent lightbulb that we use today. He developed a theory of the chemistry of surfaces—oil films on water, for example—that won him a Nobel Prize in 1932, the first such award to an industrial chemist. When Langmuir extended Lewis's ideas on the chemical bond (and got much of the credit for Lewis's ideas because of his presentation skills and popularity), Lewis developed what would be a lifelong resentment toward Langmuir, although the ill will was not reciprocated by Langmuir. In his late sixties, Langmuir began cloud-seeding experiments and claimed that he had changed the pattern of rainfall across the United States; the meteorologists rejected both his evidence and methodology.

Fritz Haber was a German-Jewish scientist who was responsible for the industrial "fixation" of nitrogen, one of the most significant technical advances of the 20th century. It permitted the production of artificial fertilizers, which have enabled the planet to feed the seven billion people inhabiting it today. He and Nernst, the leading German physical chemists of their time, had a complicated relationship, sometimes rivals and sometimes allies. Haber thought of himself as a German patriot; he converted to Christianity, served as an adviser to the Kaiser, and was known as the "father of chemical warfare" for his work in World War I. Haber was awarded a Nobel Prize for 1918, to the outrage of the Allies who had suffered under his nitrogen-based munitions and chemical weapons. When the Nazis came to power in 1933, Haber went into exile and died shortly thereafter.

Linus Pauling, an American chemist, took the physicists' new quantum theory and applied it to chemistry, extending the work that Lewis had done earlier on the chemical bond. In the 1930s, he started working in the new molecular biology, attempting to determine the nature and structure of proteins, which are some of the most important chemical constituents of living organisms. He became involved in a dispute with Dorothy Wrinch, an English mathematician with whom Langmuir was collaborating, who had developed a rival "cyclol" theory of protein structure. Pauling used his influence to block Wrinch's funding and to isolate her from the scientific community; it was not until ten years later that he was proved to be mostly right and she almost entirely wrong. He received two Nobel Prizes, the chemistry prize for 1954 for his work on the chemical bond, and the peace prize for 1963 for his work opposing atmospheric nuclear testing.

All the scientists I have selected but one, Dorothy Wrinch, were men. Women were almost systematically excluded from chemistry and physics, especially before World War I. A few prominent women—Marie Curie and Lise Meitner are examples—somehow managed to force their way in. Agnes Pockels, who could be called the first modern surface chemist, was not permitted a university education and was self-taught and self-motivated, working on experiments in her kitchen. Two women who had earned a Ph.D. in chemistry early in physical chemistry's history, Sofia Rudbeck and Clara Immerwahr, married two of the principals in this book, Arrhenius and Haber, respectively. Both had unhappy marriages; their husbands expected them to give up their scientific careers and act as conventional wives. Rudbeck left Arrhenius while pregnant with their first child, and Immerwahr killed herself. Dorothy Wrinch, who was admittedly arrogant and acerbic, suffered personal attacks more extreme than she would have had she been a man. In looking at issues of race and sex discrimination, it is difficult to find enough examples of nonwhite or even nonmale participants in physical chemistry to form a basis for general discussion. But there were a number of Jewish physical chemists, and this book uses anti-Semitism as a proxy for other forms of discrimination. Jewish scientists were expelled from universities and government institutes in Germany almost immediately after the Nazis came to power in 1933.

The scientists that I have examined were not disinterested seekers after truth. They wanted new discoveries, but they want them with their own names attached—what scientists refer to as credit for "priority," being the first to make a discovery or to develop a theory. Perhaps in no field more than science are issues of credit, priority, and prizes so important to the participants. Richards and Lewis argued over Lewis's new thermodynamic concept, Langmuir and Lewis over the chemical bond, Langmuir and Harkins over surface chemistry, and Nernst and Richards over the third law of thermodynamics. (This sort of behavior did not begin with physical chemistry, of course, but dates back at least to Isaac Newton and Robert Hooke in the 17th century.) The financial rewards for a career in science are generally modest when compared to business, medicine, or law. Of the scientists surveyed in this book, only Haber and, to some degree, Nernst seem to have been much interested in money. And scientists are usually not concerned with fame. Only the rare scientist—Newton, Darwin, or Einstein, for example—is widely known in his or her own time, and even then, the general public usually has a poor understanding of his or her achievements.

Scientists want recognition by their peers, and almost since its inception in 1901, the Nobel Prize has been the ultimate mark of that recognition. Much of this book deals with scientists' attempts to be so recognized. Van't Hoff, Arrhenius, Ostwald, Richards, Haber, Nernst, Langmuir, Urey, Seaborg, and Pauling all received Nobel Prizes in Chemistry. Lewis's failure to win a Nobel Prize is particularly instructive. While I believe I have uncovered unethical behavior by a member of the Nobel Committee for Chemistry in denying Lewis the prize, this is far from the entire story. If there were a manual on how *not* to win the Nobel Prize, Lewis could have written it. He made enemies and held grudges, isolated himself in then-remote California, and switched fields so abruptly that his work no longer seemed current by the time it was evaluated.

Like the careers of scientists, scientific ideas develop in stages. Thomas Kuhn has discussed the difference between "revolutionary" science—based on anomalous results that are unexplained by accepted theories—and "normal science"—based on work within the framework of those theories.[4] Normal science can be as ingenious as revolutionary science and as significant in the larger world (e.g., the fixation of nitrogen and the plutonium bomb). I would categorize the thermodynamic, nuclear, and surface chemistry described in this book as normal science; Arrhenius's dissociation theory, Lewis's electron-pair bond, and van't Hoff's tetrahedral carbon atom as successful revolutionary science; and Wrinch's cyclol model for protein structure as unsuccessful revolutionary science. Revolutionary science is often easier for scientific outsiders—those on the inside are more invested in the existing theories. Arrhenius and van't Hoff were outside the German organic chemistry establishment. Lewis was an outsider by nature, choosing to locate himself far from the East Coast center of American science. Wrinch was an

English woman mathematician, an outsider within American chemistry on three counts.

Physical chemistry was begun by outsiders, but by outsiders who were at least *chemists*; they shared a language and an understanding of the field with the insiders. The scientists I describe were much less successful when they tried to move into established fields in which they had little experience. Arrhenius brought the law of mass action to immunology with limited success; Lewis applied his ideas on chemical equilibrium to the wave–particle nature of light in theoretical physics and was gently rebuffed by Einstein; Langmuir made overstated claims for cloud seeding in meteorology; Wrinch, a mathematician, developed topological models for proteins that the chemists saw as contradicted by experiment; and Pauling made claims for vitamin C that mainstream physiologists and physicians rejected. These scientists did not bother to master the literature in the field they were entering and did not follow the field's accepted methodology and terminology. Entering an established field is difficult, especially if one already has a prominent reputation and is unwilling to undergo the sort of tutelage that a graduate student is forced to endure. Most mature scientists do not have the patience for it. None of the scientists were as successful as they claimed or hoped to be in changing fields, and they often did not understand why the field's participants viewed certain problems as important or what evidence they would consider convincing.

Physical chemists were more successful in attacking problems in entirely new fields such as molecular biology. As much as anyone, Warren Weaver of the Rockefeller Foundation was responsible for the creation of molecular biology as a discipline, which he called "the science of man." Using Rockefeller funds, he brought together medical chemists, biochemists, organic chemists, physical chemists, biologists, physiologists, mathematicians, and physicists. In the same way that physical chemistry had borrowed from physics, molecular biology would take what it needed from chemistry. In the molecular biology of the 1930s, it was hard to be an outsider—there was no existing field to view from the outside. Similarly, Arrhenius's model for the greenhouse effect was not the work of an outsider—there was no separate field of geochemistry or earth sciences at the time.

The sociology of science—its arrangement into specialties, each with its own practice—of course continues today. Environmental science has emerged as a separate discipline since Lewis's death in 1946, and many of the participants in the development of physical chemistry worked on problems with environmental consequences: Arrhenius's quantitative model for the greenhouse effect might have provided a warning of the consequences of industrial expansion (although Arrhenius did not intend it as such); Haber's fixation of nitrogen has allowed the planet to support billions of additional people, but with consequent increases in per capita pollution and in runoff from overfertilized fields; the Hanford nuclear plant, begun during World War II for the purification of plutonium under Seaborg's leadership, has dumped tons of radioactive material into the air and water; and Pauling and Urey identified the potentially disastrous environmental effects of atmospheric nuclear testing. Environmental concerns with nuclear power, the ozone layer, genetic engineering, and global warming are, of course, today's issues. It may be that biological "fixation" of carbon dioxide and its sequestration or conversion to methane, as envisioned by David Venter[5] and others, will lead to a new scientific area—environmental repair—that will require contributions from different areas, including molecular biology and biochemical engineering.

Cathedrals of Science

1

The Ionists

Arrhenius and Nernst

In the spring of 1884, Uppsala University's faculty and graduates gathered in their academic robes. Svante Arrhenius, a heavyset young man with few social graces, had been awarded his doctorate and had marched in the procession. The doors to the faculty chambers opened, and the professors came out to congratulate the new graduates. Arrhenius looked around for Per Theodor Cleve and Tobias Thalén, the two who had directed his studies for the past seven years. He finally saw them—a porter was helping them take off their robes and put on their overcoats. They glanced at him, looked a bit embarrassed, and left him standing there. Arrhenius never forgave this snub. Forty years later, he would recount the story, saying that Cleve and Thalén "had decided to sacrifice him."[1] His son later said that this had "opened a wound that never healed."[2]

Arrhenius would later claim that his doctoral dissertation, which he said had "received a grade only one above failing,"[3] was what earned him a Nobel Prize in 1903. But before proceeding with his story and the stories of others, it is important to examine the scientific mind-set of 1880. What were seen as important chemical problems, and how did chemists approach them? The following digression is far from comprehensive, but it gives an overview of the state of chemical theory when Arrhenius began his graduate work.

Chemistry in 1880

Modern chemistry—chemistry whose theory would be recognizable to a student today—does not have a long history. In the late 17th century, Newton had developed his theories of gravitation and mechanics by taking physical objects such as rocks or planets and abstracting them—ignoring their particular shapes or sizes and shrinking them conceptually to a single point—and then defining such concepts as "space," "force," and "mass" and formulating his three laws of motion. But abstracting chemistry proved harder. Whereas Newton's revolution in physics began in the late 17th century, it was not until Lavoisier's time in the late 18th century that two important chemical concepts emerged: (1) chemical "elements"—gold, tin, and oxygen, for example—that could not be transmuted into other elements but that could combine with each other to form chemical compounds; and (2) the principle of conservation of mass—that the total mass of reactants and products was unchanged in a chemical reaction. (Both these principles have since been qualified, as elements are transmuted into other elements in nuclear reactions, and the theory of relativity predicts minuscule changes in mass during chemical reactions.)

Some chemists tried to emulate Newton's revolution in physics by formulating mathematical laws to describe how elements formed chemical compounds. Newton had hypothesized that gravitation followed an inverse square law and was able to describe both the speed with which a rock falls and the orbits of the planets, so chemists looked for a mathematical chemical force that drove their reactions. They called this unknown force "chemical affinity." If they could identify and quantify chemical affinity, they might be able to explain why some mixtures of chemicals reacted whereas others did not; why some reactions did not go to completion but stopped part way, leaving a seemingly unchanging mixture of products and reactants; and why some acids or bases were "stronger" than others.

Around 1720, Georg Stahl and others had developed tables of *relative* chemical affinity. Stahl had no mathematical law to describe affinity, but he based his tables on the ability of one substance to displace another. For example, nitric acid would dissolve metallic silver, but if enough metallic copper were added to that solution, the copper would be dissolved and the silver would be precipitated—it would fall out of solution to the bottom of the flask. Stahl saw this as proof that copper had a higher affinity for nitric acid than did silver, because nitric acid would react with copper preferentially. In 1799–1803, Claude Berthollet saw that the progress of chemical reactions depended not only on the nature of the chemical reactants and products but also on how much of each was present—on their *masses*. He thought affinity might be identical with gravity: Newton had shown that gravity was proportional to the mass of the bodies involved, and in a similar way a large mass of a substance with weak affinity could provide a stronger chemical effect than a smaller mass of a substance with strong affinity. But Berthollet was not able to develop any quantitative laws that described chemical affinity, and after a brief flurry of interest, his ideas remained largely ignored for more than fifty years. His approach might have been neglected at least in part because chemists were distracted by the success of John Dalton's 1803 atomic theory.

Dalton proposed that each chemical element corresponded to a different *atom*, and that the atoms combined in fixed proportions, forming *molecules* to which he assigned formulas, such as carbon monoxide (CO: one carbon atom and one oxygen atom) or carbon dioxide (CO_2: one carbon atom and two oxygen atoms), but not something in between like $CO_{1.32}$. According to Dalton, the atom of each different element had a characteristic *atomic weight* (although the atomic weights Dalton originally proposed proved to be incorrect). When atoms combined to form a molecule, the *molecular weight* would be the sum of the atomic weights of its constituent atoms. The oxygen atom was eventually assigned an atomic weight of 16 and was used as a reference for the measured atomic weights of all other atoms until 1960. Determination of accurate atomic weights remained an important and laborious work throughout the 19th century.

Dalton's theory, based on combination in fixed proportions, was not immediately accepted, because substances also combined in variable proportions—examples include alcohol and water mixing in a flask, oxygen and nitrogen mixing in the atmosphere, and tin and copper alloying to form bronze; in all these cases, any amount of one substance will happily combine with the other. We now differentiate what we call chemical combinations (combinations that involve substances in fixed proportions) from physical combinations (combinations like alloys or solutions that occur in variable proportions). But this is a distinction based on definition and was not obvious at the time, so it is not surprising that Dalton's ideas met resistance. Another obstacle for acceptance of Dalton's theories was that no one had ever seen an atom, and the only evidence he had for his atomic theory was that some substances sometimes combined in fixed proportions. Nonetheless, the atomic theory proved so useful in explaining and guiding synthetic chemical work that it was eventually widely accepted.

In 1807, Humphry Davy began subjecting substances and solutions to electrical currents—"electrolyzing" them—and found that he could isolate a new element, sodium, from caustic soda (now called sodium hydroxide). He saw this as evidence that the force behind chemical affinity was electrical. In 1819, Jöns Berzelius, a Swedish chemist, developed an electrical theory known as "dualism"—that atoms were electrically charged, and that molecules were held together by the attractive forces between positive and negative atoms. For the next century, until Gilbert Lewis and others reformulated the idea of the chemical bond (described in chapter 5), variants of Berzelius's dualistic theory would be the dominant explanation for atoms combining to form molecules.

Chemists began to organize themselves by subspecialty. "Organic chemists" studied the properties of compounds containing carbon atoms, of which almost all living matter is composed. "Inorganic chemists," sometimes called "mineral chemists," studied everything else.

At first, Dalton's atomic hypothesis was taken to imply that if two substances contained the same proportions of different elements, the substances would have the same chemical properties. But by the 1830s, the idea of *isomerism* was introduced—that compounds such as ethyl alcohol and dimethyl ether, although containing the same proportions of elements (both ethyl alcohol and dimethyl ether have the same empirical formula, C_2OH_6), had very different chemical properties, which seemed to depend on the atoms' arrangement. The discovery of isomerism led to the idea of a chemical bond that linked two atoms together; different isomers were due to different structures among the same atoms, with the structures determined by the bonds.

In 1858, August Kekulé proposed the idea of the *valence* of an atom, which he defined as the number of bonds that an atom was capable of forming. He found that carbon had a valence of 4, allowing it up to four bonds to other atoms, but that some of those bonds might be double or triple bonds, which would consume two or three of the potential connections, respectively. Chemists began to draw complicated structures with atoms connected by single, double, or triple bonds, which were represented as single (–), double (=), or triple (≡) lines, respectively. They found that they could synthesize complicated organic molecules and assign structures to them by deduction. They would begin with simple molecules that had known structures and synthesize a new molecule, whose structure they could assign from the properties of the synthetic reaction. By using this new compound in another reaction, they could synthesize yet another compound, whose structure could also be assigned. Over the next century, organic chemists would build an enormous edifice upon this sort of deduction, synthesizing hundreds of thousands of compounds with assigned structures, compounds that had almost certainly never existed in nature. It might have all been a house of cards, and it was not until the mid-20th century that chemists had instruments capable of confirming that the foundations of structural chemistry were solid. But although a modern chemist can immediately understand the structures and their chemical bonds as drawn by a chemist of the 1860s, the modern chemist's understanding of the *nature* of the chemical bond is radically different, and that change in understanding is the subject of a good part of this book.

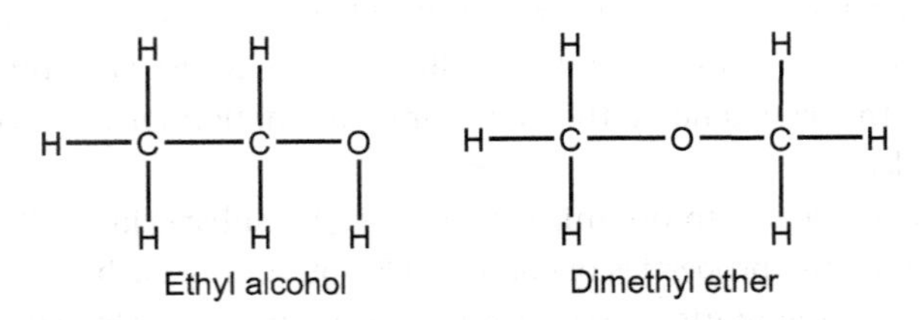

Figure 1-1. Structural isomers of C_2OH_6.

In the late 1860s, several chemists observed that the elements seemed to show a periodic similarity in their chemical and physical properties when they were arranged in order of their atomic weights. For example, members of a group including oxygen, sulfur, selenium, and tellurium often have the same valence, form similar compounds with other elements, and show regular trends in their boiling point, melting point, and color, as do members of another group including fluorine, chlorine, bromine, and iodine. Dmitri Mendeleev, a Russian chemist, is generally credited with devising the first comprehensive periodic table of the elements in 1869. He wrote the information on each of the then-known sixty-three elements on cards and laid them out on a table—much as if he were playing solitaire. He put those with similar properties in columns, ordering those in each column by increasing atomic weight. He found several holes in his table and predicted that new elements would be discovered with physical properties interpolated between the elements above and below. When several of the missing elements were found with the properties he had predicted, chemists had a new tool for rationalizing chemical behavior. A modern explanation of the periodic table is in terms of the elements' electronic structure and atomic number, but Mendeleev was working only from chemical and physical properties—the electron would not be discovered until 1897, and atomic number would have to wait until 1913.

Before the mid-19th century, textile dyes were natural products and were often too expensive for any but the wealthy to afford. Using their new knowledge of structure, chemists were able to synthesize dyes with brilliant colors

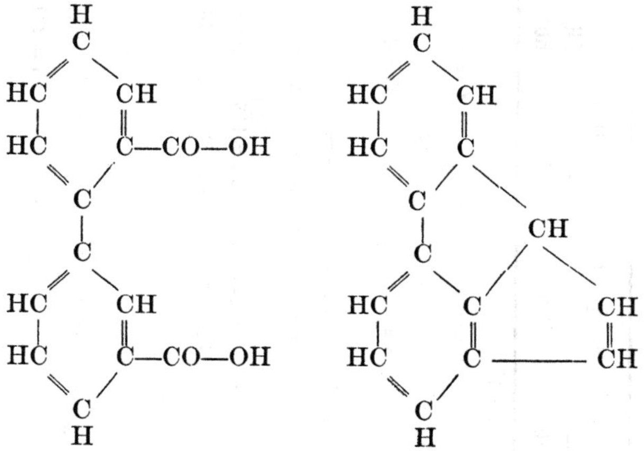

Isodiphenic Acid. **Fluoranthene.**

Figure 1-2. Chemical formulas from the first volume of the *Journal of the American Chemical Society* in 1879. Chemical notation has not changed radically since that time. Abstraction of R. Fittig and H. Liepmann, *Berichte der Deutsche Chemische Gesellschaft* 12 (1879): 163, in "On the Constitution of Isodiphenic Acid and Fluoranthene." *Journal of the American Chemical Society* 1 (1879): 169–170, on 170.

T a b e l l e II.

Reihen	Gruppe I. — R^2O	Gruppe II. — RO	Gruppe III. — R^2O^3	Gruppe IV. RH^4 RO^2	Gruppe V. RH^3 R^2O^5	Gruppe VI. RH^2 RO^3	Gruppe VII. RH R^2O^7	Gruppe VIII. — RO^4
1	H=1							
2	Li=7	Be=9,4	B=11	C=12	N=14	O=16	F=19	
3	Na=23	Mg=24	Al=27,3	Si=28	P=31	S=32	Cl=35,5	
4	K=39	Ca=40	—=44	Ti=48	V=51	Cr=52	Mn=55	Fe=56, Co=59, Ni=59, Cu=63.
5	(Cu=63)	Zn=65	—=68	—=72	As=75	Se=78	Br=80	
6	Rb=85	Sr=87	?Yt=88	Zr=90	Nb=94	Mo=96	—=100	Ru=104, Rh=104, Pd=106, Ag=108.
7	(Ag=108)	Cd=112	In=113	Sn=118	Sb=122	Te=125	J=127	
8	Cs=133	Ba=137	?Di=138	?Ce=140	—	—	—	— — — —
9	(—)	—	—	—	—	—	—	
10	—	—	?Er=178	?La=180	Ta=182	W=184	—	Os=195, Ir=197, Pt=198, Au=199.
11	(Au=199)	Hg=200	Tl=204	Pb=207	Bi=208	—	—	
12	—	—	—	Th=231	—	U=240	—	— — — —

Figure 1-3. Mendeleev's periodic table from 1871. Dmitri Mendelejeff, "Die periodische Gesetzmässsigkeit der chemischen Elemente." *Justus Liebig's Annalen der Chemie und Pharmacie* Supplement 8 (1871): 133–229, on 151.

1 H 1.00794																1 H 1.00794	2 He 4.002602
3 Li 6.941	4 Be 9.012182											5 B 10.811	6 C 12.0107	7 N 14.00674	8 O 15.9994	9 F 18.9984032	10 Ne 20.1797
11 Na 22.989770	12 Mg 24.3050											13 Al 26.981538	14 Si 28.0855	15 P 30.973761	16 S 32.066	17 Cl 35.4527	18 Ar 39.948
19 K 39.0983	20 Ca 40.078	21 Sc 44.955910	22 Ti 47.867	23 V 50.9415	24 Cr 51.9961	25 Mn 54.938049	26 Fe 55.845	27 Co 58.933200	28 Ni 58.6934	29 Cu 63.546	30 Zn 65.39	31 Ga 69.723	32 Ge 72.61	33 As 74.92160	34 Se 78.96	35 Br 79.904	36 Kr 83.80
37 Rb 85.4678	38 Sr 87.62	39 Y 88.90585	40 Zr 91.224	41 Nb 92.90638	42 Mo 95.94	43 Tc (98)	44 Ru 101.07	45 Rh 102.90550	46 Pd 106.42	47 Ag 107.8682	48 Cd 112.411	49 In 114.818	50 Sn 118.710	51 Sb 121.760	52 Te 127.60	53 I 126.90447	54 Xe 131.29
55 Cs 132.90545	56 Ba 137.327	57 La 138.9055	72 Hf 178.49	73 Ta 180.9479	74 W 183.84	75 Re 186.207	76 Os 190.23	77 Ir 192.217	78 Pt 195.078	79 Au 196.96655	80 Hg 200.59	81 Tl 204.3833	82 Pb 207.2	83 Bi 208.98038	84 Po (209)	85 At (210)	86 Rn (222)
87 Fr (223)	88 Ra (226)	89 Ac (227)	104 Rf (261)	105 Db (262)	106 Sg (263)	107 Bh (262)	108 Hs (265)	109 Mt (266)	110 (269)	111 (272)	112 (277)		114 (289) (287)		116 (289)		118 (293)

58 Ce 140.116	59 Pr 140.90765	60 Nd 144.24	61 Pm (145)	62 Sm 150.36	63 Eu 151.964	64 Gd 157.25	65 Tb 158.92534	66 Dy 162.50	67 Ho 164.93032	68 Er 167.26	69 Tm 168.93421	70 Yb 173.04	71 Lu 174.967
90 Th 232.0381	91 Pa 231.03588	92 U 238.0289	93 Np (237)	94 Pu (244)	95 Am (243)	96 Cm (247)	97 Bk (247)	98 Cf (251)	99 Es (252)	100 Fm (257)	101 Md (258)	102 No (259)	103 Lr (262)

Figure 1-4. A modern periodic table. From science.widener.edu/~svanbram/ptable_3.pdf, accessed 14 February 2008, S.E. Van Bramer, 1999. with the permission of Scott Van Bremer.

such as fuchsia, aniline blue, and imperial purple. These dyes, coupled with the textile mills of the industrial revolution, changed the way the Western world dressed and decorated. The textile industry gave chemistry a solid economic basis. By the 1870s, chemistry was largely synonymous with synthetic organic chemistry, and industrial fortunes were being built upon it, especially in Germany. Universities established chairs of chemistry—most in organic chemistry—to train new chemists for academic life and for industry.

Most chemists, perhaps in deference to the skeptical physicists, continued to refer to the atomic theory as a hypothesis. Many of the chemical papers of the mid-19th century include what seems to have been a compulsory disclaimer, saying that the chemical structures are not meant to imply the actual existence of atoms or, if atoms do exist, to imply anything about their spatial positions. But by the 1860s, many physicists had begun to accept the chemists' ideas of atoms and molecules. For substances in the gas phase, they thought of molecules as something like billiard balls, obeying Newtonian mechanics and caroming off each other and off the walls of whatever container enclosed them. Rudolf Clausius and Ludwig Boltzmann were among those who developed what was known as the kinetic theory, based on Dalton's atomic hypothesis, and they managed to derive equations that matched much of the observable behavior of matter. In 1874, two chemists, Jacobus van't Hoff and Joseph Le Bel, violated the prohibition against assigning atoms positions in space, taking two-dimensional molecular structures and moving them into three dimensions. Van't Hoff and Le Bel separately suggested that a carbon atom, with connections to four different atoms, might exist in different left- and right-handed tetrahedral forms—pyramids—that were mirror images of each other. The two molecules in fig. 1-5, for example, have the same connectivity to other atoms but cannot be superimposed upon each other—they are like left- and right-handed gloves.

Van't Hoff, one of the two who had proposed this idea, was soon to move from structural organic chemistry into efforts to apply the methods of physics to chemical investigation—to return to the ideas of Berthollet, ideas that had been largely neglected for much of the past century at least partly because of the successes of structural organic chemistry. Applying physical methods to chemistry went by many names—"theoretical chemistry" and "general

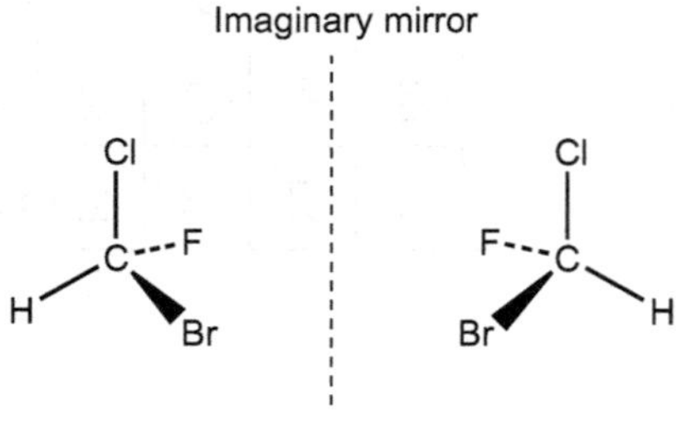

Figure 1-5. Left- and right-handed molecules may have the same bonds (connections) but are mirror images of each other, like left- and right-handed gloves. The heavy lines should be thought of as coming out of the page, and the dashed lines as going back behind the page.

chemistry" among them—but the name that would stick would be "physical chemistry."

Van't Hoff was born in 1852 in Rotterdam to a middle-class family. He began by studying organic chemistry with Kekulé in Bonn in 1872, where he found the other students unwelcoming and oppressive. He continued his doctoral work at Utrecht in Holland in 1873 and published his ideas on tetrahedral carbon while still a graduate student. Most chemists were completely uninterested, and he was unable to find work after graduation even as a high school teacher. He supported himself through tutoring until he found employment as an assistant at the local veterinary school. Hermann Kolbe, a prominent organic chemist, used him as an example of all that he thought to be wrong with modern chemistry, saying, "A Dr. J. H. van't Hoff who is employed at the Veterinary School in Utrecht appears to find exact chemical research unsuited to his tastes. He finds it more suitable to mount Pegasus (obviously loaned from the Veterinary School) and to proclaim . . . how, during his flight to the top of the chemical Parnassus, the atoms appeared to be arranged in the universe."[4] Perhaps because of this sort of reception, van't Hoff and a few others moved outside the German organic chemistry establishment and began considering Berthollet's old questions about chemical affinity. When van't Hoff was appointed full professor in Amsterdam in 1878, he quoted Kolbe's comments in his inaugural address.

Arrhenius's Dissertation

Arrhenius, whose graduation ceremony began this chapter, was an outsider in chemistry. Germany was the chemical center, and Arrhenius studied in provincial Sweden. Organic chemistry was where advancement, money, and fame were to be found, but the chemistry professor at Uppsala, Per Theodor Cleve, was a "mineral chemist"—an inorganic chemist—and was interested only in students who would work on the problems that he chose.[5] Arrhenius was not sure whether he wanted to be a chemist or a physicist, but he knew that he was not interested in endlessly synthesizing more compounds, whether organic or inorganic. He wanted to do something novel—something on the borderline between chemistry and physics that would extend chemical theory. He chose two advisers, Cleve as a chemist and Tobias Thalén as a physicist, and paid little attention to the suggestions of either of them. Because Thalén's physics laboratories were limited, Arrhenius did his experimental work at the Högskola in Stockholm, a research institute that did not grant doctoral degrees.[6] In a letter introducing Arrhenius to Eric Edlund, a professor at the Högskola, Thalén asked that he be provided laboratory facilities, but warned Edlund that Arrhenius was lazy and clumsy.[7] At the Högskola, Arrhenius worked with Edlund in physics and with Otto Pettersson in chemistry and got

Figure 1-6. Arrhenius as a young man. From www.uniarchiv.uniwuerzburg.de/images/Arrhenius%201876.jpg, accessed 14 February 2008.

along well with both of them, but they could not give him the doctoral degree that he needed. Only Cleve and Thalén at Uppsala could do that.

Arrhenius planned to try to determine the molecular weights of substances by measuring the electrical conductivity of their solutions, a topic that Cleve had specifically directed him not to pursue.[8] Molecular weight—an idea that came directly from Dalton's atomic theory—was an important parameter for any substance, and the usual method for its measurement had been the volume of the substance in the gaseous state. But there was a problem—not all substances were stable in the gas phase. When sugars were heated, for example, they decomposed before they boiled, so some other technique for measurement of molecular weight would need to be found. Arrhenius's idea for his dissertation was to measure sugars' effects on the properties of *electrolytes*—substances such as salts or acids that conduct electricity when dissolved in water. By dissolving an electrolyte in water and inserting the two terminals of a battery into the solution, its conductivity—how easily it conducts electricity—can be measured. Adding a nonconductor, such as sugar, to a solution of an electrolyte reduced the conductivity, and Arrhenius thought that the magnitude of the conductivity reduction might be related to the molecular weight of the nonconductor.

But before Arrhenius even began work, the problem of determining the molecular weights of substances like sugars had already been solved, although he appears not to have read the paper. François Raoult had found in the spring of 1882 that molecular weights of solutes (dissolved substances) such as sugars could be measured by their depression of the freezing point of the solution.[9] Unaware of Raoult's work, Arrhenius amassed conductivity data on forty-four different electrolytes at a series of different dilutions. He found that the conductivity depended as much on the concentration of the electrolyte as it did on the molecular weight of the added sugar, and he decided that he first needed to understand the dependence of conductivity on electrolyte concentration if he were to make sense of his problem. When he began his analysis of electrolyte concentration, he found the results confusing. Arrhenius eventually abandoned his original problem, the determination of molecular weights, because the dependence of conductivity upon electrolytic concentration seemed more interesting (and possibly because he had finally read Raoult's paper). Most scientists' response to confusing data is to try to find a different method to solve the original problem, but this was a case where the interest was to be found in the confusion. Arrhenius had the presence of mind to stop and examine what others might have seen as a jumble of random data. Beginning with the next chapter, we shall see other scientists, in particular Irving Langmuir, who were also willing to be distracted from one line of research to pursue an unexpected observation.

Arrhenius divided electrolytes into two classes: "strong" electrolytes such as sodium chloride (table salt) and sulfuric acid, which conduct electricity easily, and "weak" electrolytes such as acetic acid (vinegar) and ammonia, which conduct less readily. For strong electrolytes, the conductivity per weight of electrolyte changed little regardless of dilution. For weak electrolytes things were different—the total conductivity declined as the electrolyte was diluted (as would be expected), but the conductivity *per weight of the electrolyte* started off small and then increased indefinitely with increasing dilution, slowly approaching some limiting value.

The relative chemical affinity of an acid was identified with its acidic strength—whether it could displace weaker acids from their salts. For example, hydrochloric acid was considered stronger than acetic acid, because when hydrochloric acid is added to a water solution of sodium acetate (a salt of acetic acid), sodium chloride is formed and acetic acid is displaced. (This is the same reasoning that Stahl had used in 1720 in saying that copper had a higher affinity than did silver for nitric acid, as copper would displace silver.)

When Arrhenius looked at his data, he saw a strong correlation between the strength of an acid and its conductivity, and he thought that he might be able to use conductivity measurements to quantitatively predict chemical affinity beyond acids.

Arrhenius defined a concept of the "active" state of an electrolyte: *Molecules of electrolytes would conduct electricity only when they were in an active state, and this same active state was likely responsible for the electrolyte's chemical affinity.* Because conductivity could be measured, this would give a quantitative measure of affinity. He was vague about exactly what the active state comprised, saying that it is "probably a compound of the inactive part and the solvent. Or possibly inactivity may be caused by the formation of molecular complexes. Or again the difference between the active and inactive parts may be purely physical."[10] He defined an activity coefficient, which he called α, that expressed the fraction of the molecules in an active state; an activity coefficient of 1 meant that all the molecules were in an active state. He found a general pattern: a strong electrolyte like nitric acid always had an activity coefficient near 1, no matter what its dilution, but as a weak electrolyte like acetic acid was diluted, its activity coefficient increased from some fractional value and approached a value of 1 at extreme dilution. He concluded that at infinite dilution, all electrolytes, weak and strong, would be completely active.

Arrhenius's professors, Cleve and Thalén, were not impressed: his experimental work had not been very precise, as he had not controlled temperature, water quality, or concentration carefully; he had studied only a small number of compounds; he had chosen a topic on the borderline between physics and chemistry that neither of them had encouraged; his concepts of activity were vague; he had proposed a theory that contradicted the established theories with what they saw as insufficient evidence; and he had not solved the problem that he had set for himself, the determination of molecular weights by conductivity. They gave his dissertation a passing grade, but just barely—a grade of "not without praise." Such a grade would not allow him to be a *docent*, a scientist allowed to teach at the university, and his prospects seemed limited to teaching at a *gymnasium* (an academic high school). And on graduation day, Cleve and Thalén showed their contempt by walking out without even stopping to congratulate him.

The Ionists Come Together

After his humiliating graduation ceremony, Arrhenius moved back into his family home, also in Uppsala, with few prospects. Uppsala is a small city north of Stockholm, and its university—Sweden's oldest—was the center of the city's social hierarchy. Arrhenius had grown up surrounded by the pomp and academic robes of the university. His father had been a university employee, but only a rent collector—not a prestigious position when compared to the professors, whose ranks Arrhenius had aspired to join and by whom he had been rejected. Some of his difficulties with Cleve and Thalén may have been due to class differences: the professors in Uppsala were the social elite, and Arrhenius was from a lower middle-class background. Edlund, on the other hand, with whom he worked productively at the Högskola in Stockholm, was the son of peasants.[11]

At the time of Arrhenius's graduation, his father was ill. He died the following year, and Arrhenius spent much of that year caring for him and then settling his estate. Arrhenius was left with enough money that he did not need to find a job immediately, and he still believed that his theory of electrolytes was worth pursuing. Because of his dissertation's low grade, he was cut off from the laboratories at the university. Thinking that his work might find more interest beyond provincial Uppsala, he sent copies of his dissertation to several scientists whose work he respected. Only one replied positively—another chemist working outside German organic chemistry, Wilhelm Ostwald, a twenty-nine-year-old ethnic-German professor at the Riga Polytechnic in the Russian empire.

Where Arrhenius was argumentative and graceless, Ostwald was a born leader who enjoyed reaching out and encouraging others. Ostwald, like Arrhenius, had returned to the problem that Berthollet had pursued some eighty years earlier—the measurement of chemical affinity. And Ostwald was not the only one. Almost twenty years earlier, Cato Guldberg and Peter Waage, two Norwegian chemists—also outside the German chemical center—had managed to give quantitative form to Berthollet's idea that the *mass* of the products and reactants was important in determining affinity, deriving what eventually came to be known as the *law of mass action*. Guldberg and Waage derived their law not directly in terms of mass but of concentration—the number of molecules per unit volume in solution. For each reaction that had reached equilibrium—where no further spontaneous change of concentrations of reactants or products was possible—their law of mass action stated that the value of a simple mathematical formula of the concentrations of the products and the reactants would be equal to an *equilibrium constant* characteristic of that reaction. So if the concentration of any of the reactants or products were changed from outside the system—for example, by adding or removing a reactant or a product—the reaction would proceed in a direction that would return the value of the calculated formula to its equilibrium constant. But Guldberg and Waage published in Norwegian, and few were much interested in the questions they asked. Their work passed largely unnoticed, except by a few outsiders like Ostwald. Waage was in professional contact with Arrhenius's friend and chemical mentor in Stockholm, Otto Pettersson, however, and this may have influenced Arrhenius's work.[12]

Determination of chemical equilibria had practical and economic importance. Henri le Chatelier said in 1888:

> It is known that in the blast furnace the reduction of iron oxide is produced by carbon monoxide, according to the reaction
>
> $$Fe_2O_3 + 3CO = 2Fe + 3CO_2,$$
>
> but the gas leaving the chimney contains a considerable proportion of carbon monoxide, which thus carries away an important quantity of unutilized heat. Because this incomplete reaction was thought to be due to an insufficiently prolonged contact between carbon monoxide and the iron ore, the dimensions of the furnaces have been increased. In England they have been made as high as thirty meters. But the proportion of carbon monoxide escaping has not diminished, thus demonstrating by an experiment costing several hundred thousand francs, that the reduction of iron oxide by carbon monoxide is a limited reaction. Acquaintance with the laws of chemical equilibrium would have permitted the same conclusion to be reached more rapidly and far more economically.[13]

Using Guldberg and Waage's law of mass action, determination of the equilibrium constant for one set of concentrations would allow chemists to predict the equilibrium concentrations for another set. Ostwald carried out a large number of experiments in Riga confirming that the mass action law applied to the activity of organic acids and, furthermore, found that the acids' chemical affinities seemed to be related to their reaction velocities—how quickly the reactions proceeded. After exchanging letters with Arrhenius, Ostwald tried Arrhenius's conductivity approach—the organic acids he had tried were all weak electrolytes, so he thought that Arrhenius's methods should apply. Ostwald borrowed some electrical equipment from the local telephone company and used it to measure the conductivity of all the acids in his collection. The conductivity measurements corresponded almost exactly with his results on reaction velocities, so he now had two methods of estimating affinity. But he found that Arrhenius's conductivity measurements were much easier and faster than his own.

Most scientists had thought that the law of mass action would not be applicable to electrolytes because the reactions were too rapid and the equilibrium would be impossible to measure (and for strong electrolytes, in fact, there would turn out to be great difficulties in applying the law of mass action). Part of the problem was sociological—scientists studying the law of mass action formed a group that did not study electrolytes, and scientists studying electrolytes formed a group that did not have much interest in weak electrolytes, as these were relatively poor conductors. But Ostwald and Arrhenius showed that the law of mass action applied quite well to acids and bases that were weak electrolytes, such as acetic acid and ammonia. Ostwald agreed with the conclusion that Arrhenius had reached in his dissertation: "[F]or acids and bases, galvanic [electrical] activity is accompanied by chemical activity."[14] Neither Arrhenius nor Ostwald could yet explain Arrhenius's "active state" mechanistically, but they knew that it was both chemical and electrical.

Ostwald visited Arrhenius in Uppsala that fall, and the two hit it off, although it was clearly Ostwald who was the professor and Arrhenius the student. Ostwald encouraged Arrhenius to equip a small laboratory in his home. When he visited Uppsala University, Ostwald expressed his good opinion of Arrhenius to Arrhenius's former chemistry professor, Cleve, who began to revise his opinion of Arrhenius accordingly. Cleve arranged for him to be appointed a *docent* in physical chemistry (but not in physics or chemistry proper) at Uppsala, although without any salary, students, or laboratory.[15]

Ostwald visited Arrhenius because he was impressed with his scientific work, but he found him less than impressive in person. Ostwald wrote to his wife, Helene, that Arrhenius "is somewhat corpulent with a red face and a short moustache, short hair; he reminds me more of a student of agriculture than a theoretical chemist with unusual ideas."[16] But Ostwald went out of his way to encourage Arrhenius. The two of them attended a large scientific meeting in Magdeburg that September, and Ostwald graciously gave up his speaking slot so that Arrhenius could present his work. The two were now allies, dedicated to what Ostwald called the need to "reform chemistry"—to resume the study of chemical reactions rather than just to synthesize and categorize new chemicals. With Ostwald's help, Arrhenius received a travel grant for a year's study abroad. He planned to study with all the experts in electrolytes—with Ostwald in Riga, with van't Hoff in Amsterdam, and with Friedrich Kohlrausch in Würzburg.[17] Although Arrhenius had received his doctorate in 1883, he was not to hold a salaried position until 1891, when he would return to Sweden. He would spend eight years doing unpaid research in the laboratories of Ostwald and others.

While Arrhenius was in Würzburg in 1886, he met Walther Nernst, a German scientist five years his junior. Nernst was from a middle-class Prussian family; his father was a judge. Although physically unprepossessing, he had thought of becoming an actor and had written a stage play. Robert Millikan, who would later be his student, reported that Nernst spent his first two years at university as a member of one of the student fighting corps and acquired a dueling scar, but later got down to work.[18] Nernst was not at all interested in another typical student pursuit, wandering through the mountains and forests during the long university vacations. In fact, he consistently avoided physical activity. He said that he climbed a mountain once, and that that experience sufficed for his entire life.[19]

Unlike Arrhenius, van't Hoff, or Ostwald, Nernst was not an outsider but a Prussian, a citizen of the state that had led the formation of the new German

Figure 1-7. Walther Nernst ca. 1895. From www.nernst.de/, accessed 14 February 2008.

nation. By the time he arrived on the scene, a new interest in applying physical methods to chemistry was beginning to form. Before he met Arrhenius in Würzburg, he had studied chemistry and physics in Zurich, Berlin, and Graz. He did not need to fight his way into German science from the outside.

Nernst and Arrhenius impressed each other immediately. Arrhenius wrote that Nernst's work on heat conduction was "the best that any laboratory practitioner has done in a long time."[20] Nernst was fascinated with Arrhenius's descriptions of the work that he and Ostwald had been doing and resolved to go to Riga to study with Ostwald as soon as possible. Arrhenius and Nernst began spending all their spare time together. During the spring break, when the Würzburg laboratory was closed for six weeks, they went to Graz, where Nernst introduced Arrhenius to Ludwig Boltzmann, with whom Nernst had studied earlier. Arrhenius's good words about Nernst impressed Ostwald, and when in 1887 Ostwald moved from Riga to a new position in Leipzig, Germany, he offered Nernst a position as his assistant.[21] Nernst had been invited into the inner sanctum of the new physical chemistry; he said that he felt "as well qualified as if Rubens had appointed me to be one of his collaborators in his painting studio."[22]

Figure 1-8. Nernst (standing at the left) and Arrhenius (standing second from right) in Graz in 1887 with Boltzmann (seated at the table). From www.mlahanas.de/Physics/Bios/LudwigBoltzmann.html, accessed 14 February 2008.

While in Leipzig with Ostwald, Arrhenius and Nernst were not only scientific partners but also drinking buddies, often to excess, at least in Nernst's case. Arrhenius enjoyed recounting Nernst's alcoholic and "romantic" adventures. After Nernst had spent a night drunkenly banging on Arrhenius's door, Arrhenius concocted a story and convinced Nernst that he had narrowly escaped police arrest. Nernst became quite upset, and Arrhenius and his friends told him the truth only after making him sweat for a month.[23]

At about this time, Arrhenius became aware of some scientific papers that van't Hoff had submitted to a Swedish journal. Van't Hoff had moved on from organic chemistry and became interested in the same subject as had Arrhenius—the properties of solutions. Arrhenius sent van't Hoff a copy of his dissertation, and the two began corresponding. Van't Hoff was attempting to apply the new science of thermodynamics to chemical affinity.

Thermodynamics is the science of the transfer of energy between two forms, heat and work. The practical impetus for thermodynamics had been the development of the steam engine, which burned coal or wood to make heat, and then used a boiler and a piston to convert the heat into external work—driving a locomotive forward, or pumping water from a mine, for example. From the viewpoint of the steam engine, work was considered useful energy, and emitted heat was wasted energy. An important question was how much work one could extract from a given amount of heat, and the first two laws of thermodynamics, developed in the period 1824–60, addressed that question. The first law of thermodynamics states that energy, as represented by the

Figure 1-9. Ostwald, Arrhenius, and van't Hoff. Photos by Nicola Perscheid. Edgar Fahs Smith Collection, University of Pennsylvania Library.

sum of a closed system's heat and any work done by the system, is conserved: energy can be neither created nor destroyed, but only shifted from one form to another. The second law of thermodynamics states that while work can always be converted to heat, only a portion of the heat of a system, even under ideal circumstances, can be converted into work in a cyclic process such as occurs in a steam engine. The first law is sometimes stated as "You can't get something for nothing," and the second law as "You can't even break even."

There were two possible approaches to determining chemical affinity. The first was to find laws that described the interactions of atoms—an approach similar to Newton's laws of motion for classical mechanics; this approach would not begin to show real success until the 1920s. But thermodynamics offered another path. Thermodynamics does not require mechanistic knowledge of what is happening inside a system—it does not even assume the atomic theory. Its equations depend only on measurable variables such as temperature, pressure, volume, and mass. It was not immediately clear to most chemists what thermodynamics had to do with chemistry, but chemical reactions give off or take up heat, they can be made to do work (by driving a piston as in a steam engine, or by providing a voltage as in a battery, for example), and their progress depends upon pressure, temperature, and volume.

Van't Hoff and Arrhenius Make the Connection

Van't Hoff, among others, was attempting to use thermodynamics to study chemical systems. In studying large molecules, such as sugars dissolved in water, he needed a pressure measurement in order to apply thermodynamics to the molecules' properties, but sugars have a very low vapor pressure and decompose before boiling, as Arrhenius had earlier noted in what he had

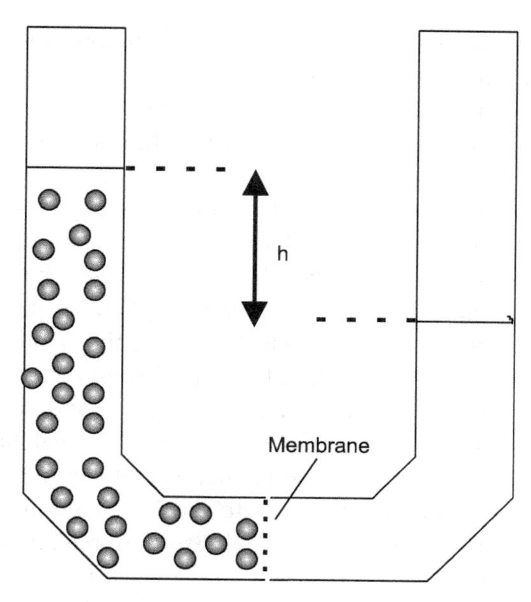

Figure 1-10. Osmotic pressure. The membrane is one through which a solvent such as water can pass, but dissolved solute particles cannot. A water solution of some salt is on the left side, and pure water is on the right. Water molecules are in higher concentration on the right side of the membrane and therefore pass from right to left more often than from left to right, causing a pressure that can be measured by the difference h in the solution levels on the two sides of the membrane. The pressure increases proportionally with the salt concentration (the number of molecules) and with the temperature (how fast the molecules are moving).

intended to be his dissertation problem. But van't Hoff found another pressure that he could use—osmotic pressure, which botanists had been studying. Osmotic pressure is described in fig. 1-10. For osmotic pressure to be observed, two things are necessary: a membrane must separate two solutions with a difference in concentration of the dissolved substance (the solute) on the two sides of the membrane—pure water on one side, perhaps, and a solution of the solute on the other side—and the membrane must allow passage of the solvent but not of the solute. Such a membrane is called *semipermeable*. Osmotic pressure can be quite large and explains, for example, how Epsom salts (magnesium sulfate) relieve swelling in a sprained ankle: a concentrated solution of Epsom salts on one side of the skin exerts osmotic pressure on the ankle's cells, pushing excess fluid out.

Van't Hoff found that osmotic pressure obeyed a law almost identical to that for the pressure of a gas: osmotic pressure increased proportionally both with the amount of the dissolved substance and with the temperature, just as the pressure of a gas increased proportionally both with the amount of the gas and with the temperature. For many substances, the equations for gas pressure and osmotic pressure even had the same constant of proportionality. But he found that the constant of proportionality for osmotic pressure was not the same for all substances, and he needed to introduce a fudge factor i into his equation for osmotic pressure, where the value of i was dependent on the particular substance.

When Arrhenius saw van't Hoff's results, he saw that for nonelectrolytes, such as sugars and alcohols, the fudge factor i was always close to 1, but for electrolytes, it was always higher than 1. He realized that if the electrolytes were breaking up into fragments, it would explain the higher osmotic pressures for electrolytes—more fragments striking the membrane would result in increased pressure. He developed a simple formula that related van't Hoff's values of i to his own activity coefficients, α, and found that the results from conductivity and osmotic pressure measurements matched closely—if he knew i for a substance, he could predict α, and vice versa. Arrhenius had been unable to define his "active state" at the time of his dissertation, but van't Hoff's experiments explained it: electrolytes breaking up into fragments were the cause of van't Hoff's i values, and because i could be calculated from his own α, his active state must consist of the same fragments of electrolytes.

But what were these fragments? Electrolytes (by definition) were molecules that conducted electricity, and the then-accepted theory held that under the influence of an electric current, electrolytes would form ions—electrically charged molecules—that would transmit electric current across the solution. But that theory held that very few, if any, of the molecules existed as ions unless there was a voltage imposed across the solution—that it was the voltage that *caused* the ions to form. And even in the presence of an imposed voltage, theory predicted very few ions would be formed. Electrical current would be able to transfer across a solution by a daisy-chain process, where one electrically charged terminal of the battery would cause a molecule in solution to become charged and form an ion, which would transfer its charge to a neighboring molecule, forming a new ion and returning to its original uncharged state. The new ion would repeat the process, and eventually the charge would make its way across the solution to the opposite electrical terminal.

Arrhenius had accepted this theory when he had written his dissertation, saying, "few hypotheses have been as generally accepted in the world of science as this one,"[24] but van't Hoff's data on osmotic pressure made him rethink things. He now believed that his "active state" consisted of fragments of electrolytes, and that his and van't Hoff's fragments were identical. And in van't Hoff's experiments, there had been no voltage or electrical current, so the fragments were not caused by applying a voltage. Because Arrhenius had identified his active state with the conduction of electricity, he reasoned that the fragments were electrically charged ions. His new idea at first seemed ridiculous to many—that the strong chemical bonds in a molecule like sodium chloride, which formed with a tremendous binding energy when created from pure sodium and chlorine, were completely ruptured in water solution even in the absence of an electrical current, and that sodium chloride in solution contained no sodium chloride at all, but only positively charged sodium ions and negatively charged chloride ions.

Van't Hoff and Arrhenius separately published their results in 1887 in the first issue of Ostwald's new journal, *Zeitschrift für physikalische Chemie* (Journal of Physical Chemistry). Ostwald later took Arrhenius's theory of electrolytic dissociation and applied the law of mass action, developing what became known as Ostwald's dilution law, which enabled chemists to predict the chemical and physical properties of solutions over a wide range of concentrations. This worked well when applied to weak electrolytes like acetic acid or ammonia, but not when applied to strong electrolytes like sodium chloride or nitric acid. This anomalous behavior of strong electrolytes—their failure to follow the law of mass action—would puzzle physical chemists for the next thirty years.

Ostwald was now installed in his new professorship in Leipzig, and he provided a place to work and publish for scholars interested in the new physical chemistry. Many German and foreign students, especially Americans,[25] got their degrees with Ostwald or spent time with him in postdoctoral study.

The Wild Army of the Ionists, and "Free Energy"

Van't Hoff was an introvert and a bit of a recluse, but Ostwald, Arrhenius, and Nernst enjoyed argument, and the four of them became known as "the wild army of the Ionists." They were ready to do battle on behalf of the dissociation theory, which was the core of the new physical chemistry, and the more rambunctious among them did their best to adhere to the image of invading Mongol hordes. In an 1889 letter to Ostwald, Arrhenius said that Nernst had decided on a particular journal as his "battlefield" to "annihilate" an opponent, but until Nernst "takes the offensive, he will keep his powder dry."[26] But in fact, the dissociation theory was accepted relatively quickly by chemists, particularly in Germany. By 1890, Edvard Hjelt, a Finnish chemist, said, "It has rarely happened that a theory has had such rapid success as the theory of electrolytic dissociation as it was formulated in 1887. Perhaps it met at first with some indifference on the part of chemists, but it has not been the target of serious opposition."[27] The theory's principal opponents were two British chemists, Henry Armstrong and P.S.U. Pickering, and the Russian originator of the periodic table, Dmitri Mendeleev. Arrhenius published a scathing attack on the revered Mendeleev in 1889. The Ionists saw themselves as evangelists: Nernst referred to his 1890 course on the new theory of solutions as an "apostolic seminar."[28] While the German chemists adopted the theory quickly, the German physicists, led by their most prominent member, Hermann von Helmholtz, were not immediately convinced, and Helmholtz opposed the creation of a chair in physical chemistry in Berlin. But in their most important forum, chemistry, the Ionists succeeded quickly, and many universities began to consider chairs in the new physical chemistry.[29]

When Arrhenius had begun his work on ionic dissociation, he had been looking for a quantitative measure of chemical affinity. Although he had been unaware of it at the time, a theoretical basis for measure of chemical affinity had been proposed when he wrote his dissertation—"free energy," which is that portion of the total energy released in a chemical reaction that can, under ideal circumstances, be converted to useful work. An American scientist at Yale, J. Willard Gibbs, had been the first to note this in 1876, but he had published almost no practical examples, so his ideas were largely ignored for about a decade. Gibbs's conclusion about the importance of free energy in chemical reactions was reached independently by Helmholtz in Berlin, who gave free energy its name in 1882. Helmholtz also noted the identity between the free energy and the amount of work that could be extracted from a battery:

> If we now take into consideration that chemical forces can produce not merely heat but also other forms of energy . . . as, for example, in the production of work by a galvanic battery; then it appears to me unquestionable that, even in the case of chemical processes, a distinction must be made between the parts of their forces of affinity capable of free transformation into other forms of work, and the parts producible only as heat. . . . I shall distinguish these two parts of the energy as the "free" and the "bound" energy.[30]

While physicists began to understand the importance of free energy in the early 1880s, chemists were slower to do so. Chemists were still looking for some property that would *correlate* with chemical affinity, as Arrhenius was doing with conductivity. Similarly, the thermochemists, Jules Thomsen in Copenhagen and Marcelin Berthelot in Paris, measured the heat given off in hundreds of chemical reactions, looking for a correlation between chemical affinity and heat. Van't Hoff later (1903) pointed out that although there is a general correlation between heat of reaction and chemical affinity, it is not universal:

> This assumption [that free energy, the maximum amount of work available, is the driving force of a chemical reaction and the measure of chemical affinity] . . . is in a measure self-evident. Whenever any change in the realm of nature can accomplish work, that is can overcome resistance, it must proceed when the resistance is removed. Now it must be noted that the accomplishment of work and the development of heat do not mean quite

the same thing. Often they go hand in hand, as in the case of explosions like gunpowder or dynamite. These materials are familiarly known to furnish by their explosions great chemical means of doing work and at the same time to develop large quantities of heat. A compound like phosphonium chloride (PH_4Cl) however, a solid body, tends to decompose at ordinary temperatures into the gases phosphine (PH_3) and hydrogen chloride (HCl) with marked absorption of heat [cooling]. Yet the decomposition products of this action may exercise a pressure of some twenty atmospheres. Here we have a case where the possibility of accomplishing work does not coincide with the capacity to develop heat, and yet where it is obviously the capacity to do work which controls the direction of the change.[31]

Free energy is related to the ability of a chemical reaction to do work, and one way for a chemical reaction to do so is to employ it as the energy source of a battery. Van't Hoff was the first chemist to make the connection between free energy and voltage,[32] but Nernst was the one who ended up with his name on the equation. While working as Ostwald's assistant in Leipzig in 1888, Nernst was following up on Arrhenius's dissociation theory and investigating the behavior of ions. Nernst reasoned that just as the rising pressure of steam can do work by driving a piston, so the rising concentration of ions can do work through the voltage in a battery, which can be used to drive an electric motor, for example. Now, this thought was not entirely original; as quoted above, Helmholtz had mentioned the idea when he had given free energy its name in 1882, and van't Hoff had suggested the same thing in 1885. In January 1888, the twenty-four-year-old Nernst managed to arrange a meeting with Helmholtz, who at age sixty-eight was known as the "Imperial Chancellor of German Science" as the president of the newly founded Physikalisch-Technische Reichsanstalt in Berlin. Nernst, who had earlier attended Helmholtz's lectures as a student, explained his ideas to him. In a letter thanking Helmholtz after their meeting, he said that their conversation would remain unforgettable to him.[33] A year later he presented an equation and experimental results showing that the free energy of a chemical reaction could be measured by making the reaction the source of energy for a battery and then measuring the voltage across the battery terminals, as Helmholtz had suggested. Many chemical reactions may be used as the energy source of a battery, although there are often practical difficulties in doing so. Two electrodes, composed of different substances, are immersed in a solution of an electrolyte. The ensuing chemical reaction will cause a voltage across the electrodes, which can be used to do work—to drive an electric motor, for instance. As an example, a car battery generates electricity from the chemical reaction

$$Pb + PbO_2 + 2H_2SO_4 = 2PbSO_4 + 2H_2O$$

By constructing a battery consisting of a lead electrode and a lead oxide electrode inserted into a solution of sulfuric acid, one could determine the free energy of the reaction by measuring the voltage. Nernst showed that there was a simple proportionality between voltage and free energy, and this relationship is now known as the "Nernst equation." Where Arrhenius had shown that conductivity measurements were strongly correlated with the vague concept of chemical affinity, Nernst went further: he knew that free energy was the theoretical measure of chemical affinity and showed how to use voltage to measure free energy.

Nernst was more focused on pursuing the problem to its logical end than was Arrhenius, who lacked Nernst's grounding in mathematics and physics, but Nernst's work on affinity was not revolutionary in Thomas Kuhn's sense of the term. Kuhn has drawn a distinction between "revolutionary" science, which focuses on explanations outside the accepted theories of the day, and "normal" science, which works within the accepted theories.[34] Arrhenius's dissociation theory was revolutionary science, while Nernst's law was normal science—it was a straightforward experimental and theoretical union of Gibbs's and Helmholtz's concepts of free energy with Arrhenius's dissociation theory.

Nernst in Germany and Arrhenius in Sweden

By 1890, both Nernst and Arrhenius were becoming well known, and academic positions in the new discipline of physical chemistry were opening up. Nernst moved to the university at Göttingen in 1891 as a lecturer with a promise that he would receive an assistant professorship. The university at Giessen then announced a new chair in physical chemistry, for which both Arrhenius and Nernst were under consideration. Nernst was clearly viewed as a young man with a future, and the Prussian minister of education, not wanting to lose him, sped up the offer of the promised associate professorship at Göttingen, which Nernst accepted. He then withdrew from consideration at Giessen, recommending his friend Arrhenius for the post. But Arrhenius longed to return to Sweden (although certainly not to Uppsala, the site of his humiliations), and he used the Giessen offer to put pressure on the Högskola in Stockholm, the institution at which he had done his research during his dissertation. In 1891, the Högskola offered him a position as a teacher in physics, including a laboratory and an assistant,[35] and in 1895 promoted him to the newly created chair of physics—although not without some internal disagreement, as he only got the promotion due to strong recommendations from German physicists and chemists.[36] He chose to forgo the big lake of science in Germany for the small pond of Sweden, although he continued to despise the Swedish "Uppsala physicists" who had snubbed him and whom he saw as having been responsible for the difficulties he had had in getting his promotion at the Högskola.

After Arrhenius returned to Sweden, he fell in love with Sofia Rudbeck, a beautiful, outgoing young woman who had graduated in sciences from Uppsala and had enrolled at the Högskola to do graduate work in chemistry and mineralogy in 1892. This was a time when Swedish women were becoming emancipated and enrolling in higher education, and the Högskola was a popular choice. Arrhenius hired Rudbeck as his assistant and was soon hopelessly smitten, sending her love poems, and she was flattered by the attentions of a brilliant young professor. By the end of the academic year, the two were engaged. From Elizabeth Crawford's biography of Arrhenius:

> Her letters to him that summer were full of endearments interspersed with descriptions of how she was learning to ride a bicycle, "flying like the wind" to go swimming in the lakes around Uppsala.... She also kept up a running commentary on how she was working her way through Nernst's [new book *Theoretical Chemistry*] giving it a second going-over to make sure she had understood everything. Amidst expressions of affection one could find... commands—those concerning appointments at the Högskola or work to be done by his graduate students, for instance. The somewhat irritated tone of his replies may have indicated that he felt she was meddling.... He [was] asked to countersign loans by his father-in-law. The prenuptial contract that Sofia insisted be drawn up stipulated that she control the assets that she had brought into the household as well as those she might receive in the future.... But, as [Arrhenius] told Ostwald later, "I was blind."[37]

Rudbeck wanted her own scientific career and was uninterested in being her husband's secretary and social companion. She was a theosophist, a teetotaler, and an antismoker and wanted Arrhenius to use the new phonetic spelling in his research papers; Arrhenius was not about to take on any of this. She moved out of their home in the fall of 1895 to a small cottage in the Stockholm archipelago. In November, she gave birth to a son without seeing her husband again. In July 1896, they divorced.

With his return to Sweden, Arrhenius essentially abandoned physical chemistry, although he welcomed to his laboratory visiting scientists and students working in the field. He took up instead what he called "cosmic physics," which would be called earth sciences today. Beginning in 1894, he was the first to construct a quantitative model of the greenhouse effect, linking variations in the carbon dioxide content of the atmosphere to climate effects. He performed elaborate calculations recording measurements every ten degrees of latitude at different seasons, used the observed reflectivity of the moon as a reference for the sun's radiative energy, and developed a model of the relationship between carbon dioxide and surface temperature: the changes in temperature would vary as the square of the carbon dioxide concentration, and the effect would be greater in the summer than in the winter. He predicted that doubling the carbon dioxide concentration would increase the mean temperature 6°C, which is not far from current estimates of 3–5°C. His interest, however, was not in warning about the perils of industrialization but in finding the cause of the ice ages. His one comment about global warming reflected his view as a resident of the far north: he estimated that the burning of fossil fuels would double the carbon dioxide content 3,000 years in the future and would "allow our descendants, even if they only be those of a distant future, to live under a warmer sky and in a less harsh environment than we were granted."[38] Apparently, he had not thought things out fully, as he was not worried that the resulting rise in sea levels would put Stockholm, the "Venice of the North," under water. Nernst later also favored increased global warming and suggested a proactive approach, proposing starting underground fires in used-up coal mines to release carbon dioxide into the air.[39]

Nernst, once established in Göttingen, married in 1892. Emma Lohmeyer's father was a professor of surgery at the university and a widower for whom Emma had kept house since she had reached age sixteen. She was twenty-one at the time of their marriage, and Nernst was twenty-eight. By all accounts, he was devoted to her and she to him, and they had two boys and three girls over the next twelve years. Despite Nernst's reputation for arrogance in the lecture hall and for abusing his staff and students, he was known for welcoming women scientists and students,[40] which was not common at the time. He was also known to have a puckish sense of humor, engaging his subject with a look of wide-eyed, deadpan innocence as he delivered his lines. Both of these qualities may have made him easier to live with than one might have expected. He loved his children and enjoyed family meals, and he took a daily trip home for lunch and a nap.[41] In his time in Göttingen, from 1890 to 1905, Nernst was at his best, both scientifically and personally. He married there, his children were all born there, he wrote *Theoretical Chemistry* there in 1893, and he was rapidly promoted through the academic ranks. In 1894, Munich invited Nernst to a professorship as Boltzmann's successor, and the Prussian education minister was able to keep him in Göttingen only by giving him a full professorship in a new chair of physical chemistry and promising him a new electrochemical institute, which was finished in 1895.[42]

After achieving his own professorship and institute, Nernst began to distance himself from Ostwald, who had become involved with his new scientific philosophy of "energetics." By 1895, most physicists acknowledged that matter was composed of atoms, although they were not convinced that the way chemists talked about atoms had much physical meaning. The energeticists took an extreme position: they not only thought that atoms were unfounded in experiment but also regarded measurements of mass itself to be of only secondary importance. They saw the essence of science to be in the transformation between different types of energy (for example, heat and work). It is surprising that Ostwald, who had built his reputation as an advocate of Arrhenius's, van't Hoff's, and Nernst's rationalization of atomic behavior, should have become the foremost spokesman for an anti-atomistic philosophy, but such was the case. In an 1895 debate in Lübeck, Ostwald and Georg Helm, a mathematician who had formalized energetics, were opposed by Ludwig Boltzmann and Max Planck, who took the atomistic view. Nernst supported the atomists from the floor, opposing Ostwald. The debate did not go well for Ostwald and the energeticists, who were generally viewed to have been trounced. After the debate, Nernst disappeared for a time from the list of authors for Oswald's journal.[43] In 1896, van't Hoff wrote to Ostwald, expressing his concern that Nernst was not showing Ostwald the proper respect:

> When I meet Nernst in person I get a very pleasant impression; but when I hear this and that about his actions, I wish he were more conscious of how much he owes you, or that he would express this consciousness better; any representative of our discipline must feel this in regard to you, but Nernst in particular, since he profited not only form your writings and organizational activities but also from personal acquaintance with you. I must try to impress this upon him in the future.[44]

The full title of Nernst's 1893 text was a slap in Ostwald's face, as well: *Theoretical Chemistry from the Standpoint of Avogadro's Rule and Thermodynamics*—Avogadro's rule was fundamental to the atomic theory. But Nernst's break with Ostwald was not complete, and the two eventually resumed correspondence.

Nernst had his practical side. While he chose problems that were theoretically interesting, he also looked for those that might pay off in prestige or in money. In 1893, he began work on an electric lamp. At the time, gas lighting was still the dominant technology, although Edison had introduced the first commercially successful incandescent lightbulb in the early 1880s. But Edison's bulb was dim, fragile, and short-lived, and there was certainly room for improvement. Nernst's lamp was based on a class of electrolytes that are insulators at normal temperatures but conductors when heated. He found that a glass containing cerium oxide, if it were initially heated externally, gave a bright white light as it began to conduct, and then required no further external heating, because the current passing through it provided sufficient heat to keep it going. He filed for a patent on his lamp in 1897 and invited the large German electric companies, Siemens and Allgemeine Elektrische Gesellschaft (AEG), to have a look. Siemens turned him down, but AEG's chairman, Emil Rathenau, was impressed and paid Nernst a million marks (worth about $4.5 million in 2007 purchasing power) for his patent. Two of Nernst's former pupils ran the development effort for AEG, and the lamp as produced was about twice the size of a modern household lightbulb, with a cerium oxide rod, a preheating element, and an electromagnetic shutoff device that prevented overheating. The lamp burned much brighter than an incandescent bulb, had a much whiter light, had lower operating costs per lumen, and was longer lasting; but it was more expensive to manufacture and required about ninety seconds' warm-up time after switching it on. It was best suited for indoor lighting for offices or stores, where the light would not need to be frequently switched on and off. It enjoyed commercial success for a time both in Europe and in America, where Westinghouse licensed the patent and built a factory in Pittsburgh for its Nernst Lamp Company in 1901. At the 1900 Paris exhibition, the German pavilion was brilliantly illuminated by 800 Nernst lamps. When Nernst visited Edison in America in 1898, he was subjected to one of Edison's standard lectures on how scientists wasted time by doing abstract research rather than focusing on the technological needs of the world, did not understand business, worked in an ivory tower, and so forth. After listening for a few minutes, Nernst shouted down Edison's ear trumpet: "How much did you get for your light-bulb patent?" When Edison said he had not been paid for his patent, Nernst shouted: "I got a million marks for mine! The trouble with you, Edison, is that you are not a good businessman!" But Nernst did note that Edison was obviously a good scientist, as he saw that he kept a copy of Nernst's *Theoretical Chemistry* on his bookshelf.[45]

After selling his patent to AEG, Nernst invested in a Göttingen hotel and restaurant, which became known as the Café N, and in a country estate. He enjoyed hobnobbing with the elite in business, politics, and science and telling stories of his experiences. He spoke of a scientific reception from about 1900 that included Lord Kelvin, the English dean of thermodynamics, then seventy-six years old, and Marie Curie, then thirty-three and considered good-looking. Curie, who had just discovered radium, told Kelvin and Nernst that she had brought a sample with her, but that its glow could not be seen in the lit room. The three of them squeezed into a dark space between double doors and waited for their eyes to adjust. Suddenly there was a knock—it was Lady Kelvin, who showed, according to Nernst, that "she was a most attentive spouse."[46]

Alfred Nobel's Will Puts Arrhenius Center Stage

While Nernst was prospering in Germany, Arrhenius was ending his unhappy marriage in Sweden. But things suddenly looked up for Arrhenius. In 1896, the Swedish industrialist Alfred Nobel died, leaving his enormous estate for the purpose of setting up international prizes in physics, chemistry, medicine or physiology, literature, and peace, directing that the physics and chemistry prizes be awarded by the Royal Swedish Academy of Sciences. The backwater of Swedish science, to which Arrhenius had retreated in 1891, had become important, and Arrhenius saw a way to return to center stage. Nobel's will was poorly drafted and confusing: he had left the bulk of his estate to a foundation that did not yet exist. The next four years, through 1900, were spent litigating the will, negotiating with Nobel's unhappy would-be heirs, setting up the Nobel Foundation, and drafting the rules for nominating and selecting the prizewinners. Arrhenius became embroiled in all of this in different roles, many of which involved him in conflicts of interest: he was a friend of the Nobel estate's attorney, Carl Lindhagen, and was consulted in the resolution of the will;[47] he was himself a candidate for the Nobel Prize in both physics and chemistry; he involved himself in setting up the rules for nomination and selection, advocating a nominating process that included scientists from countries other than Sweden, likely based on his own poor relations with the "Uppsala physicists" and on his success at achieving his professorship at the Högskola with German support;[48] he was a supplicant for funds from the estate, advocating using 25% of the prize money each year for the support of a Nobel Institute in Stockholm that would be linked to the Högskola, of which he had been appointed rector in 1896;[49] and he would be a member of the Nobel Committee for Physics and a de facto member of its Committee for Chemistry for the rest of his life, involved in awarding prizes in chemistry, physics, and even medicine from the first prizes in 1901 until his death in 1927. He used the Nobel Prizes and foundation to build a personal fiefdom. As we shall see, he would use his power to reward his friends and punish his enemies.

About 1900, the friendship between Arrhenius and Nernst ended in bitterness. When the two met in Stockholm in the summer of 1896, Arrhenius said it was "like the good old days," but that he had some difficulty in understanding Nernst's latest work on the ionic theory. When Nernst returned in 1897, he brought one of his new lamps with him to show Arrhenius. When the lamp blew all the fuses in the hotel where they were staying, Arrhenius laughed a little too hard and was not properly appreciative when the lamp finally worked. Nernst wrote Arrhenius an angry letter upon his return and began publicly attacking his work, which he followed by denigrating Arrhenius's Högskola as a scientific backwater, telling Arrhenius to his face that he was an obvious example of just how poor the place really was. In articles in Ostwald's journal, where he had resumed publication, Nernst trumpeted the superiority of his voltage measurements over Arrhenius's conductivity measurements, and Arrhenius replied by saying that Nernst's work on voltage was simply an expansion upon Helmholtz's earlier work.[50] In a 1901 letter to Ostwald, Nernst wrote,

> It is my strong belief, founded on many proofs, that a very mean fraud is being perpetrated in Sweden with regard to the Nobel estate, by which the monies will be used for completely different purposes (institutes, etc.) than those for which they were designated, and the prizes have been withheld for 5–6 years without cause. [Arrhenius] knows my view on the affair very well; but he himself is unfortunately interested in a most fatal way in

the misappropriation of the Nobel millions, because—*ipsissima verba*!—he hopes to become the director of the institutes to be built with the stolen money![51]

As young scientists, Arrhenius and Nernst had been comrades in the wild army of the Ionists; with no turf to defend, they had together taken on the scientific establishment. But as they moved into middle age, their interests were no longer aligned: Arrhenius had put chemical affinity on a semiquantitative basis, but he had then left the field to Nernst, who had surpassed him by putting affinity on a sounder theoretical and physical basis; Nernst's *Theoretical Chemistry* had become the standard text in the field, and Arrhenius had no comparable text of his own; Nernst was now at the center of German science, and Arrhenius was watching from the periphery. Arrhenius saw Alfred Nobel's bequest as his way to maintain his presence and control in physical chemistry, and Nernst, who of course wanted one of the prizes for himself, resented Arrhenius's machinations. Neither had the self-control to refrain from lashing out in public.

Arrhenius and the First Nobel Prizes

The first Nobel Prizes were awarded in 1901, and it was clear that Arrhenius thought that the Ionists (less Nernst, of course, if he had anything to say about it) were the most deserving candidates. Cleve, one of Arrhenius's scorned dissertation advisers, and Otto Pettersson, who had long been a supporter of Arrhenius, proposed a joint chemistry prize for van't Hoff and Arrhenius. But Arrhenius made it known that he did not want to receive the chemistry prize that year and supported van't Hoff. He hoped to receive the prize in physics, which was his current line of work and would have given him great satisfaction in his revenge over Thalén and the "Uppsala physicists," and he might also not have wanted to share the prize money or credit. Van't Hoff alone received the 1901 chemistry prize, and the 1902 chemistry prize went to the German organic chemist Emil Fischer. As a member of the physics committee, Arrhenius had to withdraw from consideration of his own nomination for the physics prize. By 1903, it was becoming clear that the physics committee was not about to award Arrhenius a prize, and the chemistry committee, after some argument, put forward Arrhenius for an undivided prize, which the academy duly awarded.[52]

Receiving the prize brought out the worst in Arrhenius: he wrote his own Nobel biographical notice, where he said that his dissertation had been crowned with the Nobel Prize despite its having "received only one grade above a failing one."[53] (This was misleading. He did not explain his "active state" by the dissociation theory, for which he was awarded the prize, until

1887, four years after his dissertation.) He said that he had declined professorships abroad in order to take a "modest position" at the Stockholm Högskola that was eventually transformed into a professorship despite a "last attempt to deprive him of it."[54] At the Nobel banquet, Arrhenius brushed aside an apology by Cleve and made it clear that he owed nothing to his Uppsala professors.[55]

After van't Hoff's and Arrhenius's prizes, only Ostwald remained uncrowned among the first three Ionists, but there were problems with his candidacy. He was less and less involved with physical chemistry and had retired from his professorship at Leipzig in 1906; he had embraced energetics, denying the atomic theory; and he had, in fact, not done all that much innovative research himself, preferring to take the role of leader and organizer. Nonetheless, Arrhenius managed to get him a Nobel Prize, but did so only after Ostwald announced his conversion to atomism. Ostwald was passed over in the general reports of the academy's chemistry committee in 1904, 1907, and 1908, but in 1909 the committee, based on a report written by Arrhenius that a different committee member submitted under his own name, awarded Ostwald the chemistry prize for his work on catalysis.[56] Scientists of the time were likely surprised by the award and shared the attitude of later historians, who have said that "since Ostwald had no theory of catalysis, he proposed superficial analogies," and that concerning catalysis, "no convincing answers were supplied by Ostwald."[57]

Arrhenius, Ehrlich, and Immunochemistry

While Nernst was extending physical chemistry, Arrhenius in 1900 entered yet another field—the physiological chemistry of the immune system. He became engaged in a feud with Paul Ehrlich, a professor of medicine who had developed a theory of immune reactions. The organic chemists' aniline dyes, which had built the German chemical industry and transformed the textile industry worldwide, had also found use in the biological sciences. They could be used to stain slices of tissue on microscope slides, using different dyes to selectively stain different cell types and parts. At first the selection of dyes was ad hoc, but Ehrlich's dissertation for a medical doctorate in 1878 attempted to rationalize the effect of dyes based on their chemical structure. Many of the synthetic dyes shared a common core structure but differed only in their "side chains"—the chemical groups attached to the core. According to Ehrlich, a dye would show differential biological staining when some specific (but at that time unknown) receptor on a biological cell bound chemically to a specific side chain of the dye.[58] Ehrlich continued this approach, looking for the connection between chemical structure and physiological function of drugs, using the "lock and key" analogy that had been proposed by Emil Fischer—that a chemical conformation of the drug would match with that of a receptor on the cell's surface in such a way that chemical bonds could form between the two. The opening lines of one of Ehrlich's 1891 papers could be used to describe structure-based drug design as practiced today:

> One strives for the time when insight into the *essence* of drugs is attained, and to decide the question in the first instance of the relationship between the constitution of these substances and their *therapeutic* action.... This knowledge must necessarily lead to the desired goal of the synthesis of new drugs.... [I]t will in fact be possible by means of certain combinations to eliminate nearby damaging activity without prejudicing the curative potency.[59]

Ehrlich focused on the immune system, particularly on the toxin–antitoxin relationship. When an animal's body is subjected to certain poisonous chemicals (toxins), the body in some cases produces an antitoxin that counteracts the toxic behavior and that can provide long-term immunity. This had been the basis for vaccinations for rabies, diphtheria, and tetanus, for example: injecting a patient with a weakened strain of toxin causes the body to form antitoxin, and the body then continues to produce antitoxin on its own. In 1897, Ehrlich had proposed his side-chain theory of immunology: the cell possesses many different antibodies (receptors) on its surface, and the toxin molecule possesses a side chain that binds to a specific antibody; the cell, in response to the binding by many toxin molecules to a particular antibody, makes more identical antibodies on its surface; and the cell then casts the excess antibodies into the blood, where they exist as antitoxin.[60] It was known experimentally that the toxic behavior of toxins could be separated from their binding nature: weakened toxin molecules could still bind to antitoxins but would no longer exhibit strong toxic behavior, which was the basis for vaccination. Ehrlich explained this by postulating two different side chains on the toxin molecule, a *toxophore* group that was responsible for toxic behavior, and a *haptophore* group that was responsible for binding to the antibody. But he then took this approach much further than the experimental data warranted. In an attempt to give mathematical form to his data, he proposed a large number of new hypothetical substances, which he called toxoids, toxons, prototoxoids, syntoxoids, epitoxoids, deuterotoxoids, tritotoxoids, and so forth, and then developed equations that explained his experimental results in terms of his new entities. His only evidence for most of what Arrhenius would call a "zoo" of new entities was that the more of them he defined, the better the fit to his equations.

Within the immunology community, Ehrlich took a very chemical approach, as opposed to the more process-oriented biological approaches of his rivals, but Ehrlich's approach was based on *organic* chemistry—on the idea of the specificity of chemical structure—and not on the new physical chemistry. Arrhenius entered the discussion in 1902, when he met Ehrlich in Copenhagen. Ehrlich encouraged Arrhenius to examine immune reactions from the standpoint of physical chemistry and invited him to his institute, an invitation Ehrlich likely later regretted.

Arrhenius and Ehrlich had different scientific backgrounds and styles. Arrhenius saw the toxin–antitoxin interaction as a reversible process, analogous to acid–base neutralization, which should be subject to the mass action law and to the same straightforward mathematical analysis that he had applied to the dissociation theory; Ehrlich believed the toxin–antitoxin interaction involved changes in the cell and was therefore irreversible. Arrhenius stressed the importance of using as few hypotheses as possible and rejected Ehrlich's arbitrary invention of a multitude of different chemical species. Ehrlich had worked mostly with living animals, but Arrhenius preferred to work with test tubes and Petri dishes and limited his research to the tetanus toxin, for which that approach was possible. Ehrlich knew how complex immunological reactions were—he had worked on abrin, ricin, diphtheria, and other toxins in addition to tetanus—and knew that Arrhenius's simple theories could not account for much of the observed behavior.

At Ehrlich's invitation, Arrhenius visited his Frankfurt institute several times in 1903–4 and found it nothing like his Högskola in Stockholm. Ehrlich was Jewish, as were most of the scientists on his staff, and Arrhenius found the lunchtime scientific conversations mixed with Talmudic disputes. Ehrlich was odd personally: his office was piled to the ceiling with books and journals, he had a brusque manner, and he was both reclusive and autocratic in managing his subordinates, placing written instructions on their desks every morning before they arrived and often forcing those who disagreed with him to leave the institute immediately. Arrhenius had nothing but contempt for a scientist who would put up with such lack of scientific autonomy, calling Ehrlich's subordinates "creatures" and "socially and scientifically rude persons."[61]

The scientific debate between the two is too complex for this book; interested readers should examine Arthur Silverstein's *Paul Ehrlich's Receptor Immunology: The Magnificent Obsession*[62] and Franz Luttenberger's paper.[63] But given Arrhenius's and Ehrlich's personalities, the debate was often bitter.

Once Arrhenius had gained his Nobel Prize in 1903, he went on a speaking tour of Europe and the United States to expound his immunological theories, and Nernst, who could never resist a good fight, took the opportunity to jump in on Ehrlich's side. At the first of Arrhenius's lectures in Bonn in 1904, which was a public debate with Ehrlich, Nernst stood up to attack Arrhenius's physical chemistry, saying that his data on tetanus were not supported by the law of mass action. Arrhenius tried to modify his argument on the fly, saying that the equations fit his data if the reaction of antitoxin with toxin produced an additional by-product. Nernst then pointed out that Arrhenius was doing what he had faulted Ehrlich for doing—making up new chemical entities to better fit an equation. Arrhenius continued his lecture tour, ending up at the University of California in Berkeley, where he collected his lectures in a book titled *Immunochemistry*, a word he coined that has stuck. But Arrhenius's theories had little practical effect on contemporary immunology. Arrhenius was attempting to apply quantitative methods to substances whose structure and even molecular weight were unknown—his only quantitative technique was titration of an unknown toxin against an unknown antitoxin—and he was speaking the language of physical chemistry, which few of immunology's practitioners understood. On the other hand, whatever the defects of Ehrlich's theories, he was able to make substantial practical advances, including the identification and purification of toxins and antitoxins, development of methods for keeping vaccines stable and assessing their potency, and the synthesis of Salvarsan, the first effective treatment for syphilis, in 1909.

A fundamental dictum among historians of science is that one should not judge historical science in terms of supposed "truth"—current scientific explanations—but in terms of the attitudes and knowledge of the time. Having said that, let's do it anyway: Although Ehrlich's theory of the side chain went out of fashion for forty years, it was eventually shown to be correct in its essence, as was his claim that there were two separate chemical groups involved, the toxophore for toxicity and the haptophore for binding. However, most of the Greek-named substances he invented to fit his equations have been discarded. Arrhenius was correct in his belief that toxin–antitoxin binding was reversible, but quantitative application of the law of mass action would need to wait for more detailed knowledge of the chemical entities involved.[64]

Arrhenius, Gilbert Lewis, Dorothy Wrinch, Irving Langmuir, and Linus Pauling all behaved in a similar manner when they entered an established discipline, assuming that they knew more than the discipline's current practitioners. The results were usually as poor as one would predict, and Arrhenius was more successful than most.

Arrhenius Blocks Ehrlich's Nobel Prize

Ehrlich began receiving nominations for the Nobel Prize in Medicine from its beginning in 1901, and was actively considered for the prize in 1902, 1904, and 1906. The Karolinska Institute, which awards the prize in medicine, had set forth a number of rules, and one was that the work in question must have been brought to a point where the problem had been solved entirely, so that the result constituted "a finished whole."[65] Because Ehrlich's theory was still being contested by Arrhenius, the committee had reason to override Ehrlich's staunch advocate on the committee, Karl Mörner, and to defer an award to Ehrlich.

Arrhenius, as a member of the physics committee, had little control over the prize for medicine, but in 1907, Jöns Johansson was elected as an ad hoc member of the Nobel Committee for Physiology or Medicine. Johansson was a professor of physiology at the Karolinska Institute, was a close friend of Arrhenius, and in fact had become Arrhenius's brother-in-law when Arrhenius remarried in 1905 to Johansson's sister. In the committee's 1907 deliberations, Mörner submitted a report in support of Ehrlich. Mörner had recalculated Arrhenius's data, pointing to Arrhenius's errors and to Nernst's objections, and showed that the data were inconclusive in supporting either Ehrlich's or Arrhenius's arguments. Johansson, who was not familiar with chemical kinetics, submitted a report that repeated the same points that Arrhenius had made elsewhere. Arrhenius's biographer, Elizabeth Crawford, concludes that this report could only have been written by Arrhenius, despite Johansson's signature.[66] Based on Johansson's report, Ehrlich was passed over for the prize again.

By 1908, Johansson was no longer a committee member. Despite his lack of standing, Johansson continued to oppose the award to Ehrlich, circulating memoranda to other faculty members within Karolinska, but to no avail—Ehrlich finally received the medicine prize, which he shared with Elie Metchnikoff, another immunologist. Ehrlich got in the last word against Arrhenius. From what Arrhenius had called the "zoo" of neutralization products that Ehrlich had invented, Arrhenius had used only one—toxoids—in his own work. In Ehrlich's Nobel address, he singled out toxoids, "which my honored friend Professor Arrhenius has also encountered in his numerous experiments."[67]

Arrhenius managed to delay Ehrlich's prize in medicine for seven years, but in the end he was unable to block the award. Medicine was an area in which he had little personal authority or reputation, and even the seven-year delay had been managed only with the complicity of his brother-in-law. But in chemistry and physics, Arrhenius had much more power, and he now had even more reason to resent Nernst, who had jumped in on Ehrlich's side and embarrassed Arrhenius by questioning his physical chemistry. In 1906, Nernst would make the most important discovery of his life, the third law of thermodynamics (see chapters 3 and 4). And Arrhenius would spend fifteen years as the judge of that work's suitability for a Nobel Prize.

2

Physical Chemistry in America

Lewis and Langmuir

At the July 1908 meeting of the American Chemical Society in New Haven, Connecticut, two American physical chemists met for the first time. Both were in early stages of their careers: Gilbert Lewis was thirty-three, and Irving Langmuir was twenty-seven. Lewis was already established at MIT's Research Laboratory for Physical Chemistry. His boss, Arthur Noyes, was serving as MIT's acting president, and Lewis had been named temporary director of the laboratory. Although Lewis was now doing well at MIT, he had had a difficult start as an instructor at Harvard, a position he had disliked so much that he had abruptly left for a government job in the Philippines. In 1908, Langmuir was still stuck in a low-level teaching position at Stevens Institute of Technology, a Hoboken, New Jersey, engineering school. Both Langmuir and Lewis were veterans of Walther Nernst's laboratory in Germany. Neither had liked Nernst much, and they almost certainly traded stories about his behavior. The two seemed to hit it off. Langmuir kept file cards on almost everything at this time,* and his card on Lewis reads:

> Lewis, G.N.
> Prof. Phys. Chem., Mass. Inst. Tech.
> Met at New Haven meeting ACS July 1908. I liked him + he liked me [as] well. He has A.A. Noyes place temporarily. Has studied at Leipzig + Göttingen + Harvard + was instructor at Harvard. Met him again at Baltimore Dec. 1908. Again at Boston Dec. '09.[1]

* I selected a few of the entries filed under "T" in Langmuir's card file and found cards on Langmuir's progress in improving his typing speed, the physical properties of the element tellurium, when and where he had heard performances of works by Tchaikovsky, and his tests of the performance of a thermos bottle he had purchased (Langmuir card file, Irving Langmuir Papers, Library of Congress, Washington, D.C.).

Only a few years later, Lewis and Langmuir would have reputations as the leading American physical chemists. They provide striking parallels and oppositions: To steal a line from Woody Allen, "Lewis and nature were as two," while Langmuir enjoyed climbing, hiking, skiing, and skating. Lewis was reserved, while Langmuir was outgoing and sociable. Lewis wrote beautiful prose, while Langmuir was a captivating speaker. As their careers progressed, they would both be theoretical and experimental innovators, but Lewis would have little interest in the practical applications of his work, while Langmuir would follow Edison's example and develop household and industrial products at General Electric. Lewis would be distant even from his own family, while Langmuir would have a warm family life and an interest in children's science education.

Physical Chemistry Comes to the United States: Noyes, Whitney, and Richards

In the late 19th century, becoming a serious student of chemistry meant studying in Germany. This was especially true for the new field of physical chemistry, where study with Wilhelm Ostwald or with Nernst was practically the only way in. When Lewis taught a physical chemistry course at Harvard in 1902, students were required to be proficient in German[2]—all the important chemical literature was in German, and a chemist without German would have been like a rabbi without Hebrew. Before Lewis and Langmuir studied with Nernst, an earlier group of Americans had studied in Germany and had returned to the United States to preach the gospel according to the Ionists. In all, forty-four Americans spent at least some time in Ostwald's laboratory between 1889 and 1904.[3] We shall examine three of them: Arthur Noyes, Willis Whitney, and Theodore Richards.

Arthur Amos Noyes received his bachelor's degree in chemistry in 1886 from MIT, then known as Boston Tech, and immediately joined its faculty as an instructor. MIT was not nearly as prestigious as it is today—it had only a small endowment and was in the shadow of its Cambridge neighbor, Harvard, and of the other Ivy League schools; MIT focused on training chemists and engineers for industry rather than on faculty research. Noyes took a sabbatical from MIT to study in Germany in 1888, received his Ph.D. under Ostwald at Leipzig in 1890, and then returned to the MIT faculty. Upon his return, he soldiered on with his research as best he could despite MIT's lack of support.

Willis Whitney was only a few years behind Noyes, receiving his bachelor's degree from MIT in 1890, joining the MIT faculty as an instructor, going off to earn his Ph.D. under Ostwald in 1896, and then returning to the MIT faculty. Like Noyes, Whitney wanted to do research. In 1898, Noyes and

Whitney jointly developed an industrial process for the recovery of alcohol and ether vapors in the production of photographic paper. The motivation was not environmental—no one was much concerned with industrial pollution in 1898—but economic, based on the savings from the recovery and reuse of the solvents. Noyes and Whitney licensed their process to the American Aristotype Company for a quarter of the company's savings from its use. This amounted to more than $1,000 per month for each of them, more than five times their salaries as professors. When American Aristotype later merged with Eastman Kodak, they exchanged their interest in the patent for stock in Eastman Kodak.

Noyes then had more than enough money, and he used it to give himself what he wanted—complete freedom to do basic research in physical chemistry. He entered into a partnership with MIT, promising to pay part of the costs for a new Research Laboratory for Physical Chemistry, of which he would be the head. Noyes's idea was that the laboratory would be dedicated exclusively to basic research, that its faculty would have no undergraduate teaching duties, and that the laboratory would eventually be staffed by eight to ten researchers. The MIT administration resisted at first, but Noyes managed to get additional financial support from the Carnegie Institution, which finally convinced MIT to go ahead with the project. In 1903, Noyes's new laboratory was funded and built, and Whitney joined as a senior staff member.

But Whitney had his own plans. General Electric had been formed in 1892 from the merger of Thomas Edison's Edison General Electric Company and the Thomson-Houston Company; its new headquarters were in Schenectady, New York. Its technical director, E.W. Rice, wanted to establish a basic-research laboratory. At a time when industrial laboratories were focused on straightforward product improvements and on process monitoring, this was a radical idea. Rice attempted to recruit Whitney as GE's research director in 1900, but Whitney demurred. He was happy working with Noyes at MIT, had plenty of money from his share of the sale of their patent, loved teaching, and doubted that he would find intellectual stimulation in industrial research. Rice talked him into accepting a half-time appointment, and in 1904 Whitney decided that he liked GE enough to work there full time. GE established the first basic-research industrial laboratory in America, and its success would make it a model for other companies worldwide.

Theodore Richards received his Ph.D. in chemistry from Harvard in 1888, joined the Harvard faculty, and made his name by determining accurate atomic weights. This was still no simple matter, even ninety years after Dalton had proposed the atomic theory: the atomic weight of oxygen was by definition 16, and the weight of all other elements depended, directly or indirectly, on the measurements of the weights of oxides. But any measurement could be quite indirect, involving a chain of intermediate compounds, and it was essential that all compounds in the chain be pure and that moisture be completely excluded. Richards developed accurate methods of weighing, assaying, purifying, and drying chemical samples, and used multiple pathways for measuring each element's weight, always searching his own work and that of others for errors or discrepancies. His latest and most accurate table of atomic weights was published in each issue of the *Proceedings of the American Academy of Arts and Sciences*, of which he was a fellow. In 1894, Harvard sent him off to Europe to learn the new physical chemistry from Ostwald and Nernst. He returned to Harvard as an assistant professor, where he applied the same painstaking care to thermochemical and electrochemical measurements that he had applied to atomic weights, designing calorimeters and electrodes that were precision instruments. His work must have impressed the Germans, because they offered him the position of professor of physical chemistry at Göttingen in 1901, a singular honor for an American. Harvard reacted quickly in order to keep him, promoting him to full professor, promising him $750,000 for a new research laboratory, and reducing his teaching load. He stayed at Harvard.

Noyes, Whitney, and Richards were of similar background. They had all been born within the two-year period 1866–68 and were from middle-class New England Yankee families. By the first years of the 20th century, Americans who had studied with Ostwald, including these three, had built a number of centers for physical chemical research in America and had begun training and hiring American research chemists,[4] among them Lewis and Langmuir, the subjects of this chapter.

Gilbert Lewis

Gilbert Lewis was born to Frank and Mary Lewis in Weymouth, Massachusetts, in 1875—another Yankee physical chemist. The Lewis family moved to Nebraska when Lewis was nine years old. There was a land boom in Nebraska at the time, and Gilbert's father Frank, a lawyer and later a banker, apparently decided to try his luck there by joining a business called "Western Investments." He had advanced ideas: he later wrote a book, *State Insurance*, which proposed that the government be responsible for providing a basic social safety net. Gilbert was the middle child of three, including an older sister, Polly, and a much younger brother, Roger. Lewis's mother, Mary, saw herself as responsible for the children's education even before the move to Nebraska. Lewis was home-schooled entirely through elementary school and attended high school only briefly. In addition to his mother, his aunt Miriam, who later taught school in Salt Lake City, may have tutored him. He was reading by the age of three and learned not only the basics but also Latin, Greek, French, German, history, and algebra. Gilbert's sister, Polly, who shared the same education, later founded a private school in Cleveland. Her educational philosophy was apparently controversial; the school closed in a "possibly scandalous situation," according to Lewis's son and biographer.[5]

Throughout his life, Lewis would show the effects of his home schooling: He was a skilled writer (see the epigraph at the beginning of this book) and a punster with a quick, dry sense of humor. He had a wide and eclectic knowledge and, like Svante Arrhenius, felt confident that he could understand and contribute to any field—he would eventually publish not only in chemistry but also in theoretical physics, anthropology, geology, biology, and economics. But he had missed the socialization of the playground, and it seems to have affected him. Like Arrhenius, he would be quick to take offense at any slight, real or imagined. But where Arrhenius's response to slights was usually a vengeful head-on attack, Lewis's response was to leave the scene—to take his ball and go home—although he would throw asides into his scientific publications to skewer the offending party. He could be gregarious and social with those he trusted, but he would always be on guard in the wider world.

Lewis began his undergraduate education at the University of Nebraska but transferred to Harvard when his family returned to the Boston area in 1892, and he graduated from Harvard with a bachelor's degree in chemistry in 1895. He had little interest in descriptive chemistry (he received a D in advanced organic chemistry in his senior year),[6] but he excelled at mathematics and physics, and at both experimental and theoretical chemistry. After graduation, he taught for a year at Phillips Andover Academy and then returned to Harvard in 1896 as a Ph.D. candidate under Richards, who was only seven years his senior.

Richards was not always an easy mentor. He found it difficult to give his junior collaborators the freedom to learn from their mistakes: "[A]ssistants who are not carefully superintended may be worse than none, for one has to discover in their work not only the laws of nature, but also the assistant's insidious if well meant mistakes. The less brilliant ones often fail to understand the force of one's suggestions, and the more brilliant ones often strike out on blind paths of their own if not carefully watched."[7] Lewis would later be known for giving graduate students almost complete freedom in their research during his time at the University of California at Berkeley, perhaps to distance himself from Richards's methods.

Richards also differed with Lewis over matters of science. A primary and consequential difference concerned mathematics:

> Throughout his career, Richards was of two minds regarding the value of mathematics in the physical sciences. On the one hand, he recognized that it was sometimes a necessary tool. On the other hand, he never regarded it of sufficient value to merit an important place in the training of chemists, and in his private work often expressed a suspicion of the work of those scientists who made significant use of it.... How much mathematics did

Richards know? Not very much, at least by the standards of 20th-century physics. His papers on chemical thermodynamics and the compressible atom depict a chemist struggling to handle the elementary calculus. Nor do they reveal a truly expert grasp of physical concepts.[8]

Robert Oppenheimer took Richards's physical chemistry course at Harvard in 1925 and said that "Richards was afraid of even rudimentary mathematics."[9] Richards excelled at the experimental end of physical chemistry, but he did not have the mathematical background to contribute to theory. Unlike Richards, Lewis had an excellent grasp of mathematics, physics, and especially thermodynamics. He had read J. Willard Gibbs, the American pioneer of theoretical thermodynamics, and was familiar with the work that the European thermodynamicists were doing. Within a short time after beginning his graduate work in thermodynamics under the direction of Richards, Lewis understood far more about the subject than did Richards. As a graduate student, he wrote two unpublished works on thermodynamics: "Equilibrium, and the Equilibrium of Supersaturation and Superheating" and "A Theory of Gravitation."[10] Where Newton had been unwilling to decide on a mechanistic explanation for the inverse-square gravitational law, the young Lewis was fearless in this second manuscript. His theory of gravitation was essentially a rediscovery of Georges-Louis Le Sage's "push" theories of gravitation from the 18th century[11] (of which he was unaware at the time), but was based on thermodynamic reasoning. It appears that Richards may have viewed Lewis's gravitational theory as "a blind path" by one of his "more brilliant ones." Lewis later wrote:

> I went from the Middle-west to study at Harvard, believing that at that time it represented the highest scientific ideals. But now I very much doubt whether either the physics or chemistry department at that time furnished real incentives for research. In 1897 I wrote a paper on the thermodynamics of the hohlraum [the paper on gravitation mentioned above] which was read by several members of the chemistry and physics departments. They agreed unanimously that the work was not worth doing, especially as I postulated a pressure of light, of which they all denied the existence. They advised me strongly not to spend time on such fruitless investigations.[12]

Richards may have also squelched Lewis on his theories of chemical bonding, in which Lewis would eventually make what many now view as his greatest contribution. The electron was discovered by J.J. Thomson in 1897, and there was immediate speculation that it provided the key to understanding the chemical bond—Jöns Berzelius had postulated his dualistic electrical theories of bonding almost a century earlier, and Thomson now identified the negatively charged electron as the fundamental subatomic unit of electricity. Michael Kasha, Lewis's last graduate student in the 1940s, reports reading a handwritten paper by Lewis titled "The Electron and the Molecule" dating from Lewis's graduate student days; unfortunately, the paper has not been preserved.[13] As we shall see from his later comments, Richards likely regarded Lewis's speculations about the electron as yet another "blind path."

Despite their differences, Lewis and Richards came to an agreement regarding the subject of Lewis's doctoral research. Lewis's 1899 dissertation consists of two parts: a careful experimental study of the properties of zinc and cadmium amalgams as electrodes, and an entirely theoretical study of an equation he had derived for the determination of free energies.[14] The first part of Lewis's dissertation was a straightforward extension of the sort of work Nernst had been doing connecting voltage and free energy. By developing better electrodes, Richards and Lewis made possible more accurate measurement of the voltage of reactions involving zinc and cadmium, and voltage could be used to measure free energy, as Nernst had shown. When this work was published in the scientific literature, the authors were Richards and Lewis, in that order.[15]

Figure 2-1. Lewis as a young man. Courtesy of the College of Chemistry, University of California, Berkeley.

The second part of Lewis's dissertation was published under Lewis's name alone.[16] Lewis was attempting to improve upon what Nernst had done with his "Nernst equation"—the equation that allowed free energy to be determined from voltage. Lewis was trying to find a way to determine free energy from heat measurements, which were easier to make and more universally applicable than voltage measurements.

Lewis's Dissertation and Early Career

The 19th-century physicist Lord Rayleigh once observed, "Perhaps... a young author who believes himself capable of great things would usually do well to secure the favorable recognition of the scientific world by work whose scope is limited, and whose value is easily judged, before embarking on greater flights."[17] If Lewis ever read this advice, he certainly paid no attention. The second part of Lewis's Ph.D. dissertation and maiden solo flight into publication was titled "The Development and Application of a General Equation for Free Energy and Physico-chemical Equilibrium." Like Arrhenius, Lewis wanted a dissertation that would change chemical theory. Lewis's dissertation claimed to "express the relations deduced from [the first and second laws of thermodynamics] in a single equation, the most convenient and general possible, which may serve to systematize a part of our present knowledge."

The dissertation and its subsequent publication were not well received by Wilder Bancroft, professor of chemistry at Cornell and editor of the *Journal of Physical Chemistry*, who wrote, "The author declares a single equation which should enable one to predict anything; but which does not lead the author to anything new.... What we need in physical chemistry is a closer adherence to facts and less approximation theory."[18] Lewis did not forget Bancroft's comments: during Lewis's long career, he was never to publish in Bancroft's journal.

Lewis hoped that his equation would lead to a new method for determining free energy—which he understood to be the true measure of chemical affinity, even if Richards did not.[19] And a new method was certainly needed, because free energies were difficult to measure. At the end of the 19th century, there were only two known methods for determining the free energy of a reaction: calculating from the equilibrium constant, and using the Nernst equation.

Because free energy determined the point of chemical equilibrium, one could work backward from equilibrium to calculate free energy: let the reaction reach equilibrium, measure the concentration of all components, repeat this last step at increasing dilutions, extrapolate all the concentrations to infinite dilution to obtain the equilibrium constant, and then use a simple known equation to calculate the free energy from the equilibrium constant. This method was arduous and fraught with difficulty. It was necessary to work at very low dilutions, to ensure that the reaction had actually reached equilibrium, and to ensure that the equilibrium was not disturbed by the act of extracting and measuring the concentrations of the components.

The second way to determine free energies was from the Nernst equation, which equated free energy with the possible external work that could be done by making the reaction the energy source of a battery. Unfortunately, this method for measuring free energies also had limited applicability. One had to be certain that the cell's electrodes were not contaminated and that only the desired reaction was occurring. Lewis would later say: "Of all the chemical reactions which we meet in our thermodynamic calculations, comparatively few may be studied by the simple measurement of [voltage]. This is due to the difficulty of finding a galvanic cell in which a given reaction occurs, and occurs with such ease as to permit an approach to complete reversibility."[20]

So while at the end of the 19th century free energies were known to be the key to determining chemical affinity, they were elusive experimentally. But measuring heats of reaction was easy—just let the reaction take place in a calorimeter (a sealed vessel immersed in a water bath) and measure how much the temperature of the water increased or decreased. If only a way could be found to determine free energies directly from heat measurements! Lewis's new equation suggested that it could be done. To determine the free energy of reaction, one could make a series of measurements of each reactant's and product's *specific heat*—how much heat was absorbed in changing the temperature by one degree—all the way from the reaction temperature down to absolute zero.

Lewis's method was theoretically correct but was not a practical suggestion. Low-temperature chemistry was limited in 1900. It was relatively easy to get down to the temperature where liquid nitrogen boils, but no one knew much about how specific heats behaved below this point. But measuring low-temperature specific heats was possible at least in principle. There was another problem with Lewis's method that was more fundamental: his new equation included an "integration constant," the value of which he saw no way to determine. Knowing the free energy curve as a function of temperature allowed calculation of the curve for heat of reaction, but that did not work in reverse—knowing the curve for heat of reaction did not allow calculation of the free energy curve, because the unwanted and unknown integration constant showed up in the resulting equation. Although Lewis had not known it at the time of his dissertation, Henri le Chatelier, a French mining engineer, had first identified the integration constant as the crux of the problem of determining free energies from heat measurements in 1888: "It is very probable that the constant of integration is a determinate function of certain physical properties of the substance in question. The determination of the nature of this function would lead to a complete knowledge of the laws of equilibrium. It would permit us to determine *a priori*, independently of any new experimental data, the full conditions of equilibrium corresponding to a chemical reaction."[21] We shall see in chapter 3 how Nernst managed to determine the integration constant in 1905, coming up with what would eventually become known as the third law of thermodynamics. But when Lewis stated his new (but then unsolvable) equation in 1899, he was unable to apply it practically and could use it only to analyze the approximations that were implicit in the ad hoc thermodynamic equations in use at the time.[22]

Despite their differences in style and approach, Richards and Lewis were on good terms at the time that Lewis completed his doctoral work. Four cordial letters from Lewis to Richards have been preserved, indicating that Richards was helping Lewis by reviewing the proofs of his first solo publication on the free energy equation.[23] Upon completing his dissertation, Lewis spent a year as an instructor at Harvard, where he continued his work on the theory of thermodynamics. He submitted the first of his papers on the concept of what he would later call "fugacity," but that he called "escaping tendency" in this first paper.[24] He thought that the Ionists had been sloppy in the way that they had applied thermodynamics to chemistry, and fugacity was a key concept in his effort to put chemical thermodynamics on a firmer mathematical footing. He defined fugacity as the tendency for a substance to move from one phase into another, for instance, for water to move from liquid water into steam or vice versa. He believed that fugacity was a concept as fundamental as temperature—just as a difference in temperature expressed the tendency for heat to flow, so a difference in fugacity would express the tendency for mass to flow from one phase to another.

In late 1900, Lewis followed in Richards's footsteps, going off to Europe to study with Ostwald in Leipzig and Nernst in Göttingen. He planned to return to Harvard as an instructor in chemistry in 1902. When he entered Ostwald's laboratory in Leipzig, he was welcomed and encouraged. In the same letter to Richards that accompanied his second fugacity paper, Lewis described an account of his presentation of his concept of fugacity to an audience in Leipzig that included Ostwald and Ludwig Boltzmann, the physicist who had led the way in using statistical arguments to explain thermodynamics:

> In addition to my regular work I have been kept busy by being called in to [*sic*] demand as a German lecturer. Notwithstanding the common impression that Germans are devoid of a sense of humor I was asked to speak before the physico-chem. Seminar [*sic*]. In a moment of temporary aberration I consented and talked for three quarters of an hour on a theory of surface tension which has grown out of my thermal pressure theory [fugacity]. The discussion ran over to the next meeting and occupied the whole hour. Ostwald and particularly Boltzmann taking the most prominent parts. It was bad enough to give a prepared address before half a hundred Germans but it was kindergarten play compared with carrying on an extemporaneous discussion.

It was not a grand success, one German told me that he understood my German perfectly but not my physical chemistry and I thought a good many went away with a very hazy idea of what we had been talking about. Later I reviewed my thermal pressure theory [fugacity] before the physical colloquium and again Boltzmann insisted on continuing the discussion to two hours, much to the disgust of others who were billed to speak. It required all my nerve to face him with the proposition that the kinetic energy of a molecule is not a measure of its temperature[25] and other such heretical ideas. His final remark was that although he had no more objections at the moment that he could raise, he was not convinced. Some of the other men, Wirdeburg among others, supported me on different parts of my theory although I don't know that any of the professors swallowed it whole.[26]

Richards Claims Credit for Lewis's Ideas

While in Leipzig, Lewis continued to correspond with Richards and in January 1901 sent him a draft of his second paper on fugacity. Richards kept Lewis's letters from this period and also kept a heavily edited pencil draft of one of his own letters to Lewis. This letter from Richards to Lewis, following shortly after Lewis sent Richards the draft of his second paper on fugacity, is what sparked a rift between the two. Richards's draft is rubber-stamped "MAR 16 1901":

> Your paper . . . gives mathematical form . . . to the feeling and intuition which I have had for two years or more. I am very glad that you have come to agree with me about the importance of *pressure*. . . . That you did not thoroughly take my point of view was evident, because I remember you told me that you did not see why I published the paper on the "driving tendency." . . . That paper on "driving tendency" (of which I sent you a copy) was my first vague attempt to express the thought which you have expressed so much more adequately in this new paper. I remember that in my lectures I even spoke of an "outward tendency" as the pressure between phase and phase. The really great gain which it seems to me you have made is the making of this point of view applicable rigorously for conditions which are not perfect—a state of affairs which I only hinted at.
>
> The only lack which I have discovered thus far in your paper is a possible lack of care in tracing the origin of your ideas. . . . Especially the important influence of van't Hoff is to be perceived in your work, and to be perfectly frank, I think that something is due to the pressure point of view also. I have been hesitating about calling your attention to this, for obvious reasons; but I feel that it is only fair to you to point it out.[27]

By "the pressure point of view," Richards meant his own work. Accompanying the draft of the letter to Lewis is an additional note written in

Richards's hand, apparently the acknowledgment that he was requesting that Lewis insert into his paper:

> Recognition of the importance of the conception of "escaping tendency" is partly due to Richards, who named it "outward tendency" in his lectures during the year 1899–1900. He applied it to the equation of van't Hoff, and in a paper called the "Driving tendency of physico-chemical reaction" points out that it is probably the determining factor in both physical and chemical change. His definition of it was however vague and amounted to little more than an intuition of the parallel significance of pressures of all kinds; the conception was applied by him definitely only in the case of ideal conditions. The present paper gives [construction?] and applicability to the idea.[28]

Richards's letter put Lewis in a difficult position. Lewis took the letter as a claim by his dissertation director—the chairman of the Harvard chemistry department where he planned to teach and the person upon whom he was relying for his introduction to the world of physical chemistry—to share credit for his concept of fugacity. Furthermore, the acknowledgment that Richards had requested Lewis to add as a footnote was untenable on two counts: first, Richards's concept of "outward tendency" had nothing to do with fugacity; and second, Lewis could not call Richards's ideas "vague" without insulting him, even though the ideas were in fact vague and Richards had requested the use of the term.

The paper on "driving tendency" to which Richards referred was one he had recently published.[29] This paper is a mathematical muddle whose publication Lewis had been right to question, although Richards did thank Lewis in a footnote for reviewing the mathematics. In his letter to Lewis claiming a share of the credit for the idea of fugacity, Richards was confusing what he had called in his paper the "driving tendency" or "reacting tendency" or "outward tendency" with Lewis's concept that he called "fugacity" or "escaping tendency." It is possible that Lewis changed the name of his own concept from "escaping tendency" to "fugacity" between the publication of his first and second fugacity papers to avoid confusion with Richards's paper, which was published between Lewis's two papers.

It is unlikely that Richards was intentionally misappropriating Lewis's ideas. Richards was not an intellectual thief, and his claims were likely due to his lack of understanding of Lewis's ideas and of thermodynamics in general. But Lewis was quick to write a response to Richards disputing Richards's claim for shared credit for the concept of fugacity. In a letter from Lewis to Richards from Rome dated 3 April 1901, he wrote:

> I thank you for your frankness which is but another name for friendly consideration and I will answer, if I may, with equal candor. I should be sorry

to have you think that I lack a full appreciation of my immense indebtedness to you in this matter. I can express but inadequately how strongly I feel that what I know of chemistry as well as what I have accomplished and may accomplish, I owe in the largest measure to your kindly assistance and to those conversations with you which have determined the course of my work. In the line of thought of which the present paper is an item these conversations had an influence even greater than normal upon my ideas since we had both thought a great deal upon the subject independently.

It is always difficult to trace the evolution of an idea. One is often at a loss to remember the source of the most trivial thought of the moment and it is quite impossible to distinguish the mesh of ideas which for several years converging, intertwining, diverging finally produce a theory which is woven strongly enough to hold.[30]

Lewis then proceeded to give detailed accounts of conversations with Richards and with others, including citations from his laboratory notebooks to make the case that Richards was not the source of his ideas. He concluded:

I have written much more than I intended and can hardly believe that it will interest you but I have taken your letter as an excuse for recalling some facts which I would have soon forgotten, and if you will keep this letter I may care to see it at some future time. I think I have written enough to show that I can hardly be charged with ever ignoring the "pressure point of view" whether you mean osmotic pressure or external pressure, and moreover that in the development of any theory of thermal pressure and escaping tendency I owe far more to the helpful and encouraging communications which I had with you than to your papers or your lectures. Of course you have forgotten most of these conversations as you were much occupied at the time but I think they considerably influenced the ideas of both of us.

I should like to have you insert the following as a footnote:... "In the further extension of this theory an analogy will be seen between the conception of fugacity and the driving tendency of chemical reaction as used by T.W. Richards.... It is a pleasure to recall how much I owe to the many conversations full of assistance and encouragement which I had with Professor Richards during the early development of the theory of fugacity (escaping tendency)."

Or if you care to suggest any changes in or substitute for this note I should be very glad to put it in when I read the proof, as I think I had better.[31]

While Lewis expressed this with some courtesy, the letter suggests that Lewis was putting Richards on notice that he would not accept Richards's arguments that he deserved any credit for the concept of fugacity; that Lewis had clear documentation that this was the case; that he intended to limit his acknowledgment to thanking Richards for general conversations; that he had influenced Richards's ideas, as well; and that he intended to carefully read the proof of the paper, which was to be submitted to the *Proceedings of the American Academy of Arts and Sciences*. Richards published regularly in that journal and was a fellow and officer of the academy, and Lewis likely wanted to ensure that the acknowledgment was printed exactly as he had written it. While Richards must have been disappointed with Lewis's reply, he apparently accepted this, since the footnote that Lewis suggested was published verbatim in the second of Lewis's fugacity papers.[32] I have found no later correspondence between Richards and Lewis from Lewis's time in Germany among Richards's or Lewis's papers. It is likely that the two stopped corresponding after this unpleasant exchange.

As Arrhenius had felt oppressed by what he saw as unfair treatment by his advisers Per Theodor Cleve and Tobias Thalén, Lewis felt put upon by Richards; as Nernst had been exasperated by Ostwald's anti-atomistic philosophy of energetics, Lewis was frustrated by Richards's failure to understand and embrace his ideas on fugacity; and as Nernst broke with Ostwald, Lewis broke with Richards. But Lewis was in a much more difficult position than Nernst had been: when Nernst broke with Ostwald in 1895, he was firmly established with a full professorship and an institute, but Lewis was only an instructor at Harvard and would soon be returning there to work under Richards. Furthermore, Nernst regarded Ostwald's energetics as misguided but not directed at him personally, which was certainly not the case with Lewis's beliefs about Richards's claims on fugacity. Nernst broke with Ostwald openly, opposing him from the floor in the debate on energetics, but Lewis could not break openly with Richards, because he needed to return to Harvard.

Although Richards did not press his claims on fugacity, Lewis never trusted him again. This would become a pattern for Lewis, who discovered (and made) enemies often, and never forgave a slight. By the end of his career, he would manage to estrange himself from many scientists, including, as described in following chapters, Richards, Nernst, Langmuir, and Harold Urey.

Lewis and Nernst

After finishing in Ostwald's lab during the time of the above exchange of letters with Richards, Lewis was ready to move on to Nernst's laboratory in Göttingen. In a letter dated 30 March 1901, Ostwald wrote to Richards, "I have great esteem for Dr. Lewis for his insights and broad thinking that will lead to great accomplishments."[33] While Ostwald was a welcoming man who thought it was his duty to encourage younger scientists, this was not Nernst's reputation. As we will see shortly, Langmuir recorded Nernst's rudeness in some detail, and Lewis, who was much more sensitive to slights than was Langmuir, must have been miserable in Göttingen. No correspondence has been preserved from Lewis's time with Nernst, and Lewis published nothing from the period. But something must have happened there: Lewis and Nernst developed an obvious enmity that would last the rest of their lives and that may have cost Lewis a Nobel Prize (see chapter 7). The groundwork for Nernst's animosity might have been laid even before they met. Lewis had left Harvard for Germany in 1901, intending to work in Ostwald's and in Nernst's laboratories for half a year each, just as Richards had done in 1895. While Lewis was working in Ostwald's laboratory, he wrote Richards saying that he had changed his mind and would spend the second part of the year with Jacobus van't Hoff rather than Nernst:

> But I have concluded to spend the next semester with van't Hoff. It was not very easy to decide between Berlin and Göttingen but van't Hoff was here at our Christmas feast and I broached the idea to him. He has only a small private laboratory and takes but a half dozen students. He said that he might have a place vacant and if so would reserve it for me in case I cared to work along the lines which are receiving attention there, as his laboratory is only equipped for this work. He asked me to write to him and the other day I received a reply stating that he would keep a place open for me.
>
> I liked van't Hoff immensely and the thought that I should be able [to] see a great deal of him in so small a laboratory, while Nernst [*sic*] students say that they receive almost no attention from him decided me although I might have a more congenial subject for investigation with Nernst.[34]

A few weeks later Lewis changed his mind:

> My plans for next semester have become unsettled. I have just received a postal from van't Hoff saying that he is to leave Berlin soon after the beginning of the semester and remain absent until the latter part of July. He is going to America to speak at the Chicago Univ. anniversary 16 June, and will remain a fortnight in America. I hope he will receive the L.L.D. [an honorary degree] at Chicago. I knew some time ago that he thought of going but he refused even to say something about it then. I am debating whether it will be worth going to Berlin. I should like much better to spend a semester there than in Göttingen, but it seems to be not worthwhile if van't Hoff is only to be there half the time. I think if Nernst has a place for me I had better go there.[35]

It is possible that Nernst later learned of Lewis's conversations with van't Hoff. If so, he would not have appreciated being Lewis's second choice.

Lewis Adrift at Harvard

In 1902, Lewis returned from Germany to Harvard as an instructor in the chemistry department, where Richards was his senior. The University of

Göttingen had offered Richards a professorship in 1901, and Richards had used the offer to negotiate a reduction in his own teaching load at Harvard to three lectures a week;[36] it is likely that Lewis was forced to pick up the lectures Richards was giving up. Lewis was assigned a teaching load of three courses for 1902. He taught one course jointly with Richards, so the two of them must have been forced to have regular conversations. Lewis presented his ideas on valence and the electron—essentially the ionic chemical bond—during that course, as Lewis documented by later printing his 1902 lecture notes.[37] Richards had this to say about valence theory in his class notes from that year: "Twaddle about bonds: a very crude method of representing known facts about chemical reactions. A mode of represent[ation] not an explanation."[38] Lewis was becoming fed up with Harvard and with teaching. He found the faculty uninterested in his work and ideas, and he felt that his teaching load was so heavy that he had no time for research. He published nothing from his time in Germany or from his time at Harvard after returning from Germany, the longest break in publication in his life.

After being accused of appropriating Richards's ideas (at least as he saw it), Lewis was likely avoiding any but the most necessary discussions with Richards after returning from Germany. In 1904, he abruptly left Harvard to take a position as Superintendent of Weights and Measures in the Philippines. He was twenty-nine years old and unmarried, and he might have wanted a year of adventure in his life. This was shortly after the United States acquired the Philippines in the Spanish-American War, and he might also have been motivated by patriotism—this was a time when many Americans believed in "manifest destiny." During his time in the Philippines, Lewis took up smoking cheap "Fighting Bob" cigars—"Regular 10¢ value, now 5¢"—which he later upgraded to another Philippine cigar, the Alhambra Casino, that he would chain-smoke for the rest of his life. But it was not all fun. He had a laboratory as part of his position, and he resumed experimental work in physical chemistry. He published three papers from that time, the first publications since his fugacity papers. And he nursed his resentment against Nernst. Lewis said that he took only one book with him to the Philippines—Nernst's *Theoretical Chemistry*—and spent the next year finding all the errors.[39]

Richards seems to have held no lasting grudge against Lewis from their dispute over fugacity in 1901 or from Lewis's abrupt departure from Harvard in 1904, although the physical organic chemist Louis Hammett later said that when he was a Harvard student in 1915, Richards mentioned G.N. Lewis as a bright young fellow but did not seem to entirely approve of him.[40] When Lewis returned to Boston from the Philippines in 1905, Richards was one of three on a National Academy of Sciences committee that appointed Lewis to lead a collaborative effort to gather information on free energies.[41] In 1924, Richards would nominate Lewis for the Nobel Prize in Chemistry.[42]

If Richards forgave all, Lewis did not. Although he sent Richards a cursory congratulatory letter when Richards received the Nobel Prize in Chemistry for 1914 for his work on atomic weights,[43] he never forgave Richards for his claim for priority on fugacity. In the 1940s, Lewis was still complaining about Richards, saying, "Imagine that—stealing from a graduate student!" to his own graduate student, Jacob Bigeleisen.[44] After Richards's death in 1928, Sir Harold Hartley asked Lewis for information on Richards, probably for a scientific obituary. In his reply, Lewis mentioned that Richards was "conscientious," "accessible," "meticulous," "pleasant," and "a good conversationalist," and then said, "I seem to be running on without saying anything worthwhile" and referred Hartley to Richards's other students.[45] The editor of *Science* wrote to Lewis, as Richards's most prominent student, asking him to write a scientific obituary for Richards.[46] Lewis did not comply, and may not have replied; in any case, the obituary that appeared was written by another of Richards's students.[47]

Lewis Established at MIT

Lewis returned to Boston in 1905 and took a position at MIT in Arthur Noyes's Research Laboratory for Physical Chemistry. Lewis did reapply to join the faculty at Harvard, but withdrew his application once his appointment at MIT was made permanent.[48] He blossomed at MIT. He had no teaching duties, and Noyes encouraged his research. In 1907 and 1908, while Noyes served as acting president of MIT, Lewis served as acting head of the laboratory. He continued working to put chemical thermodynamics on a rigorous footing, extended his theoretical and experimental work, headed up the National Academy of Sciences' collaborative effort to collect data on free energy of reactions, and took on his first Ph.D. student, Merle Randall.

Lewis's goal was to simultaneously make thermodynamics accessible to chemists and make it rigorous. The theory behind thermodynamics had followed two separate paths, a more mathematical one by the physicists and a more utilitarian one by the chemists. Lewis attempted to reconcile the two approaches, to preserve the work that had already been done by chemists but to add the necessary rigor. While at MIT, he laid out the approach that he would follow throughout his career: develop exact equations and clearly identify and separate any approximations. In 1907, he announced his approach to thermodynamics in a paper that, like his dissertation, had an all-embracing title: "Outlines of a New System of Thermodynamic Chemistry."[49] From the paper's introduction:

> In the rapid development of theoretical chemistry, in which the two laws of energy have played so important a role, two thermodynamic methods have been widely used. The first, employed by Gibbs, Duhem, Planck, and others,

is based on the fundamental equations of entropy and the thermodynamic potential. The second, employed by such men as van't Hoff, Ostwald, Nernst, and Arrhenius, consists in the direct application to special problems of the so-called cyclic process.

... It must be admitted that it is the second method to which we owe nearly all the advances that have been made during the last thirty years through the application of thermodynamics to chemical problems, and which is now chiefly used by investigators and in the text-books of physical chemistry.

Yet the application of this method has been unsystematic and often inexact, and has produced a large number of disconnected equations, largely of an approximate character. An inspection of any treatise on physical chemistry shows that the majority of the laws and equations obtained by the application of thermodynamics are qualified by the assumption that some vapor behaves like an ideal gas, or some solution like a perfect solution....

... [T]he old approximate solutions of thermodynamic chemistry will no longer suffice. We must either turn to the precise, but rather abstruse, equations of entropy and the thermodynamic potential, or modify the methods which are in more common use, in such a way as to render them exact.

The latter plan is the one followed in the present paper, the aim of which is to develop by familiar methods a systematic set of thermodynamic equations entirely similar in form to those which are now in use but which are rigorously exact.[50]

Telling van't Hoff, Ostwald, Nernst, and Arrhenius by name that they were responsible for "old approximate equations" that were "unsystematic," "inexact," and would "no longer suffice" was not a politic thing to do. It is evidence of what one might choose to call either Lewis's courage or naiveté, and Lewis would repeat this sort of behavior frequently without much thought for its effects on his career.

By this time, Lewis was becoming known as a leading American thermodynamicist, and his new call to arms was likely taken more seriously than his earlier attempts at reformulation. But his paper was only what he had called it in its title—an *outline* of a new system of a chemical thermodynamics. He had not yet done the experimental work to back up his claims, and the paper was more a preview of coming attractions than a report of research performed. He had earlier introduced his concept of fugacity, which may be thought of as an idealized pressure, and in this paper he introduced his most important thermodynamic concept, *activity*. His choice of the term "activity" was unfortunate, as it was already in general use with a vague meaning at the time—Arrhenius had used the term "active state" almost a quarter of a century earlier. But Lewis meant something quite specific by activity. He thought of it as an idealized concentration—the concentration a dissolved substance would exhibit if it behaved according to the ideal solution laws. Almost all the approximate thermodynamic equations that had been developed by physical chemists became exact by substituting Lewis's fugacity for pressure and his activity for concentration. But without ways to determine the activity of a component as a function of concentration, this would have been only a reformulation of the problem, and nothing would have been gained. It would have been a case of not solving the difficulty but baptizing it, as Lewis later quoted Henri Poincaré to have said.[51] But Lewis was to spend the next fifteen years developing methods to make activity determinable experimentally.

For Lewis, activity would be the route to the solution of the problem that the Ionists had left unsolved—the anomalous behavior of strong electrolytes.[52] As Arrhenius had shown, strong electrolytes (for instance, sodium chloride, sulfuric acid, and potassium hydroxide) are virtually completely dissociated at any concentration when dissolved in water solution, whereas weak electrolytes (acetic acid and ammonia, for example) are only partially dissociated at any actual dilution. Solutions of weak electrolytes follow the law of mass action, but solutions of strong electrolytes do not, and there was no adequate explanation for this discrepancy. This was considered the most pressing problem in physical chemistry and would remain central to Lewis's work through 1921, when he would define his concept of "ionic strength," which would give a practical method for determining the activity of strong electrolytes (see chapter 7).

Lewis had gone through some difficult times with Richards and Nernst (due in good part to his own making), but by the time he met Irving Langmuir in New Haven in 1908, he was already a presence in American physical chemistry. In 1908, Langmuir, six years Lewis's junior, was stuck in the same position at Stevens as Lewis had found himself at Harvard six years earlier—working for what he saw as an oppressive boss, with no time for research, and with no one who seemed to appreciate his ideas.

Irving Langmuir

Irving Langmuir was born in 1881, the third of four sons of Charles and Sadie Langmuir. Irving's father was an executive with New York Life Insurance, and the Langmuirs traced their ancestry back to Scots who had immigrated to Montreal in the 1840s and later to the United States. They lived in Fort Greene, a leafy neighborhood of brownstones in Brooklyn, just across the East River from Manhattan. Irving's oldest brother Arthur was a chemist and helped Irving construct a home chemistry laboratory, where Irving spent his time setting off the usual minor explosions. When Irving was twelve, his father was sent to Paris to run New York Life's European operations. Irving attended French schools, became fascinated with opera and classical music, and began the intense mountain climbing, skiing, and skating that he would continue for the rest of his life. He began keeping a detailed log of each mountain he climbed on his vacations, with elevations and times of ascents and descents. In 1898, when Irving was seventeen, the family planned to return to Brooklyn. Irving and his younger brother, Dean, went first to start the school year, but as the rest of the family prepared to return, Irving's father fell ill and died after a few days' illness. Irving's mother was left with enough money that she could continue to indulge her interest in world travel, and her correspondence with Irving shows that she spent much of the rest of her life on steamships and in hotels. Irving corresponded with her regularly until her death in 1936. His candid letters to his mother are often the most revealing documents among his papers.

Upon graduation from high school, Langmuir entered the program in mining engineering at Columbia. He later wrote, "The course was strong in chemistry, it had more physics than the chemical course, and more mathematics than the physics course—and I wanted all three." During the summers, he hiked and bicycled in the Catskills. One of his professors was impressed with him and asked, "If you could do what you most wanted to do, what career would you choose?" Langmuir answered, "I'd like to be situated like Lord Kelvin—free to do research as I wish."[53] Kelvin was the famous English scientist who had been engaged in developing the first two laws of thermodynamics and had been in charge of laying the first sub-Atlantic telegraph cable, among other accomplishments.

Langmuir and Nernst

In 1903, after receiving his bachelor's degree from Columbia, Langmuir headed off to Germany for graduate work, unsure of whether he would choose Göttingen and Nernst or Leipzig and Ostwald. Students in German universities were not required to stay at one place—they could attend classes at any university, eventually submitting a dissertation at one. Göttingen was the great center of German mathematics; Felix Klein, David Hilbert, Carl Runge, and Hermann Minkowski, all of whom were extraordinary mathematicians, were professors there. Leipzig, on the other hand, had Ostwald. Langmuir went to Göttingen first and attended a lecture by Nernst, whose German he was surprised to find he could understand. "It is probably because Nernst speaks rather slowly and very distinctly. He certainly seems to be a splendid professor,"[54] Langmuir wrote in a letter to his mother. He sent her a carefully constructed table comparing the advantages and disadvantages of studying in Göttingen with Nernst versus studying with Ostwald in Leipzig, including not only the science, but also theater, music, the countryside, and the local German accent.[55] When Langmuir visited Leipzig, he found that Ostwald had virtually retired, that the university was full of students from the United States (which was not what he wanted for his German experience), and that the scientific lectures seemed inferior to Göttingen's, so he chose Nernst and

Göttingen.[56] Within a short time, Langmuir gained an almost unaccented German fluency to go with his excellent French. (Lewis could read and speak German, but his accent was "incredibly awful."[57]) Langmuir took off that summer for mountain climbing in Berchtesgaden, and by fall had settled himself in Göttingen. By the end of November he had received cross-country skis that he had ordered from Munich, and he taught himself to ski and was soon teaching the German students, as well. During the summers, he climbed some of the most challenging peaks in the Alps. His brother Arthur recalled, "These climbs were made without a guide, but using a guidebook listing the various handholds on the cliffs and in the chimneys. Near the top of one of these chimneys, while holding on by his fingertips, it was necessary to swing out into space with a drop of 500 feet in order to secure a foothold for further advance." In June 1905, Langmuir's diary shows that he hiked 52 miles in a single day at an average speed of 3.4 miles per hour, ascending 4,940 feet in the process.[58] Some years later, Langmuir and his friends were observing a group of seals from a sailboat when Langmuir decided to see if he could approach closer. He took a black poncho, draped it over his head, cut two eyeholes and a nose-hole, tied a black canvas strip around his neck, jumped over the side, and dog-paddled over to the seals, who allowed him to join the group, although he said that they viewed him with some suspicion.[59] There is no account from Lewis of any exercise during his time in Germany. This might have been the one point on which he agreed with Nernst: climbing one mountain would have been enough to suffice for an entire life.

After a semester of coursework at Göttingen with Nernst, Klein, and Minkowski, Langmuir passed the qualifying exams and began his doctoral research with Nernst in the spring of 1904. The *Arbeit*, or research problem, that Nernst suggested to Langmuir was related both to Nernst's lamp and to the "fixation of nitrogen"—the synthesis of other chemical compounds starting with atmospheric nitrogen. Nernst wanted Langmuir to examine chemical equilibrium in gas reactions, in particular, the formation of nitric oxide from a mixture of nitrogen and oxygen. The solution of this problem had great economic potential, as nitric oxide could be converted to fertilizer or to munitions. Plants for manufacture of nitric oxide by this process had already been constructed at Niagara Falls and in Norway. In a series of letters to his mother, Langmuir wrote:

> [Nernst] proposed two ... subjects, one which I liked rather well so I am reading up on it a little.... Nernst gave me a short article to read, written by himself, on the subject of the new Arbeit.[60] I read it over and could not understand it thoroughly I thought, but after going over it carefully and comparing my notes with other students, I have come to the conclusion that Nernst must have made a couple of foolish little mistakes. This makes it very awkward for me, for Nernst is not the sort of man that likes to be told

he is wrong, and if I tell him I don't understand it he will think I don't know anything. But of course, before I can undertake an Arbeit whose object will be to prove the formulas he has deduced I have got to understand what those formulas mean to be sure that they are right. The Arbeit Nernst proposes is to study chemical equilibrium in spaces where there is a large difference of temperature between one part and another. He has not told me anything definite for me to do yet, but said I might study the dissociation of water in the region around the filament of a Nernst lamp and then later apply what I find that way to the production of oxides of nitrogen from the air by electric arcs and also Nernst filaments.[61]

The reaction of nitrogen and oxygen to form nitric oxide would only take place at high temperature, so Nernst's idea was to allow the reactant gases

Figure 2-2. Irving Langmuir at about age 19. Courtesy of the Chemical Heritage Foundation.

to come into contact with each other on a hot surface (initially the filament of his lamp, but later a hot platinum wire, which might have the added advantage of acting as a catalyst and increasing the reaction's velocity). As the nitric oxide formed, it would move from the hot wire to a cooler region. Nernst hoped that as the products and reactants cooled, the reaction might slow enough that their amounts would not change before they could be measured. Nernst initially showed great enthusiasm for the problem. Langmuir wrote: "This week I have just begun the experimental work. Nernst has been so interested in this work himself that he has set up the apparatus himself in the laboratory. Nernst has admitted a couple of mistakes that I wrote about before, but I think there are still a couple more which I will have to convince him of."[62] And the following week:

> This week I have been working on my Arbeit night and day practically. I had no idea I would find it so exceedingly interesting. I have really been very lucky in getting such an Arbeit and particularly in having Nernst so much interested in it himself. As I wrote last week, he set the apparatus up in his own lab and for the first three days I worked with him most of the time. Weeks before last I had not seen Nernst at all, but was spending all my time reading all the literature he had given me and in working over the part that I had had the argument about so that I could know what I was talking about when I saw him again Monday. When I saw him he said that he had started laboratory work on my Arbeit on his own account as he wanted a few general results for a paper he was going to publish soon. He had set up the apparatus and had made only one preliminary experiment. So I made the rest of the experiments of that set. They consist of leading a current of air over the filament of a Nernst lamp and then determining the quantity of nitric oxide formed.... Nernst then went to work and passed air through his iridium furnace heated to 2000° (or even 2100° in one experiment) and cooled the gases down from this temperature very rapidly by passing them into a capillary quartz tube. This gave him 0.8% of the air changed to nitric oxide where my results had so far only shown 0.3%, although the filament had a temperature of 2200°C.
>
> Last Saturday I started to set up apparatus for experiments myself and have everything ready now to begin tomorrow morning. Nernst came around to see me every little while so that I might get all the apparatus I need, while most of the other students have to wait their chances to see him once a week.[63]

The equilibrium point of the reaction that Langmuir was studying moves toward increased production of nitric oxide as the temperature is increased. It unfortunately turned out that unless the nitric oxide was cooled very rapidly, it would quickly dissociate back into nitrogen and oxygen when it left the wire, which is probably why Langmuir's results using the hot wire showed only 0.3% yield, while Nernst's results with an iridium furnace and more rapid cooling showed 0.8% yield. Water vapor and carbon dioxide did not have the same problem with a rapid reverse reaction, so they would eventually become the subject of Langmuir's dissertation.[64] But their formation had no economic significance, and Nernst soon lost interest in Langmuir's project.

At the end of 1904, Nernst announced that he was moving from Göttingen to Berlin. He advised Langmuir to stay behind and finish his work with his successor, who was yet to be determined.[65] By the spring of 1905, Nernst was dividing his time between Göttingen and Berlin and was not viewed as helpful by the students. Langmuir wrote:

> Nernst is rather hard to get along with. The students try to avoid him every way possible when they are working in the lab and nearly everyone is glad when he doesn't make his daily rounds.
>
> The number of curses laid upon his head by the Americans alone are more suitable for the blackest villain than for Nernst. The trouble is principally in his manner and his lack of sympathy with a man in difficulty.
>
> He comes around to say, *"Ja, Herr ___, haben Sie neue Zahlen."* [Well, Mr. ___, do you have any new numbers? ("Zahlen" meant not just "numbers," but quantitative results that Nernst might be able to publish.)] Well it so happens that you were just able to start getting the apparatus ready that morning, so that it will take at least a week to get the apparatus all constructed and ready for producing "*Zahlen*" but nevertheless Nernst will be around every morning and ask you for more results and moreover never once take the least interest in any fine idea you may have in a new form of the apparatus.
>
> He takes absolutely no interest in the experimental part of the work except on paper. A little sketch of a new apparatus might please him, but the apparatus itself, although it might be most skillfully carried out would not interest him in the least.
>
> When you do have *Zahlen* he sometimes seems a little pleased to give you some suggestions which are often valuable but sometimes are absurd showing that he really does not know just what you are doing or at least that he was thinking of something else at the time.
>
> Very often you give him results and he stands there thinking about them or something else and you mention something which gives you difficulty or prevents you getting the results he expected and he suddenly says, *"Ach, ich habe keine Zeit, Adieu"* [Alas, I have no time. Adieu], and turns his back and runs out of the room.
>
> Other times he loses his temper over trifles and gets furiously mad and says the most insulting things. For instance, one day he came into the lab where Merriam one of the American students works and found it cold because the door was open. Well he began talking about that and asking Merriam how he Merriam expected him to talk over results in such a cold room. Well Merriam didn't care much whether results were talked over or not but Nernst stayed there five minutes just boiling over and getting hotter all the time until when he did go the room was plenty warm enough.[66]

This sort of thing must have pushed Lewis over the edge during his stay with Nernst. Langmuir resented Nernst's rude behavior but had the self-confidence to realize that it was not directed at him personally—Nernst treated all his subordinates that way.

Nernst's successor at Göttingen was Friedrich Dolezalek, who had been Nernst's student only a few years earlier. Langmuir got on well with Dolezalek: "I have handed in the first part of my Arbeit to Prof. Dolezalek and he seems to think that it is quite good, especially a mathematical theory which I got up to explain some results. But he says and I agree with him that the subject is an unsatisfactory [one] to work on, as it is far too complicated to understand thoroughly."[67] While he was writing up the dissertation, Langmuir continued to work on deriving equations that would describe the behavior of the gases, and he made what he considered a real breakthrough. From Langmuir to his mother:

> I am in the happiest state of mind that I have been in for years. I think just a couple of hours ago I have made a purely theoretical discovery. . . .
>
> Nernst first had me make experiments with the filament of a Nernst lamp in air. He said that it would illustrate his theory that there is bound to be a temperature at which the amount of NO [nitric oxide] which is measured in the cold air corresponds to that of the temperature of the filament. This theory of Nernst's, which he applied to other phenomena, explosions for example, I have now definitely proved to be wrong, both by experiments and by theory.
>
> When Nernst saw that the experiments did not check up with the theory he lost all interest in my Arbeit and tried to get me to start immediately on another problem, the dissociation of water vapor on a white hot platinum wire. However I was very much interested in the first problem and notwithstanding Nernst's impatience and the very tedious experiments I kept it up a month or so longer. I am very sorry now that I did not spend more time on them.[68]

Langmuir included his new equations in his dissertation but was too pressed for time to return to his experiments on nitric oxide. Dolezalek gave him the highest possible grade on his dissertation. Langmuir published both his experimental results on carbon dioxide and water[69] and his theoretical results[70] after he returned to America.

Young scientists find a certain pleasure in proving older scientists wrong, and Langmuir and Lewis were no exception, especially when it came to Nernst. From Langmuir's point of view, showing that Nernst's theory was wrong (at least to his own satisfaction) was one of the high points of the graduate career, just as Lewis said that he spent his year in the Philippines finding all Nernst's errors. But Langmuir had the good sense not to explicitly point out Nernst's errors in his publication. Lewis would spend the next twenty years publishing Nernst's errors.

Langmuir Adrift at Stevens

When Langmuir completed his dissertation in 1906, he needed a job. His oldest brother, Arthur, an industrial chemist, had several times tried to push him into industry, but Langmuir resisted. Langmuir had corresponded with his other older brother Herbert in 1904, agonizing over whether to pursue a career of pure research in an academic setting or to enter the world of business and try to make his fortune. Herbert had advised him that he had the ability to do great things, and that he should forget about money and pursue his love for research.[71] Upon receiving his doctorate, Langmuir took Herbert's advice and accepted a position at what he thought was an up-and-coming scientific institution, Stevens Institute of Technology in Hoboken, New Jersey. Stevens was a new engineering school, and Dr. Stillman, who ran the chemistry department, offered Langmuir a position as instructor in chemistry at $900 per year. Stillman was head of a two-man department, and Prof. F.J. Pond served as the other faculty member. It soon became clear, at least from Langmuir's point of view, that his job was to do all the dirty work, preparing and supervising the laboratory sessions. "Professor Stillman," he wrote to his brother, "is a rather old man and he never did a bit of work except to walk around the laboratory two or three hours every afternoon (spending much of the time telling them funny stories). The students had no lectures on their work and understood nothing they did; they simply followed the directions in their text book and as a result got rotten results." Langmuir tried to organize lectures on the lab experiments, but found that the students resented what they regarded as extra work. He found the students rowdy and inattentive in class. He asked the college president for a raise and to his surprise got it, to $1,200 per year, but was unable to get the school to agree to give him an assistant. In 1908, he organized a modest research program of his own and published a paper on gas reactions. In 1909, Stillman was planning to retire and Pond would be taking his place. Langmuir wanted two things: a pay raise and a promotion to assistant professor. Pond was to be in England that summer visiting the college president, Alexander Humphreys, and Langmuir sent him a letter with his requests for money and advancement and asked him to cable back an answer. In order to keep the return cable short, he wrote Pond:

> First question: In the case of Dr. Stillman's retirement, will you recommend to Pres. Humphreys that I receive a salary of eighteen hundred dollars ($1800) a year?
>
> Second question: In the case of Dr. Stillman's retirement, will you recommend to Pres. Humphreys that I receive the title of Assistant Professor?
>
> If you will simply cable "Yes, Yes" I will understand the first word to apply to the first question and the second word to the second question.[72]

Pond soon answered: "NO NO RECOMMEND $1400."[73] Langmuir was to be left with no promotion and with only a modest increase in salary.

Langmuir Established at General Electric

Despite his brother Arthur's advice to enter industrial chemistry, Langmuir feared that it would be a life of drudgery. Langmuir had run into a classmate from Columbia who was now on the staff of General Electric in Schenectady working for Willis Whitney and was enjoying the work. Langmuir wrote to Whitney, who invited him to GE to give a talk and to take a tour of the laboratory. He then offered Langmuir a summer job, which Langmuir accepted. Given the situation at Stevens, industrial employment was looking more attractive.

Langmuir was amazed at the facilities and the scientists that he met at GE—it seemed like his dream of scientific research come true. The most pressing problem at GE's research laboratory was the improvement of the incandescent lamp (the ordinary household lamp), which was then not much changed from the lamp that Edison had developed in 1879. It used a carbon filament, since carbon was a conductor of electricity with a high melting point (3,527°C). The bulb was evacuated to a high vacuum so that the filament would not be oxidized. But the bulb's light was dim unless the current was increased, and when that was done the bulb blackened quickly. Furthermore, the carbon filament was fragile. Other lighting systems, including the Nernst lamp, Nikola Tesla's fluorescent lamp, carbon arc lamps, and gas lights with thorium mantles were all actual or potential competitors to the incandescent lamp.

Langmuir's first success at GE was developing what is essentially the modern incandescent lightbulb, work he began that summer and would continue for three years, and it is instructive to examine this work in some detail because it shows Langmuir's approach to research. Most scientists use experiments in one of two ways: either to prove an already-developed theory, or to solve a particular problem, stopping when the problem is solved or when the results do not seem to be leading to a solution. Langmuir's approach was different. In his work at GE, he always focused on the practical, but he would allow himself to become interested in confusing results along the way—much as Arrhenius had been distracted by his strange results on the conductivity of electrolytes, which had pulled him away from the problem of determining molecular weight. Langmuir believed that it would not be possible to make practical advances in GE's products until he understood the applicable chemistry and physics in detail.

GE was searching for a better lamp filament than carbon—one that would have a high melting point but that would be less fragile and would avoid the blackening problem. Tungsten has the highest melting point of any metal (3,422°C) and seemed to be the best candidate for a new filament. But tungsten had its own problems—it was hard and brittle, and it was difficult to draw into a thin wire. William Coolidge, who like Whitney had studied with Ostwald and had been on the MIT faculty before joining GE, had just succeeded in drawing tungsten into a filament, but he found that the tungsten almost always became brittle and eventually crumbled when alternating current was applied. But of the hundreds of tungsten filaments he had made, three gave excellent results, maintaining their flexibility. Whitney let Langmuir pick his summer research project, and he decided to investigate the difference between Coolidge's good and bad tungsten filaments.

Langmuir thought that perhaps GE's problem with brittle tungsten filaments was related to the adsorption of gases onto the tungsten. (Adsorption is the adherence of one substance to the surface of another.) If he were to heat the lamps to a high temperature, any adsorbed gases might be emitted from the filaments, and by comparing the gases that were emitted from the good filaments with those emitted from the bad filaments, he might be able to determine why the tungsten had become brittle in the bad filaments. He set up an apparatus to trap any gas that might be emitted and heated both sorts of lamps.

The experimental results seemed to make no sense at all. Both the good and bad lamps emitted gas—lots of it. In fact, after a few days, each lamp had emitted a volume of gas that, at atmospheric pressure, occupied 7,000 times the volume of the filament itself. That much gas could not be coming from the filament! Langmuir continued his work on the original problem—the brittle tungsten filaments—but started a second investigation into the gases that appeared when the lamps were heated. He told Whitney that he thought he was wasting GE's money on an impractical problem. Whitney told him not to worry, but to work on whatever he found interesting. (After all, GE was in the electric lamp business, and it would be a good thing to know more about the behavior of gases in the lamp.) Langmuir found that the emitted gas was not coming from the filament, but from gas adsorbed on the bulb. He still had not made any progress on the original problem of the difference between the good and bad tungsten filaments, but Whitney was impressed enough to offer Langmuir a permanent position at GE at the end of the summer. He accepted and did not return to Stevens.

If the bulb of an incandescent Edison lamp contained oxygen gas, everyone knew that the carbon filament would burn just as coal (which is itself carbon) does in a furnace, so Thomas Edison's solution had been to remove the oxygen by evacuating the bulb to as high a vacuum as possible. The general belief at GE when Langmuir arrived was that if only a better vacuum could be produced, the blackening of the bulbs would disappear—that the blackening was due to some reaction between residual gas and the carbon filament—perhaps the reaction was with the remaining oxygen, or perhaps

Figure 2-3. The General Electric laboratory, about 1912. Left to right: J.A. Orange, Leonard D. Dempster, Irving Langmuir, George Hotaling, Willis Whitney, William Coolidge. General Electric Research and Development Center, courtesy of the AIP Emilio Segrè Visual Archives.

with some other gas. A perfect vacuum is a negative idea—an enclosure that contains nothing at all. A perfect vacuum is not actually attainable, but a "good" or "high" vacuum in a lamp approaches as closely as possible to this ideal, removing as much gas as possible. Langmuir had already discovered that baking the bulbs to remove adsorbed gases greatly improved the vacuum, but the blackening persisted. Langmuir believed that the easiest way to determine if residual gases were the problem was not to try to further improve the vacuum in order to remove more gas, but to add particular gases one at a time and then to study the effects of each gas on the lamp. Rather than work on improving the vacuum, he would spend the next three years studying the effects of methodically *degrading* the vacuum. He later wrote:

> When it is suspected that some useful result is to be obtained by avoiding certain undesired factors, but it is found that these factors are very difficult to avoid, then it is a good plan to increase deliberately each of these factors in turn so as to exaggerate their bad effects, and thus become so familiar with them that one can determine whether it is really worth while [*sic*] avoiding them. For example, if you have in lamps a vacuum as good as you know how to produce, but suspect that the lamps would be better if you had a vacuum say 100 times as good, it may be the best policy, instead

> of attempting to devise methods of improving the vacuum, to spoil the vacuum deliberately in known ways, and you may find then either that no improvement in vacuum is needed, or may find just how much better the vacuum needs to be.[74]

This sort of approach would be typical of Langmuir: to step back from the immediate problem, to refuse to accept the common belief, and to investigate the underlying chemistry and physics.

Investigating gases inside lightbulbs was something that he already knew how to do. His experimental setup in Nernst's lab in Göttingen had been essentially a lightbulb—a vacuum enclosure with a hot filament (a platinum wire) into which he had introduced gases and observed their effects. Within this experimental apparatus, he had been able to control and measure almost everything. He had been able to control which gases were present, the gas pressure, the temperature of the wire (by controlling the current), and the temperature of the bulb. He had been able to measure the temperature of the wire using its electrical resistance, which varied as a function of the temperature. He reasoned that at low pressure, the temperature of the gas inside the bulb was the same as the temperature of the lamp's bulb: on average, the gas molecules struck the bulb much more often than they struck the tiny filament or each other, so he could control the temperature of the gas by simply controlling the bulb temperature. He could even immerse the bulb in liquid air in order to reduce the gas temperature dramatically.

Langmuir used almost the same apparatus at GE to investigate the effects of individual gases within the lightbulb. GE had much better vacuum systems than he had used in Nernst's laboratories, and tungsten had a much higher melting point than did the platinum he had used there. So while he used the same experimental setup as in his dissertation with Nernst—a lightbulb as his test platform—he could use a much wider range of temperatures and pressures.

He began working on introducing gases one at a time. When he introduced hydrogen gas into the bulb, he observed something surprising. He measured the heat loss from the filament as it was heated. Up to 2,070°C, hydrogen behaved the same way as did nitrogen or mercury vapor. But at higher temperatures the heat loss increased dramatically, and the hydrogen gas began to disappear quickly. Where was the hydrogen going, and what was causing the heat loss?

Hydrogen gas occurs as H_2, where two hydrogen atoms are joined by a single bond. Langmuir suspected that when the hydrogen molecule came into contact with the hot filament, it dissociated into two hydrogen atoms. He thought that the heat loss he observed was caused by the energy consumed in breaking the bond between the two atoms. But if that were the case, when the hydrogen atoms moved some distance away from the filament, he would have expected that they would have been quite reactive and ready to recombine. As the bond between the hydrogen atoms reformed, this would release the heat that had been used to break the same bond earlier, so that there would be no net heat loss within the bulb, and as the molecular hydrogen gas reformed, there would be no net loss of gas. But this was not what was happening.

Where was the hydrogen going? Was it stuck on the tungsten filament? Langmuir heated the filament almost to the melting point of tungsten and observed no hydrogen emitted. That left only the bulb. When he baked the bulb almost to the melting point of the glass, the hydrogen finally emerged. He reasoned that the hydrogen molecule had been broken into hydrogen atoms on the tungsten filament, and that the atoms then shot off from the filament into the space within the bulb. But the bulb was at high vacuum, so the hydrogen atoms had almost no chance of colliding with anything else before they struck the surface of the bulb, where they stuck. Because the hydrogen showed up when he heated the bulb sufficiently, he now knew where it had gone.

He found something else: if he kept adding hydrogen, the heat loss eventually dropped back to its normal value, and no more hydrogen gas was taken up. It appeared that the bulb could only take so many hydrogen atoms, and after that there was no room for more.

Langmuir developed a theory to explain the data: the hydrogen atoms formed a layer exactly one atom thick on the surface of the bulb. The surface of the bulb was like a checkerboard with one hydrogen atom per square. From the measured volume of hydrogen that could be taken up and the measured surface area of the bulb, he calculated the size of the checkerboard's squares—they were spaced every 2.8 hundred-millionths of a centimeter. Once all the squares on the surface were filled, a second layer did not form. There was no more room, and any remaining hydrogen atoms would bounce around inside the bulb until they ran into other hydrogen atoms, at which point they would recombine to form molecular hydrogen and release the heat that they had consumed when they had formed by the rupture of the molecular hydrogen bond on the filament's surface. This explained why, when enough hydrogen was added, the hydrogen stopped disappearing and the heat stopped being taken up. This was an astounding claim—that there was a layer on the surface of the bulb exactly one hydrogen atom thick, or about one hundred-millionth of a centimeter, but Langmuir's theory seemed to fit the data.

The idea of a layer one molecule or atom thick was one that would form the core of Langmuir's work for much of the next forty years—what would be known as *surface chemistry*. Chemistry is the science of the organization of matter and of its transformations from one form to another. Chemistry had always been thought of as three-dimensional—what happens within a flask, for example. Langmuir would suggest that many chemical phenomena occurred in two dimensions rather than three, as was the case with the hydrogen atoms on the surface of the bulb. Other examples of this sort of behavior would include oil films on water, polymer films, protein films, and catalysis by metals. His work in this area would eventually earn him a Nobel Prize in 1932, but in 1912 it still had not led to building a better lightbulb.

GE's experimental tungsten lamp could give a bright, white light if the temperature were increased, but this led to increased bulb blackening, just as it did for the carbon-filament lamp. The common belief at GE was that blackening was due to a reaction between residual gases and the filament. By investigating gases one at a time, Langmuir found that only water vapor caused blackening, but that there was not enough water vapor present even in the then-produced Edison bulb to account for the observed blackening. By process of elimination, Langmuir had shown that bulb blackening in both the carbon and tungsten lamps was not due to reaction with any gas but to direct *sublimation* of the filament itself to the bulb's surface—evaporation straight from solid to vapor in the same way that snow can disappear from a sunny hillside without passing through the liquid state. The vacuum that GE was using was already good enough to eliminate the effects of gas reactions; to fix the blackening, something else would need to be done. This was the first result Langmuir had been able to produce that had immediate relevance to lightbulbs, but it was a negative result—he had established what was causing the blackening, but he did not yet know how to prevent it. At one point, after nearly three years of research, Langmuir went to Whitney and said, "I'm having a lot of fun, but I don't know what good this is going to do the General Electric Company." Whitney replied, "That's not your worry, it's mine."

Langmuir continued to experiment with other gases, including nitrogen, and it was here that he finally found the solution to the problem of bulb blackening. Like hydrogen, the nitrogen molecule also consists of two identical atoms, but they are joined by a much stronger triple bond, N≡N, rather than the single bond in H–H. At the temperature of the tungsten filament, nitrogen was largely inert—it did not appreciably dissociate into atomic nitrogen or react much with the tungsten filament. But the blackening of the bulb decreased sharply if he filled the bulb with nitrogen—and the higher the pressure of nitrogen, the less the blackening occurred. Based on these observations, Langmuir calculated that, although tungsten still evaporated from the filament when the bulb was filled with nitrogen, the straight path from the filament to the bulb was no longer available to the evaporated tungsten atoms. In the presence of significant amounts of nitrogen, the atoms of tungsten were likely to collide with inert nitrogen molecules before hitting the bulb. After a series of such random collisions with nitrogen, the tungsten atoms had a good chance of colliding with the filament before they struck the bulb, redepositing themselves there—the inert nitrogen atoms provided a screen around the filament to keep most of the tungsten close to home. The bulb could run hotter and brighter with less blackening when nitrogen filled the bulb than in a vacuum. Adding argon, an inert gas that constitutes almost 1% of the earth's atmosphere but that had been discovered only in 1895, worked even better than pure nitrogen—nitrogen did react to a small extent with the tungsten that had evaporated, forming a brown tungsten nitride residue, but argon did not. Langmuir went to Whitney and suggested that the conventional thinking was wrong—the solution to a better lightbulb was not a better vacuum, but rather no vacuum at all: fill the bulb with an inert gas and the lamps would run much better.

As a summer employee, Langmuir had been asked to figure out why some tungsten filaments were stable while others became brittle and crumbled. He had not solved this problem—others at GE eventually did by adding thorium dioxide to the tungsten—but had become distracted by the behavior of different gases interacting with a hot tungsten filament, and Whitney had supported him in what many might have thought to be idle experiments. At the end of what seemed to be an unfocused multiyear research project, Langmuir's discoveries revolutionized the electric lamp business. Throughout his career, Langmuir would show unwavering loyalty and respect for Whitney, who had not asked for "*Zahlen*" each morning as Nernst had. Whitney's characteristic greeting was "Are you having fun?" GE's deep pockets and Whitney's patient commitment to fundamental research would become a model for other companies around the world.

There was one drawback to Langmuir's new gas-filled lamps. The inert gas transferred heat from the filament to the bulb by both conduction and convection, as in a hot wind, whereas in a vacuum the only mechanism for heat transfer is radiation, as in the heat that reaches the earth from the sun. The purpose of a lightbulb is to emit visible light, and heat is just wasted energy. So for the same filament temperature, more energy is wasted in a gas-filled bulb than in a vacuum bulb. But the gas-filled bulb did not need to be run at the same low filament temperature to avoid blackening, and the emitted visible light increased dramatically with increased temperature. Langmuir found that by increasing the diameter of the filament and coiling it tightly, he could further suppress the evaporation of tungsten by reducing the filament's effective surface area, and that by designing the bulb's shape to direct the gas convection currents appropriately, he could direct most of the tungsten that did strike the bulb to areas where it was less visible. The gas-filled, tungsten-filament lightbulb was much less fragile that Edison's vacuum, carbon-filament bulb, it consumed less energy for the same amount of light, and it had a longer life. It is essentially the same bulb that GE sells today. Its introduction made enormous profits for GE, saved consumers millions of dollars in electricity costs every day, and made Nernst's electric lamp obsolete. Langmuir was, of course, a hero at GE. For the rest of his career, GE encouraged Langmuir to work on whatever he wanted. At age thirty-one, he had achieved his ambition—to be like Lord Kelvin, free to do research as he wished.

3

The Third Law and Nitrogen

Haber and Nernst

In November of 1892, Fritz Haber, a twenty-four-year-old German-Jewish chemist, stood at the baptismal font of the largest Lutheran church in Jena. He recited the Apostles' Creed:

> I believe in God, the Father Almighty,
> Maker of heaven and earth,
> and in Jesus Christ,
> His only Son, our Lord:
> Who was conceived by the Holy Spirit,
> born of the Virgin Mary,
> suffered under Pontius Pilate,
> was crucified, died, and was buried.
> He descended into hell.
> On the third day He rose again from the dead.
> He ascended into heaven
> and sits at the right hand of God the Father Almighty,
> from there He shall come to judge the living and the dead.
> I believe in the Holy Spirit,
> the Holy Christian Church,
> the communion of saints,
> the forgiveness of sins,
> the resurrection of the body,
> and the life everlasting.
> Amen.

The pastor poured water over Haber's head, saying, "I baptize you in the name of the Father, and of the Son, and of the Holy Spirit. Amen." Haber was now a Christian and, so he thought, a true German. When the Nazis would come to power forty-one years later, he would find that not all Germans would think it was that easy to stop being a Jew.

Fritz Haber

Haber was the physical chemist who would have the greatest economic and social impact on the world. His solution to the problem of the "fixation of nitrogen"—the conversion of atmospheric nitrogen into ammonia—would eventually lead to an end to famine in China and India and would, for better or for worse, permit the world's population to quadruple in the period from 1900 to today. At the end of World War I, Haber received the 1918 Nobel Prize in Chemistry, an award that angered the Allies because his process had allowed Germany to produce not only fertilizer but the munitions that had prolonged the war for four years, while twenty million had died. And the anger would turn to outrage when it became known that Haber had led Germany's poison gas efforts—he became known as the "father of chemical warfare."

Haber was born in Breslau, Prussia (now Wroclaw, Poland), in 1868, just three years before the German nation was to be united in Bismarck's Second Reich. His father was a successful dealer in wool and dyes, running a business that he had inherited from his father and grandfather. The Haber family was Jewish, but it had not been observant for some time, as can be seen from Haber's first name, Fritz, and his father's first name, Siegfried. These are as Germanic as possible, and the family celebrated Christmas with a tree, although not with a crèche. Haber's mother died three weeks after his birth, and throughout his life Fritz had a difficult relationship with his father, who seemed to blame him for his mother's death. For his first five years, Fritz was raised by his aunt. When his father remarried and had three daughters by his new wife, Fritz developed a loving relationship with his stepmother, but his father remained distant. Father and son were temperamentally mismatched—where his father was conservative and controlling, Fritz was quick, risk-taking, unfocused, and a bit of a loudmouth. As Fritz moved into his teenage years, his father seemed to consider him a drinker and a lout. After Fritz finished his studies at the *Gymnasium* with good but not outstanding results, he wanted to go to university to study chemistry, which his father opposed as a waste of money and time. It was only through the intervention of his uncle that Fritz was able to convince his father to pay for university.

Breslau had its own university, but Haber wanted to get away and chose the Technische Hochschule in Charlottenburg, a neighborhood in Berlin, Prussia's and Germany's capital. There he attended lectures by Hermann von Helmholtz in physics and by August von Hoffmann in chemistry, perhaps the two most prominent German scientists of the day. But they were old, and he found himself dissatisfied as they mumbled their way through their lectures. He moved to the university at Heidelberg, where Robert Bunsen taught him spectroscopy (although also well past his prime at seventy-six), and Leo Königsberger introduced him to calculus. Haber tried to be the stereotypical Heidelberg student and managed to acquire a dueling scar on his jaw. The source of the scar seems to be unclear, as there were stories that Haber's came from an ordinary fight[1] or that he walked through a glass door while drunk.[2]

Military duty was a three-year obligation for Prussian young men, but a university student who was willing to pay all his own costs, including horse and equipment, could serve only one year. Haber's father agreed to pay, and in 1888 Haber joined a field artillery regiment in Breslau. He applied for a commission as a reserve officer. In Prussia, an officer's uniform carried status. But he was turned down for a commission, and he knew why—he was a Jew. No Jew had ever received a reserve commission in Prussia except in the medical corps. But this was not the way the new German system was supposed to work, at least in theory. With the unification of the German Reich in 1871, all legal restrictions on Jews had been abolished. They could worship, work, live, and intermarry as they liked. But while German anti-Semitism was then nothing compared to what it would become under the Nazis, it still flourished, sometimes with academic dressing. Heinrich von Treitschke, a prominent philosopher and historian at the University of Berlin, ranted, "Year after year, from the inexhaustible cradle of Poland, a throng of ambitious young trouser merchants pushes its way across our eastern border, intent on conquering Germany's stock exchanges and newsrooms with their children and their children's children," and ended his speeches with the refrain, "The Jews are our misfortune."[3] Haber would spend his life trying to distance himself from his Jewish origins and to prove himself a true German.

After his military service, Haber returned to Berlin and completed his doctoral dissertation in organic chemistry with a mediocre grade. He was completely dissatisfied, writing to a friend, "One and a half new compounds, produced like rolls in a bakery, all in the trash, and besides that a bunch of negative results . . . plus results that I can't publish at all out of fear that some competent chemist will look at them and prove that I don't know what I'm talking about. You learn to be modest."[4] In 1891, during his last year at university, Haber met Richard Abegg, who was to become a close friend. Abegg was interested in the new physical chemistry and would eventually study with Svante Arrhenius in 1893 and be Walther Nernst's assistant in 1894. Both Abegg and Haber wrote to Wilhelm Ostwald in 1891, applying to study in his laboratory; Abegg was accepted, but Haber was not.

When Haber graduated from university, he had few prospects. His father arranged some opportunities in chemical plants in Germany, Poland, and Hungary, in each of which Haber spent a few months. While he felt isolated in these industrial sites, he was impressed with the efficiency and power of the plants, which harnessed modern chemistry on a massive scale. Throughout his life, he never had much interest in theory unless it could be put into practice. Of his time at the soda-ammonia factory in Poland, he wrote: "Life here is quite monotonous. I am learning things about production and am gaining an eye for technical realities and questions.... Social life is nonexistent.... As to the soda factory—a splendid and energetic intelligence, sustained by large amounts of money, dominates here.... I have reason to be very happy about the opportunity that came my way to work here."[5]

In 1891 he made what may have been the first of many visits to sanatoria and spas for "nerves." Throughout his life, he was bombastic and excitable and talked too much. He would repeatedly work himself to the point of collapse and would then go off for a rest cure. Between his attempts to ingratiate himself with German society and his difficult family life, he must have been under constant psychological strain. It is difficult to know how much this strain was behind Haber's chronic illness. Hannah Arendt, who was herself a Jewish exile from Germany in 1933, has written on the psychological state of assimilated Jews in Germany: "Assimilation, in the sense of acceptance by non-Jewish society, was granted [Jews] only as long as they were clearly distinguished from the Jewish masses.... Jews who heard the strange compliment that they were exceptions, exceptional Jews, knew quite well that it was this very ambiguity—that they were Jews and yet presumably not *like* Jews—which opened the doors of society to them."[6] But even Jews who converted were not allowed to simply disappear into society: they were valued by liberal Germans precisely because they *were* Jews and thus proved the enlightened nature of those who socialized with them. Educated Jews were forced to choose between being pariahs and standing outside German society, or being parvenus and constantly monitoring the moods and sensitivities of society. Haber chose the latter. Arendt continues:

> The social destinies of average Jews were determined by their eternal lack of decision. And society certainly did not compel them to make up their minds, for it was precisely this ambiguity of situation and character that made the relationship with Jews attractive. The majority of assimilated Jews thus lived in a twilight of favor and misfortune and knew with certainty only that both success and failure were inextricably connected with the fact that they were Jews. For them the Jewish question... haunted their private lives and influenced their personal decisions all the more tyrannically. The adage, "a man in the street and a Jew at home," was bitterly realized.[7]

In 1892, Haber gave academic life another try, moving to Zurich for a semester that he found unrewarding. He then joined his father's firm, essentially as a technically trained salesman and purchasing manager. The family business, which was based on the import and processing of organic dyes, was being hurt by the new synthetic dyes that were being produced by the German chemical industry. When a cholera epidemic struck Hamburg in 1892, Fritz saw an opportunity for the family business to make a quick financial score. At the time, the common belief was that cholera was caused by an airborne "miasma" that could be eliminated by cleaning surfaces with chloride of lime, a powder that is now sold in solution as household bleach. Fritz figured that there would be a run on chloride of lime and convinced his father to buy a large amount for resale. When the epidemic subsided, his father found himself stuck with a product for which there was no demand. He fired Fritz.

Haber took an unpaid assistantship at a provincial university in Jena in 1892 and again applied to study physical chemistry with Ostwald. This time he got as far as an interview but was again rejected. It was while in Jena that he decided to convert to Christianity. His motives were apparently not religious—he had not been an observant Jew before his conversion and was not an observant Christian thereafter. He might have been motivated by simple personal ambition, believing that his difficulties in advancing his career were due to anti-Semitism and that he could help himself by converting. It was certainly true that being Jewish did not help a career. Supposedly one of his contemporaries said that before age thirty-five he was too young to be a professor; after age forty-five he was too old; and in between he was a Jew. But being Jewish was not an insurmountable obstacle to academic advancement, and many of Haber's contemporaries, including his future friend Richard Willstätter, made successful careers as academic chemists without converting. Haber might also have seen conversion as a way to reject his father as his father had rejected him, and Haber's second wife later said that his father saw his son's conversion as a betrayal of his ancestors.[8] And finally, the spirit of the times—patriotism for the new German nation—might have been a motive for Haber's conversion. Theodor Mommsen, a liberal German classicist, answered the anti-Semite Treitschke with what was seen as a liberal solution to the "Jewish problem"—assimilation. The new German nation had been formed by uniting Prussians, Bavarians, Schwabians, and others, and Jews were welcome, too. But for Mommsen, there was a condition—the Jews should give up their separateness, should destroy the barriers that divided them from their fellow German citizens. What he meant by "barriers" was the Jewish religion; he considered being Christian part of what it meant to be culturally German.[9] Haber wanted to be a German, and even as late as 1926 he said that he did not regret the decision to convert.[10]

Haber Established

In 1894, at the age of twenty-eight, Haber achieved his first paying academic job—assistant at the Karlsruhe Technical Institute. There were two chemical institutes within Karlsruhe, one headed by Carl Engler and the other by Hans Bunte, both of whom maintained close ties with the chemical giant BASF, whose business was concentrated in the dye industry and whose plant was at nearby Ludwigshafen. Haber had arrived at Karlsruhe without any guarantee of a job, but Engler sent him to Bunte, who hired him. Haber completed his *Habilitation*, a dissertation past the doctorate that is required for those who wish to become a *Docent* and to advance in the German academic system. His dissertation was on the decomposition of organic compounds with heat, a problem with immediate practical applications to the fuel industry. He sent a copy of the dissertation to Ostwald, who reviewed it positively in print but did not respond with an invitation to his laboratory.

Haber must have realized that Ostwald was not going to be the answer to his desire to learn physical chemistry, so he decided to do it on his own. While working as an assistant in Karlsruhe in 1897, Haber became interested in electrochemistry, the field that had sparked Arrhenius's dissociation theory and that Nernst was then developing in Göttingen. Hans Luggin, a young colleague on the Karlsruhe staff who had studied with Arrhenius in Stockholm, helped Haber by guiding his reading. By 1898, Haber had completed his first research project on the electrochemical reduction of nitrobenzene, and he was ready to present his work at the meeting of the German Electrochemical Society at a session chaired by Ostwald. As always, Haber spoke confidently. The next day, a more senior scientist, Prof. Jacques Löb from Bonn, vehemently attacked Haber's work, and Haber ably defended himself. More important, Ostwald also took Haber's side, a story Ostwald would recall in his autobiography.[11] Although Haber was an entirely self-taught physical chemist, he then wrote a book on electrochemistry that was well received, and he was promoted to the rank of extraordinary professor at Karlsruhe, one step from a full professorship.

Although he was a Prussian, Haber was in many ways more of an outsider than Arrhenius, Ostwald, or Jacobus van't Hoff. By the time he attempted to break into physical chemistry, it was already an established discipline, with scientists from all over the world fighting for places with Ostwald and Nernst. Haber was Jewish, had a mediocre academic record, and was rejected three times by Ostwald. He became a recognized physical chemist through will and intelligence. There is a story that a group of chemists were arguing over a few beers about the lecturing skills of the great physical chemists. When someone asked Haber what he thought, he replied, "I wouldn't know. I never attended a physical chemistry lecture."

Marriage to Clara

By March 1901, Haber was established as an academic chemist and was ready to marry. He had long been interested in Clara Immerwahr, a young scientist from Breslau whom he had met ten years earlier and who was now working for his friend, Richard Abegg, who was on the faculty at Breslau. Haber wrote to the two of them, inviting them to meet him at a scientific meeting in Freiburg. Like Haber, Immerwahr was from Breslau's Jewish community and had converted to Christianity. She was the first woman to receive a doctorate from the university there, where she had studied under Abegg's direction. After she received her degree, Abegg had hired her as his assistant, and he and Clara were close friends. Her letters to him, which have been preserved, show a strange intimacy. Although she referred to him in the third person as "Herr Professor," she showed a girlish flirtatiousness and revealed her doubts and aspirations. Abegg was married, and there is no evidence of a romantic relationship with Clara. Throughout the miserable years of her marriage to Haber, she would continue to correspond with Abegg until his death in a ballooning accident in 1910. (Abegg had also been the confidant of Arrhenius's unhappy first wife, Sofia Rudbeck.) Clara and Fritz were married in August, and the marriage was almost immediately unhappy. Clara was a shy, sensitive woman who had given up a scientific career to manage a professor's household. Fritz thought only of his work and wanted a wife who would host dinner parties and coddle him through his nervous bouts.

After their child, Hermann, was born in June 1902, Haber left almost immediately on a four-month trip to America to gather information on the American electrical industry as the representative of the German Electrochemical Society. His report, which was presented in a speech and then published, was welcomed by the Germans, but many Americans viewed it as industrial espionage, for he disclosed things that the Americans thought they had revealed in confidence.

Haber was becoming frustrated at Karlsruhe. Max Julius Le Blanc had been given the new professorship in physical chemistry at Karlsruhe in 1901, a chair Haber thought should have gone to him. He made Le Blanc's time in Karlsruhe miserable, correcting his errors in seminars and sniping behind the scenes. In 1906, when Ostwald retired at age fifty from his professorship in Leipzig after a spat with the university's administrators, Le Blanc accepted the chair there, possibly in order to get as far from Haber as possible. There were two chief candidates for the now-vacant chair in Karlsruhe: Haber and Fritz Förster, who was already a full professor at Dresden. The faculty commission that evaluated the two found them equal scientifically, but voted three to two in favor of Förster based on his "personal qualities."[12] This might have been anti-Semitism, but might also have been based on what the commissioners had seen of the way that Haber had treated Le Blanc. In any case, Engler, the director of the chemistry institute, knew Haber's merits and fixed things, and his name appeared at the top of the list before the vote was sent to the education minister. At age thirty-seven, Haber was a self-taught full professor of physical chemistry. He was beginning to be seen within Germany as a rival to Nernst.

Nernst and Haber shared an interest in determining the free energy of gas reactions, particularly those of nitrogen with hydrogen or oxygen—the "fixation of nitrogen," which they knew had an enormous potential payoff. Nernst had initially set this as a dissertation problem for Langmuir—the equilibrium of nitrogen and oxygen—as described in chapter 2. When Langmuir's initial results had not been promising, Nernst had lost interest in that particular approach, but he continued to ponder the larger problem. The yields of these reactions of nitrogen were very low—nitrogen was not very reactive, and extraordinary means would be required to increase the yield of nitrogen compounds to an economically attractive level. Increasing the yield meant shifting the equilibrium point of the reactions, and that meant understanding the behavior of the reactions' free energies, a subject of interest to both Haber and Nernst.

Nernst in Berlin, and the Third Law

While Haber was finding his place in physical chemistry, Nernst was at the top of his game. In 1905, he moved from Göttingen to Berlin, the capital of the German empire. During the middle of this move, he came up with the solution to the problem that had puzzled Gilbert Lewis in his dissertation—the determination of the integration constant in what Lewis had called his "general equation for free energy and physico-chemical equilibrium," discussed in chapter 2. Lewis had hoped to find a way that free energies could be determined from measurements of heat, which would be much easier and more universal than Nernst's method of voltages, but he had found himself stuck with an integration constant whose value he did not know how to determine. By the early 1900s, Lewis, van't Hoff, Haber, and Nernst all had concerned themselves with this problem, knowing that its solution would allow them to determine the equilibrium point of a chemical reaction.

In December 1905, Nernst would be the first to announce a solution to the determination of the integration constant—he said that the answer came to him while he was giving a lecture. His achievement would eventually come to be known as the third law of thermodynamics. But the source of his solution would remain a subject of dispute for years, as Theodore Richards claimed that in 1902 he had done something that was the equivalent of determining the integration constant, although he was working on a different problem at the time. Using reliable published sources, Richards collected the heats and the free energies of a number of reactions as a function of temperature to support a paper he was writing on his theory of the compressibility of atoms.[13] His theory is not important in itself; in fact, it shows his fundamental misunderstanding of thermodynamics. He drew the conclusion that free energy was *not* the true measure of chemical affinity, a question that had been settled for twenty years to the satisfaction of nearly everyone else working in the field. But the data he collected was suggestive of how the integration constant might be determined. A competent thermodynamicist who examined his data might have been led to Nernst's ideas.

Richards himself did none of the experimental work to determine the data that he presented but carefully chose published data from scientists whose work he respected. All the data had been available at the time of Lewis's 1899 thesis (which Richards had supervised but apparently not understood), and if Richards had thought then to suggest to Lewis that he examine the same data, or if Lewis had thought of collecting it himself, Lewis might have then seen a way to determine the integration constant in his equation. Some of the data were quite old, and Richards corrected the results with the latest atomic weights, which were his experimental specialty. He plotted the free energies and heats of reaction as a function of temperature, extrapolating the measured data down to absolute zero using known thermodynamic formulas. What was remarkable in the graphical presentation of the data was not that the free energy and heat of reaction were equal at absolute zero—this was clear from the mathematical equation defining the free energy—but rather how rapidly they approached each other as the temperature approached absolute zero. A later diagram of Lewis's illustrates the general pattern that was there to be seen in Richards's data—the pattern that Richards missed:

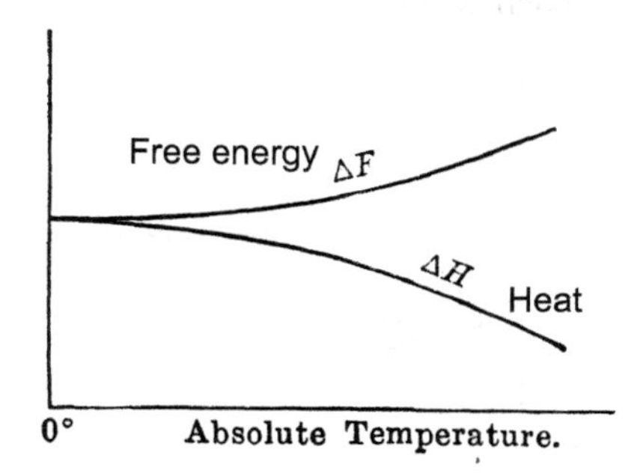

Figure 3-1. Graph showing the pattern in Richards's data. The two curves tail into each other as they approach absolute zero. Gilbert N. Lewis and Merle Randall, *Thermodynamics and the Free Energy of Chemical Substances*. New York: McGraw-Hill, 1923; 437.

It appears that in the reactions that Richards studied, the two curves tail into each other and become parallel to the *x* axis—mathematicians say "become asymptotic with zero slope"—at low temperatures. For any chemical reaction that behaved in this manner, it was a matter of a few simple mathematical steps to take the equation from Lewis's dissertation and come to the conclusion that his integration constant could be set to zero. The mathematics of Nernst's discovery was easy—the key to the discovery was to assume, as Nernst did, that this sort of asymptotic behavior was a general case for the free energy and heat curves for all chemical reactions. Richards did not make this leap.

Had Richards looked around, the person who could have helped him understand what his curves might imply was just down the hall. Lewis had returned to Harvard as an instructor in the chemistry department by 1902, and he certainly understood the importance of the behavior of the two curves. Richards had some awareness that Lewis's equation applied to the situation—he quoted it in his paper but then failed to apply it. Richards had the data, Lewis had the equation and theoretical insight, and if the two of them had spent an afternoon on the problem, it might have all come out differently.

But after being accused of appropriating Richards's ideas on fugacity (chapter 2), Lewis was likely avoiding any discussions with Richards. Apparently, the two were no longer exchanging their papers for mutual review; there is no acknowledgment in Richards's 1902 paper thanking Lewis for reviewing the mathematics as there was in his earlier 1900 paper. Lewis may not have even have seen Richards's data before 1904, when he became fed up with Richards and Harvard and left for the Philippines, where he was to be found when Nernst announced the solution of the problem. In any case, Richards and Lewis missed their opportunity at discovering the third law.

Van't Hoff and Haber had independently made stabs at the problem of determining free energies from heat measurements in 1904 and 1905, respectively. Both had examined Richards's data but failed to see the implications of the asymptotic behavior of the two curves. Van't Hoff had discussed two possible behaviors of the curves as they approached absolute zero:

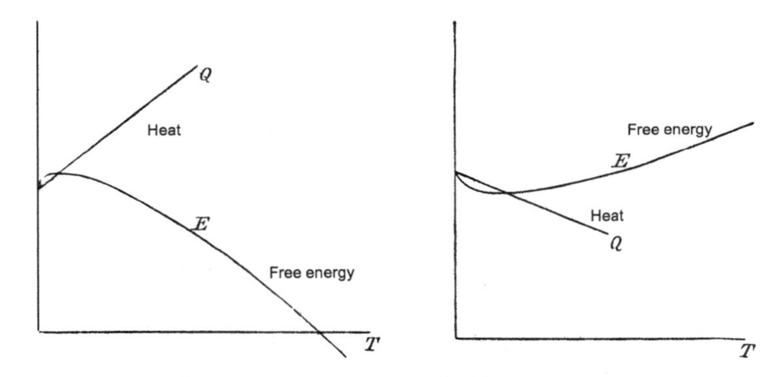

Figure 3-2. Van't Hoff's ideas on how free energy and heat of reaction might behave as they approached absolute zero temperature. He did not consider the case where they approached each other with zero slope, which is what led Nernst to the third law. Jacobus van't Hoff, "Einfluss der Änderung der Spezifischen Wärme auf die Umwandlungsarbeit," in *Festschrift der Ludwig Boltzmann*. Leipzig: Verlag von Johann Ambrosium Barth, 1904; 233–247, on 237.

But van't Hoff had neglected to examine the important and actual case, where the curves approached each other asymptotically.[14] Haber attacked the problem by going off on a tangent, trying to develop a new "chemodynamic" temperature scale defined such that at one degree above absolute zero, the heat of reaction and the free energy would be identical. He hoped that by investigating the experimental behavior of gases, he would be able to determine the magnitude of a single degree on his new temperature scale and then determine the integration constant.[15]

In December 1905, Nernst said "the realization pressed itself forward" that the free energy and heat of reaction curves were asymptotic at low temperatures for all solids and liquids. He published his claims in early 1906.[16] His realization was not entirely correct in the form that he stated it, but it was the necessary insight. He called this his "heat theorem," although it would have been better called his "heat postulate," as it was a simple assertion, and it would not be put on a theoretical foundation until Einstein applied Planck's quantum hypothesis to the problem in 1907. Nernst's heat theorem was based on intuitive grounds, and was possibly based on examination of the figures that had been previously published by Richards, although he did not cite Richards's paper. There is, of course, nothing unethical about drawing conclusions from the published work of others, but if Richards's data were

Figure 3-3. Walther Nernst in 1906, at the time of the heat theorem. Photo by Nicola Perscheid.

in fact the basis for Nernst's ideas, he might have been more forthcoming about his sources.

Before Nernst's announcement of the heat theorem, he and Richards had been on cordial terms. Richards had studied in Nernst's laboratory in 1895, and the two had maintained a collegial scientific correspondence. While Nernst did not mention Richards's work in his original 1906 paper on the heat theorem, he did mention it briefly in his Silliman lectures at Yale in October of 1906:

> [I]n 1902 a very interesting paper was published by Th. W. Richards on "The Relation of Changing Heat Capacity to Change of Free Energy, Heat of Reaction, Change of Volume, and Chemical Affinity". . . . I am very glad to be able to state that our formulae agree qualitatively in many cases with the conclusions of Richards. I do not wish to enter here into a discussion of the difference in the quantitative relations.[17]

At first, Richards did not feel that Nernst had behaved badly: Richards nominated Nernst for the Nobel Prize in Chemistry in both 1909 and 1910.[18] But, as we shall see in the next chapter, Richards would eventually become convinced that he, not Nernst, was the one who deserved credit for the third law. It is possible that after enough of his colleagues looked at his data and asked him, "How could you have missed it?" he came to believe that the important point was his collection and organization of the data, and that the conclusion that Nernst had drawn was trivial and obvious. But both van't Hoff and Haber had missed it, as well, after examining Richards's graphs, so it may not have been as obvious as it seemed to many who looked at the solution retrospectively. Richards reacted to Nernst's heat theorem in the same way that he had to Lewis's fugacity. In both cases, Richards did not know enough mathematics or thermodynamics to fully understand either the problem or its solution, but he knew that his own work was earlier and that it was related to the problem, and so he came to believe that he deserved credit.

For those substances to which it applies, the third law of thermodynamics can be stated a number of different ways: that the integration constant in Lewis's equation may be set to zero, that the free energy and heats of reaction approach each other asymptotically at very low temperatures (Nernst's preferred formulation), or that the reaction's change in *entropy*—a measure of the randomness due to heat or atomic disorganization—is zero at absolute zero. Unlike the first two laws of thermodynamics, which were developed in examining how steam engines produce work, the third law's practical impetus was chemical—the desire to obtain free energies from heat measurements, and Henri le Chatelier, Lewis, Richards, van't Hoff, and Haber had all attacked the problem. Richards would become increasingly bitter about what he came to view as Nernst's theft of his idea, and he seems never to have grasped the essence of the insight that he had missed: that the two curves approach each other asymptotically, not just that one had positive slope near absolute zero and the other negative, which is what he had claimed. He did not mention the missed opportunity of collaborating with Lewis, and Lewis did not discuss whether he had missed seeing the third law in what he would later call "the curves presented by Richards [that] very nearly imply the generalizations which were later to be embodied in the third law of thermodynamics,"[19] nor did Lewis mention when he had first seen Richards's data. In a later letter, Lewis said of Richards, "He had something pretty close to the third law of thermodynamics without realizing what he had."[20] Lewis likely kicked himself, too, over the missed opportunity. And to the end of his life, Haber regretted that it had been Nernst rather than himself who had discovered the third law.[21] But it was Nernst to whom the realization had "pressed itself forward."

The Fixation of Nitrogen

The third law was a breakthrough in chemical theory, and Nernst would immediately apply it to the practical problem of the fixation of nitrogen. In 1898, William Crookes, a British chemist, gave a talk to the British Academy for the Advancement of Science, of which he was then president. His talk was intentionally alarming, just as talks on global warming are today. His message was Malthusian—that Britain faced famine in the not-too-distant future, and that the only hope was to be found with the chemists. The population of England and Wales had more than tripled during the 19th century, and other European nations had shown a comparable increase. Yet the amount of arable land had increased very little, so it had been necessary to increase the yield per acre, which had been done through the use of imported fertilizers. Cereal grains, such as wheat, maize, oats, and rye, deplete the soil of its essential elements, chief among which is nitrogen. The nitrogen in the soil had to be replaced, which had been done traditionally by crop rotation and by spreading manure. Crop rotation involved planting legumes, such as beans or peas, which are able to utilize nitrogen from the air. This replaced the soil's nitrogen, but it also reduced production of more profitable cereal grains. And spreading manure only recycled a portion of the nitrogen that had been extracted from the soil. As the cities grew, the recycling of nitrogen was less efficient than it had been for subsistence farming. When food was exported from farms to cities, the nitrogen was lost in sewers or privies and not returned to the soil. Crookes pointed out that additional nitrogen was already being introduced from outside the closed system so that Britain could grow enough grain to feed its people. One source of nitrogen was the industrial revolution itself. Nitrogen-containing ammonia was a by-product of processing coal into coke for use in Britain's steel industry. In addition, thousands of tons of nitrogen were imported each year from guano-covered islands off Peru and from nitrate beds in Chile. But the nitrogen from processing coal was not sufficient by itself, and the nitrogen from South America was what we would call today a "nonrenewable resource"—it would inevitably run out at some point, and Crookes predicted that when that happened, famine would result. What was true for Britain was, of course, true for all of Europe. But Crookes did see hope, saying, "It is through the laboratory that starvation may ultimately be turned into plenty."[22]

There was no shortage of nitrogen, only of nitrogen chemically bound or "fixed" to other atoms. Nitrogen is the most abundant element in the air, forming nearly 80% of the atmosphere, but it is present in air in an inert form—as nitrogen gas, N_2, which cereal grains are unable to process. Crookes's hope was that a way could be found to chemically combine atmospheric nitrogen with hydrogen, carbon, or oxygen to form compounds that then either could be used directly as fertilizer or could easily be transformed into fertilizer. Ammonia (NH_3), urea ($(NH_2)_2CO$), nitric acid (HNO_3), and nitric oxide (NO) were all likely targets as products of fixation, but the only known methods for their synthesis from atmospheric nitrogen required so much energy as to be economically infeasible. The challenge was to find a reaction that not only would work in the laboratory, but also would have both high yield and low energy costs so that its nitrogen-containing products could be sold at a profit. Haber was the one who would eventually come up with such a reaction, but it is worth looking at the alternative processes before going to his solution.

In the first years of the 20th century, there were three industrial sources of fixed nitrogen. First, as already mentioned, there was coal. About 1–1.4% of coal consists of nitrogen compounds, which are a residue of the vegetable and animal protein from which the coal was formed by nature. When coal is burned, the nitrogen compounds go up the chimney and are lost. If coal is heated in the absence of air, however, as was done in the production of coke for the steel industry, about 12–17% of the nitrogen could be recovered in the form of ammonia if a properly designed oven were used. Extraction of nitrogen from coal was important, but even at best, this was going to account for less than a third of the needed supplemental nitrogen.

Another approach was known as the *cyanamide* process. Ammonia could be synthesized from atmospheric nitrogen starting with calcium oxide in a three-step process. It involved the intermediate synthesis of calcium carbide (CaC_2), a compound reactive enough to combine with atmospheric nitrogen to form calcium cyanamide ($CaCN_2$):

$$CaO + 3C = CaC_2 + CO$$
$$CaC_2 + N_2 = CaCN_2 + C$$
$$CaCN_2 + 3H_2O = CaCO_3 + 2NH_3$$

But even when this process was perfected, formation of the carbide required so much energy that it was uneconomical.

The third industrial process simply united nitrogen and oxygen into nitric oxide using an electric spark in what was called the *electric arc* process:

$$N_2 + O_2 = 2NO$$

This was the chemical reaction that Nernst had assigned Langmuir to investigate for his dissertation, and in which Nernst had lost interest when Langmuir's results did not back up his theory (chapter 2). When this reaction reaches equilibrium, there are actually two chemical reactions occurring simultaneously—nitrogen and oxygen are converted to nitric oxide in one reaction, and nitric oxide is converted to nitrogen and oxygen in the other. The equilibrium point is that point at which both reactions are occurring at the same rate—reactants and products are still being consumed and formed, but there is no net change. Equilibrium applies in everyday situations—the traffic load on a highway, for example, is at equilibrium when the same number of cars are entering and exiting. The equilibrium point of a chemical reaction determines the product yield and can take any value from 0% to 100%; it can be moved one way or the other by changing the reaction's temperature and pressure. In this case, the objective was to move it to the right, forming more nitric oxide.

For electric arc formation of nitric oxide, increasing the temperature would move the equilibrium point in the desired direction. But it required temperatures above 3,000°C if much nitric oxide were to be formed—when Langmuir and Nernst had investigated the reaction at 2,100°C, they had gotten less than 1% nitric oxide. And once nitric oxide was formed, there were still difficulties—it had to be cooled very rapidly or much of it would revert to nitrogen and oxygen, as Langmuir had seen in his experiments. Just as for the cyanamide process, the energy requirements of the electric arc process were economically prohibitive. Both cyanamide and electric arc plants were built, but only in places where cheap hydroelectric power was available. Haber had visited an electric arc plant at Niagara Falls during his visit to America in 1902, but the company closed shop in 1904 because its products were too expensive to be profitably sold.

An obvious approach was to try to make ammonia directly from its elements, nitrogen and hydrogen:

$$N_2 + 3H_2 = 2NH_3$$

The problem with this approach was that the yield was extremely low—the equilibrium point of the reaction lies very far to the left under ordinary conditions, so when nitrogen and hydrogen are combined, virtually no ammonia is formed. Unlike the situation with nitric oxide, the equilibrium point of the formation of ammonia moves to the right—more ammonia—with a *decrease* in temperature, which offered some promise of avoiding the energy costs associated with nitric oxide. In his Nobel lecture, Haber included a table he had calculated (using Nernst's new heat theorem) showing the predicted equilibrium concentration of ammonia as a function of temperature and pressure (table 3-1). From the table, the way to increase the output of ammonia was obvious: high pressure, low temperature.

But it was not that easy. Even the best steel might fail at pressures as high as 200 atmospheres. And operating at low temperatures was good advice for moving the equilibrium point of the reaction in the desired direction, but not for getting the reaction to occur expeditiously. Many chemical reactions do not approach thermodynamic equilibrium quickly. For example, the equilibrium point of the reaction of hydrogen and oxygen to form water lies very far to the right—toward the formation of water—yet mixtures of hydrogen and oxygen can sit for millions of years with no appreciable water formed. Many reactions require some sort of push to get started. An analogy is a car on a hill: take the brake off and put it in neutral, and the car will roll down the hill and reach equilibrium at the bottom. But if the car starts at the bottom of even a small dip in the hill, it will need a push out of the dip before it will begin rolling. The height to get out of the dip is analogous to a reaction's *activation energy*, the energy barrier that must be overcome for the reaction to occur. For a mixture of hydrogen and oxygen, the push can be supplied by a spark or flame—a spectacular example of this is the dirigible *Hindenburg*'s explosion in 1937. An electric spark also worked for the formation of nitric oxide from nitrogen and oxygen—hence the name "electric arc process"—because of the reactive nature of oxygen. But a spark would not work for the formation of ammonia from nitrogen and hydrogen, because nitrogen is much less reactive than oxygen. One way to overcome the activation energy is heat—run the reaction at a higher temperature, shaking all the molecules. This was the approach Haber would take in his early research, when he would produce a

Table 3-1. Haber's predicted concentration of ammonia at equilibrium, as a function of temperature and pressure

Temp., °C	at 1 atm	at 30 atm	at 100 atm	at 200 atm
200	15.30%	67.60%	80.60%	85.80%
300	2.18%	31.80%	52.10%	62.80%
400	0.44%	10.70%	25.10%	36.30%
500	0.13%	3.62%	10.40%	17.60%
600	0.05%	1.43%	4.47%	8.25%
700	0.02%	0.66%	2.14%	4.11%
800	0.02%	0.35%	1.15%	2.24%
900	0.01%	0.21%	0.68%	1.34%
1,000	0.00%	0.13%	0.44%	0.87%

From Fritz Haber, "The Synthesis of Ammonia from Its Elements, Nobel Lecture 20 June 1920," in *Nobel Lectures, Chemistry 1901–1921* (Stockholm: Elsevier, 1966).

very small amount of ammonia at 1,000°C. But for the formation of ammonia, high temperatures had the drawback of pushing the equilibrium point in the worng direction—producing less ammonia—and of course high temperatures would also consume energy and make economic payoff unlikely. Another way to move the reaction along is through use of a *catalyst*—a compound that would not take part in the reaction directly, but that would reduce the activation energy. To return to the analogy of a car on a hill, using a catalyst would be like taking a shovel and reducing the height of the dip so that a smaller push would get the car rolling. A catalyst would not change the reaction's equilibrium point, but it would make the reaction come to equilibrium more quickly. For many gas reactions (including the formation of ammonia, as we will see shortly), metals act as catalysts; for example, in an automobile's catalytic converter, platinum, rhodium, and palladium act as catalysts to convert the carbon monoxide and nitrogen oxides in the exhaust to compounds less damaging to the environment. When metals are used as catalysts in gas reactions, the gases generally adsorb onto the surface of the metal, where they are able to react more easily. As the products are formed, they leave the catalyst's metal surface and make room for the next set of reacting molecules.

In 1900, Ostwald announced that he had succeeded in making ammonia from hydrogen and nitrogen at high temperature and pressure using an iron catalyst. He filed for a patent and notified the chemical giants Hoechst and BASF. BASF immediately negotiated an agreement with Ostwald to investigate his process and turned the investigation over to a young researcher, Carl Bosch, who reported that the ammonia that Ostwald had produced came not from atmospheric nitrogen but from contaminants in the catalyst Ostwald was using. Ostwald was irate and told BASF that "when you entrust the task to a newly hired, inexperienced chemist who knows nothing, then naturally nothing will come out of it."[23] But Ostwald eventually realized that Bosch was correct, and he withdrew the patent application. Working with Haber, Bosch would eventually be the one to make ammonia synthesis a commercial reality and in 1931 would be the first chemical engineer to receive the Nobel Prize in Chemistry.

Henri le Chatelier tried synthesizing ammonia in 1901 but abandoned it when his test apparatus exploded under high pressure. In 1904, Haber was approached by the Margulies brothers, two industrial chemists from Vienna who hired him as a consultant to undertake the problem. He began by examining the equilibrium between hydrogen and nitrogen at 1,000°C in the presence of iron, as Ostwald had done, but devised a system that would immediately remove ammonia as it was formed. Removing the ammonia shifted the equilibrium toward making ammonia. Haber was able to achieve some very small amount of ammonia synthesis—about 0.01%. He realized that the synthesis would work commercially only at very high pressure and at a lower temperature, which meant finding a better catalyst than iron. He gave the Margulies brothers a discouraging report and published his results in 1905.

Nernst Attacks Haber, and Haber Attacks Ammonia

In 1906, Nernst was busy applying his new heat theorem to data published by others in an effort to validate his claims. He found that Haber's 1905 results on ammonia were at variance with his predictions and wrote Haber about the discrepancy. Nernst then repeated Haber's experiments and found significantly more ammonia than Haber had—in fact, even more than his own heat theorem predicted. Haber, having just achieved his professorship in physical chemistry at Karlsruhe, was unhappy at being reproached for sloppy work by Nernst and immediately repeated his own measurements. He found more ammonia than he had originally detected (but less ammonia than Nernst had found), and the amount he found almost exactly agreed with Nernst's predicted theoretical value. One would think that Nernst would have been happy with this confirmation of his new heat theorem.

But Nernst, being Nernst, was not. At a scientific meeting in 1907, he attacked Haber's results, saying that they were once more incorrect, and that the commercial synthesis of ammonia was impossible: "It is very unfortunate that the equilibrium is more displaced toward the side of very low ammonia formation than the strongly inaccurate figures of Haber had formerly led us to assume, since one had inferred from them that it might be possible to synthesize ammonia from nitrogen and hydrogen. Now, however, the conditions are much less favorable, the yields being about a third what was thought previously."[24] Svante Arrhenius, whose rivalry with Nernst was unabated, endorsed Haber's results, but Nernst's unprovoked attack damaged Haber's reputation as one of the leading young physical chemists, and Haber decided that he could not allow it to stand unanswered. Nernst had given up on synthesizing ammonia, but Haber and his collaborator, Robert Le Rossignol, an English chemist working with him at Karlsruhe, began to repeat the experiments again, this time at 30 atmospheres pressure, which increased the yield by a factor of 28. Haber began to believe that ammonia synthesis might actually be commercially feasible, but only if a catalyst could be found so that the temperature could be reduced. When he looked at the trade-offs, it seemed to him that operating at 200 atmospheres and 600°C was the point to shoot for, which he predicted would give an 8% ammonia yield at equilibrium (see table 3-1). By removing the ammonia as it was formed and recycling the unreacted nitrogen and hydrogen, ammonia might then be produced in a continuous fashion.

About the same time, Haber had made some improvements in the electric arc process for the production of nitric oxide and found that BASF was interested. Carl Engler, Karlsruhe's director who had pushed through Haber's professorship, was a member of BASF's board of directors. Engler urged BASF to engage Haber to work on the fixation of nitrogen, and Haber signed a contract in 1908 promising him 10% of the profits from any such process that he developed. Although BASF's interest was in the electric arc production of nitric oxide, Haber believed that the most promising approach was the direct synthesis of ammonia from nitrogen and hydrogen, and that was where he put his efforts. In 1909, he found a catalyst that would work—osmium, a rare element of which only 100 kg had been isolated in the entire world. Osmium was being used as an electric filament by a lighting company, the Auergesellschaft, for which Haber was a consultant, which was why he had osmium on hand to try.

Chemistry was only part of the problem. This was perhaps the most demanding technical chemical project ever undertaken in an academic setting. The design of a reactor that would remove ammonia as it was formed, would recycle the unreacted material, and would work at 200 atmospheres pressure was largely the work of Le Rossignol, aided by Karlsruhe's technician, Friedrich Kirchenbauer. Le Rossignol and Kirchenbauer designed a new conical valve that could withstand the high pressures. As the hot gases left the reaction chamber, they were used to preheat the incoming gas for the next cycle so that no energy was wasted. The ammonia was removed as it was formed, and the unreacted nitrogen and hydrogen were fed back into the reaction chamber. In late March 1909, Haber and Le Rossignol put the new reactor and the osmium catalyst together and fed in the nitrogen and hydrogen. As the ammonia was produced, it was chilled below its boiling point and fed drop by drop into a collection flask. Haber ran through the institute, calling to everyone, "Come on down! There's ammonia!"

Haber wrote to BASF, telling them the news and encouraging them to buy up the world's supply of osmium, but he found them skeptical. BASF was interested in the electric arc process and was not at all sure that the pressures required by Haber's ammonia synthesis were feasible. Engler interceded, and BASF agreed to come to Karlsruhe for a demonstration. Three attended from BASF: Heinrich von Brunck, the chairman of BASF's board; August

Figure 3-4. The Haber-Le Rossignol apparatus for the synthesis of ammonia from nitrogen and hydrogen. Deutsches Museum, Munich.

Bernthsen, BASF's research director; and the same Carl Bosch who had debunked Ostwald's patent application nine years earlier. Bernthsen skeptically asked what the required pressure would be. Haber shaded the truth, saying, "at least 100 atmospheres." Even this value, half the pressure in the demonstration he was about to see, appalled Bernthsen, who said, "100 atmospheres! Why just yesterday seven atmospheres blew up one of our autoclaves!" But Bosch, who said that he knew what the steel industry was capable of, said, "I think it can work," and von Brunck committed to funding the process's development.

Continuing his work on catalysts, Haber found that uranium served as well as osmium and, although expensive, was much more available. BASF was trying to get its patents filed and to keep things as secret as possible, but Haber, over BASF's objections, announced his results at a scientific meeting in Karlsruhe on 10 March 1910, talking about the "extraordinary need for bound nitrogen, mainly for agricultural purposes and to a much smaller extent for the explosives industry."[25] Given that the huge arms buildup that would lead to World War I was well under way, this was an understatement. Nitrogen was as essential to the manufacture of explosives as it was to fertilizer—the explosive element in gunpowder is potassium nitrate, or saltpeter. As Timothy McVeigh showed with his fertilizer-truck bomb in Oklahoma City, there is no essential difference between nitrate-based fertilizer and a weapon. Haber was well aware that the German government was worried about being cut off from the supplies of South American nitrates by a British blockade during a war and saw his work as vital for the national defense.

Haber's announcement was immediately viewed as a great discovery, and he was inundated by offers from chemical companies. BASF had been unable to get him to forgo the announcement, but he had agreed to suppress as many of the details as possible. But the word was out—Haber had done what had seemed to be the impossible.

Haber Becomes an Adviser to the Kaiser

Two months after Haber's talk, the Prussian ministers called a meeting of industrial moguls and made a proposal that had been endorsed by Kaiser Wilhelm himself—that the German industrialists follow the model of the Rockefeller and Carnegie families in the United States, funding scientific institutes that would keep German science and the German nation at the forefront. Kaiser Wilhelm had modestly agreed that these might be called "Kaiser Wilhelm Institutes." Leopold Koppel, an industrialist who controlled the Auergesellschaft, the lighting company that had supplied Haber with osmium, saw an opportunity. He had been happy with Haber's work as a consultant and had offered him the position of research director at his firm, which Haber had declined. Koppel offered to fund a Kaiser Wilhelm institute of physical and electrochemistry in Berlin if two conditions were met: Haber would be the director, and the institute would be able to apply for patents, which would then be available for license. Koppel, of course, expected an inside track on the patents. He was quite explicit about his motives—he wanted to fund physical chemistry "because this has the most direct impact on the industry to which I am personally closest."[26] Although he had agreed to fund the institute itself, the salaries, including Haber's, would be paid by the national government, something that would become important in 1933 when the Nazis would fire all Jewish civil servants (see chapter 6). At the announcement ceremony for the new institutes, the biochemist Emil Fischer gave the keynote address. When Fischer mentioned Haber's achievement of extracting nitrogen from the air for fertilizer, the Kaiser nodded his head.[27]

The organic chemist Richard Willstätter, who was to become Haber's closest friend, agreed to head another Kaiser Wilhelm institute next door to Haber's, and he built his home across the street from Haber's. Like Haber, Willstätter was a Jew who felt himself a German patriot, but he had not converted to Christianity. Willstätter wrote that Haber's appointment to director of the institute "completed [Haber's] transition from great researcher to great German."[28] Haber had now achieved everything he had aimed for. He had acquired a family (although his wife Clara felt stifled and ignored); he was universally acknowledged as a great physical chemist; he had a brand-new Kaiser Wilhelm Institute for Physical and Electrochemistry in Dahlem, a suburb of Berlin; he would soon be wealthy through the ammonia patent; and he was as German as was possible: he was now an adviser to the Kaiser—a *Geheimrat*. He was a scientific leader in the new German nation.

Making Ammonia Industrially: Carl Bosch

Turning Haber's laboratory synthesis of ammonia into a commercial ammonia plant was no easy task. Carl Bosch's statement, "I think it can work," was ambitious, because no one had yet built a plant that would operate at a pressure this high. When Bosch made that statement in 1909, he was thirty-five years old. He was the son of a master plumber and had received a secondary education in a technical *Oberrealschule* rather than an academic *Gymnasium*. After graduation, he spent a year in a smelting company, where he apprenticed as a molder, mechanic, carpenter, and locksmith. He then went to the Technische Hochschule in Charlottenburg (the same school that Haber had attended a few years earlier) and studied as a metallurgist and a mechanical engineer, gaining practical experience in the steel mills. Only then did he study organic chemistry, earning a doctorate with Johannes Wislicenus at Leipzig in 1898, and then staying for another year as his assistant. In his application to BASF for employment, he listed not only his academic credentials but also his practical apprenticeships.[29]

The industrial synthesis of ammonia came to be known as the Haber-Bosch process, and Bosch deserved to have his name on the marquee. The process had begun with theory and moved step by step toward practice: Haber had begun working on the problem in earnest only after a theoretical dispute with Nernst over Nernst's predictions from his heat theorem; using theoretical predictions, Haber had decided on the optimum temperature and pressure; by trial and error, Haber had found catalysts that would make the reaction possible at the target temperature, and Haber, Le Rossignol, and Kirchenbauer had come up with a test rig that would produce a few drops of ammonia, showing the process's potential. But a commercial process would need to produce tons of ammonia every day. Bosch was a problem solver, and there would be plenty of problems in making the ammonia synthesis work. Bosch saw three principal difficulties: finding a low-cost supply of pure hydrogen and nitrogen feedstocks, finding a better and cheaper catalyst, and designing reaction converters that could handle 200 atmospheres pressure.

The hydrogen and nitrogen that were to be used had to be very pure, or the impurities introduced would poison whatever catalyst was used. Pure nitrogen could be obtained from the fractional distillation of air, but pure hydrogen was more of a problem. BASF had ready supplies of coal, and Bosch decided to produce hydrogen by reacting glowing coke (pure carbon, made by heating coal in the absence of air) with water vapor according to the equation

$$C + H_2O = H_2 + CO$$

Finding a catalyst was a bigger problem. Haber had come up with osmium and uranium as catalysts, but neither was suitable for a commercial process. Both were expensive, osmium was too rare, and uranium was too sensitive to contamination by oxygen and water. Bosch assigned Georg Stern the job of testing catalysts, and Stern designed a high-pressure apparatus into which cartridges containing 2 grams of a potential catalyst could be quickly inserted. Stern had thirty of these test stations operating simultaneously, testing every metal sample that he could lay his hands on. He found that a one-off sample of iron ore from a mine in Sweden worked as well as osmium or uranium. It seemed that trace amounts of other materials were the key to its success as a catalyst, and Stern began examining these. He eventually identified not only catalytic promoters but catalytic poisons that must be rigorously excluded, including sulfur, phosphorus, and chlorine. So Stern found after a great deal of work that the most successful catalyst was iron, with which Ostwald had begun in 1900. The difference was that only the right sort of dirty iron would work.

Bosch had two test reactors built, which he housed in concrete enclosures as a safety precaution. After 80 hours of operation, both of them burst. When the engineers examined the converters, they found that the metal had become brittle and had fractured. Because the reactor needed to withstand such high pressures, it had been designed with the strongest steel available—high-carbon steel. Bosch discovered that the hydrogen had diffused into the steel walls and then reacted with the carbon in the steel to form methane, a gas. When the methane formed within the steel, it was at 200 atmospheres pressure, and it quickly caused the walls to disintegrate. Bosch's solution to the problem was a two-tube reactor—a thin inner tube made of soft, carbonless steel enclosed by a carbon-steel jacket that could withstand the pressure. While the hydrogen could diffuse into the inner lining, it could not damage it, because there was no carbon there to form methane. Whatever hydrogen diffused through the inner tube to the carbon-steel outer jacket was allowed to escape through small holes drilled in the jacket.

Bosch's philosophy was to measure wherever possible. Pressure, temperature, and gas sensors for products, reactants, and contaminants were placed at each stage in the process, so that he could know exactly what was going on within the reactor. In May 1910, just ten months after Haber's demonstration in Karlsruhe, Bosch had a small pilot plant in continuous operation, and by December 1910, he had a larger pilot plant producing 18 kilograms of ammonia per day.

The Patent Suit: Haber and Nernst Join Forces

Haber and BASF had, of course, filed for a number of patents and had kept what Bosch was doing technically as quiet as possible. But BASF's competitors could see what Haber's discovery would mean and feared that BASF would monopolize the new technology. BASF's rival Hoechst filed a claim that Haber's principal patent—on the catalytic synthesis of ammonia from nitrogen and hydrogen—was void, on the grounds that it had been done earlier by Nernst in 1907. Hoechst had retained Ostwald as a consultant for the trial, and he gave an opinion that the patent was simply an extrapolation from Nernst's low pressures to higher pressures, which was easily predicted by applying the laws of thermodynamics without the need for experimentation. According to Ostwald, everything Haber had done was just engineering.

If BASF were to prevail, it would need a consultant of its own, and who better than Nernst? Nernst no longer had a horse in the ammonia race, as he had abandoned research in ammonia in 1907 when he had concluded that the "the strongly inaccurate figures of Haber had formerly led us to assume . . . that it might be possible to synthesize ammonia from nitrogen and hydrogen." In return for 10,000 marks per year for five years (a total of about $225,000 in 2007 purchasing power), he agreed to a consulting contract with BASF.[30] Two days before the patent trial, the BASF side judged its case to be weak and believed that Hoechst would likely prevail. But when the trial began, Nernst carried the day. Richard Weidlich, Hoechst's attorney, apparently did not know before the trial that Nernst was now on the BASF payroll and had this to say after the trial:

> It was very doubtful what the decision . . . would be. . . . During the oral arguments, to my surprise, Nernst appeared arm in arm with Haber, and after Bernthsen [the BASF research director] had given only a brief reply to my thorough justification for our suit, my principal witness, Nernst, whose fundamental contributions to the ammonia synthesis I had just praised, made a passionate speech in which he explained that his work had no technical relevance and his results were of scientific interest only. Only Haber had created the prerequisites for a technical success by investigating new pressure ranges. If chemical entrepreneurs had not had any interest in his, Nernst's, work, that was understandable. However, had such an entrepreneur rejected Haber's process, then he would have had to have been blind. The decisive effects of Nernst's speech were evident, and I, too, was impressed, so that I whispered to Müller-Berneck [his colleague], who was accompanying me, that we could go home.[31]

On 12 March 1912, BASF won the case, and Hoechst was ordered to pay all costs of the suit. And the feud between Haber and Nernst was settled as well, as Nernst had retracted his nasty remarks (granted, for payment) and had in the process made Haber a rich man. Haber had come to an agreement with BASF that he would receive 0.015 marks per kilogram of ammonia produced over the fifteen-year life of the patent, which would amount to several million marks by the mid-1920s.[32] (The German mark of the time had a value of about $0.23, equivalent in purchasing power to about $4.50 in 2007.) Nernst, with his lamp, and Haber, with his ammonia synthesis, had both proven that it was possible to be an academic scientist and to become rich, too.

With the patent case against Hoechst settled, BASF was ready to proceed to commercial production by building an enormous ammonia plant at Oppau. The plant was almost entirely self-contained. The factory really needed only one input other than air and water—coal, which was used for heating, coking, and making hydrogen from reaction of coke with water vapor. The Oppau plant was producing 10 metric tons of ammonia a day by the end of 1913 and double that in 1914, about 70% of it used for fertilizer. But on 28 June 1914, everything changed: Gavrilo Princip, a Serbian nationalist, assassinated Archduke Ferdinand of Austria-Hungary. World War I erupted on 1 August 1914, and the ammonia from Oppau was no longer scheduled for fertilizer, but for munitions.[33]

4

Chemists at War

Haber, Nernst, Langmuir, and Lewis

On 22 April 1915, Fritz Haber, now a captain in the German army, looked out from the German trenches at Ypres and watched as his Pioneer regiment opened the valves on 6,000 cylinders of chlorine gas. The gas fed through manifolds to nozzles pointed toward 7,000 meters of the Allied lines. As it was released, the chlorine formed a cloud 10–30 meters high, white at first from the water that condensed as the cold gas escaped, but then yellow-green. The cloud advanced with the wind at the pace of a slow walk. The soldiers in the French trenches were Algerians. They were used to artillery, to machine guns, to mortars, and even to the tear gas that the Germans called *T-Stoff*, but this was something new. Fritz Haber had organized the first attack with lethal gas.

Immediately before the war, Haber had been nursing his health problems, this time gallstones and nervousness, at a spa in Karlsbad.[1] Walther Nernst and his wife Emma had just returned from a lecture tour in South America, where Nernst had studied Spanish and Emma had checked his homework just as she had done for the children.[2] But when the war began, both Nernst and Haber immediately volunteered for the military, although Nernst was now fifty and Haber forty-six. And both of them, as advisers to the Kaiser, began applying what they knew—chemistry—to the German war effort. The United States would not enter the war for almost three years, and then both Gilbert Lewis and Irving Langmuir would apply themselves to the Allied war effort.

The first three chapters of this book have presented scientists as acting with a number of motivations—a desire to understand nature; personal and corporate aspiration to wealth; ambitions for advancement, recognition, and awards; and personal rivalry. This chapter introduces another element—militaristic patriotism. Science was more important in World War I than in any previous war, and scientists—particularly chemists—were called to participate. World War I remains the only international conflict in which large-scale chemical attacks with lethal gas were a standard tactic, although use of chemicals in combat has certainly not disappeared—as evidenced by the use of the defoliant Agent Orange by the U.S. military in Vietnam and of poison gas by Saddam Hussein against internal dissidents in Iraq, to name only two of many cases. The story of chemists' participation in gas warfare is disturbing, and I have not attempted to make it any less so. I have tried to describe their actions dispassionately, using the same tone as in describing their peacetime work. I use the same terminology that they would have used—an "improved" poison gas that "works better" is one that is more lethal and can be delivered more effectively. The scientists involved in chemical warfare saw themselves as solving technical military problems—attempting to kill the enemy or to defend the soldiers of their own nation—and used the same sort of reasoning that they had applied to the problems of a better electric lamp or to the synthesis of ammonia. Their methodology did not change substantially from peacetime to wartime, nor did their motivations: greed, ambition, and personal rivalry were still in evidence. The principal change was a substitution of patriotism for a desire for scientific understanding. There was a war on, and scientists did not always have the time to stop and examine everything in detail.

Choosing a Poison Gas

Nitrogen, oxygen, and chlorine fall in three adjacent columns of the periodic table (see fig. 1-4 in chapter 1), and they also fall in that order in increasing reactivity to most organic compounds. In its gaseous form, nitrogen (N_2, or N≡N) is held together by a triple bond that takes almost twice as much energy to break as does the double bond in oxygen (O_2, or O=O). Because the bond must be broken before a gas can react with other compounds, nitrogen is much less reactive than oxygen—which is why Haber's forcing nitrogen to unite with hydrogen to form ammonia had been such an achievement. Chlorine (Cl_2, or Cl–Cl) is held together by a single bond that takes only about half the energy to break as does the double bond in oxygen, making chlorine more reactive still. Once the chlorine bond is broken, the resulting chlorine atoms are themselves very reactive, pulling on the electrons in almost any other substance. When chlorine comes into contact with unprotected human tissue, it reacts immediately, burning the skin or the eyes if the exposure is prolonged or concentrated. When chlorine is inhaled, it corrodes the lungs, which fill with fluid. There is no antidote to chlorine poisoning—with moderate exposure, the body may heal itself, but if the exposure is severe, the victim drowns in his or her own fluid. One soldier described it this way: "It produces a flooding of the lungs—it is an equivalent death to drowning only on dry land. The effects are these—a splitting headache and terrific thirst (to drink water is instant death), a knife edge of pain in the lungs and the coughing up of a greenish froth off the stomach and the lungs, ending finally in insensibility and death. The color of the skin from white turns a greenish black and yellow, the color protrudes and the eyes assume a glassy stare. It is a fiendish death to die."[3]

Haber was the one to advise the use of chlorine. It had several advantages: it was commonly available in cylinders as an industrial chemical, it is quite poisonous, and it is heavier than air, so it sank into the trenches. Haber watched as the first chlorine cloud approached the French lines. The Algerian soldiers who had not succumbed ran in terror. If one could get clear of the cloud, it could be outrun, and it in fact provided a visual screen against the German machine guns, but it was easy to become blinded and disoriented. When the gas had dissipated sufficiently, the German soldiers (who had not been issued protective masks but only wet cloths to hold over their faces) advanced tentatively for four miles into the abandoned French positions. The Germans were almost as surprised by the effects of the gas attack as were the French and stopped at that point to regroup. Night was falling, and they had accomplished their immediate objective. For the rest of his life, Haber would claim that this had been the missed opportunity for Germany to win the war. If the generals had only listened to him and had been ready to advance in force, he believed that they could have shattered the Allied lines and forced a negotiated peace.

When war had been declared on 1 August 1914, the Kaiser had presented himself as the personification of the Reich and of the German *Volk*. He had been called out onto the balcony of his palace time after time that evening to thunderous ovation, while he proclaimed, "I shall lead you forward to glorious times." The German General Staff had planned for a quick war, invading neutral Belgium on the way to a wheeling attack across France. If France could be forced to surrender quickly, as it had in the Franco-Prussian War in 1870, the General Staff believed that France's allies, Britain and Russia, would be compelled to negotiate. It almost worked. But on 6 September 1914, the French counterattacked at the Marne River, just southeast of Paris, and General Alexander von Kluck broke off his advance to meet the attack. On 9 September, von Kluck was forced to retreat across the Marne River, and the western front bogged down into four years of trench warfare.

Nernst and Haber, Companions in Arms

Nernst, who was interested in automobiles and owned several, volunteered as a military driver and was sent on 21 August with his car to carry documents from the General Staff to von Kluck's army during its advance. He stayed with von Kluck through the battle of the Marne and came close to being captured in the retreat. When he returned home, he told his closest friends and family that the war was lost—he believed that Germany could not win a long war. Nevertheless, he put all his efforts into the war over the next four years, trying to stave off defeat even if he no longer believed victory was possible. His two sons, Rudolf and Gustav, were immediately called up, and Rudolf was killed in the war's first year.[4]

The Germans had bet everything on a quick victory and were completely unprepared for a longer war. Trench warfare consumed an enormous amount of artillery and small-caliber ammunition on a daily basis, and the Germans had enough munitions on hand for only six months. The British navy deprived the Germans of imports from outside Europe, and the German High Command had no concept of what the economy would need to do to keep the army functioning in the field. Three weeks into the war, a meeting was called at the German war ministry at the instigation of Walther Rathenau, an industrialist who had succeeded his father Emil as the head of AEG, the company to which Nernst had sold the patent for his electric light. Rathenau had convinced the army chief of staff, General Erich von Falkenhayn, that something must be done quickly. Rathenau was put in charge of the materials department of the war ministry. Since the generals were now running the country, this meant that Rathenau was running the economy. If there were to be any hope of Germany winning (or even of surviving more than a few months), substitutes would need to be devised for essential materials that were unobtainable, a continuing supply of munitions would need to be found, and new weapons would need to be developed. Rathenau called in the scientists and put Haber in charge of the chemistry effort—the "Haber office," as it came to be known.

The first priority was munitions, and the Haber-Bosch process would be the key to providing them. When the Germans seized the Belgian port at Antwerp, they got a brief respite when they found 20,000 metric tons of nitrates in a warehouse. But that was only one month's supply, and they would need to produce the same amount each month on their own, because the British navy had now cut them off from the Chilean nitrate beds. If the Germans were to use Haber's process to make munitions, the ammonia produced at BASF's Oppau plant would need to be converted to nitrates. While this was not a trivial problem, it was something that could be done. Wilhelm Ostwald had developed a process to oxidize ammonia to sodium nitrate using platinum as a catalyst, and BASF found that a less expensive iron catalyst would also work. BASF negotiated to build a plant that could convert the Oppau ammonia to nitrates, and promised that it could have the plant ready for production in six or seven months. Haber was, of course, involved in the discussions, but he had conflicting interests: the Kaiser's government, to which he was a *Geheimrat* or adviser, and BASF, to which he was a consultant, and from which he received royalties on ammonia made using the Haber-Bosch process. The government would have been happy had any ammonia—including ammonia from the production of coke or from the cyanamide process—been used to feed the new BASF plant for producing nitrates. But Haber and BASF both wanted to see that the ammonia used for conversion to nitrates came exclusively from BASF and the Haber-Bosch process. In the meetings with Rathenau and others, where he was acting as an adviser to the government, he did disclose that he had a financial interest, but he was not explicit about just how large his interest was. He did not patriotically offer to forgo royalties on ammonia used to produce ammunition for the Reich. In 1918 alone, he would receive royalties equivalent to about $4 million in purchasing power today.[5] In return for agreeing to build the new nitrate plant, BASF was granted rights to exclusive supply of the input ammonia.[6] The ammonia production at Oppau was increased fourfold to 80 metric tons a day, and a massive new ammonia plant was built far to the east at Leuna. It would be BASF and the Haber-Bosch process that would prolong the war for four long years, feeding the artillery batteries and the machine guns that froze the Allies in place along the western front.

The Manifesto of the 93

The Allied press portrayed the Germans as barbarians, particularly around the German invasion of neutral Belgium, printing stories of German soldiers bayoneting Belgian babies and indiscriminately raping Belgian women. It was certainly true that the Germans had planned the invasion of neutral Belgium—this was central to their Schlieffen plan for the invasion of France. While the Allied press exaggerated the extent of German atrocities in Belgium, there was substantial truth to the stories; as many as 6,500 Belgian civilians and prisoners of war were executed by the Germans, with the atrocities worst in Louvain.[7] In October 1914, ninety-three German intellectuals, many of them scientists, attempted to refute what they saw as Allied propaganda and issued a "Proclamation to the Civilized World":

> As representatives of German Science and Art, we hereby protest to the civilized world against the lies and calumnies with which our enemies are endeavoring to stain the honor of Germany in her hard struggle for existence—in a struggle that has been forced on her.
>
> The iron mouth of events has proved the untruth of the fictitious German defeats; consequently misrepresentation and calumny are all the more eagerly at work. As heralds of truth we raise our voices against these.
>
> *It is not true* that Germany is guilty of having caused this war. Neither the people, the Government, nor the Kaiser wanted war....
>
> *It is not true* that we trespassed in neutral Belgium. It has been proved that France and England had resolved on such a trespass, and it has likewise

> been proved that Belgium had agreed to their doing so. It would have been suicide on our part not to have been beforehand.
>
> *It is not true* that the life and property of a single Belgian citizen was injured by our soldiers without the bitterest defense having made it necessary....
>
> *It is not true* that our troops treated Louvain brutally. Furious inhabitants having treacherously fallen upon them in their quarters, our troops with aching hearts were obliged to fire a part of the town, as punishment. The greatest part of Louvain has been preserved....
>
> *It is not true* that our warfare pays no respects to international laws. It knows no undisciplined cruelty. But in the east, the earth is saturated with the blood of women and children unmercifully butchered by the wild Russian troops, and in the west, dumdum bullets mutilate the breasts of our soldiers....
>
> *It is not true* that the combat against our so-called militarism is not a combat against our civilization, as our enemies hypocritically pretend it is. Were it not for German militarism, German civilization would long since have been extirpated....
>
> We cannot wrest the poisonous weapon—the lie—out of the hands of our enemies. All we can do is proclaim to all the world, that our enemies are giving false witness against us....
>
> Have faith in us! Believe that we shall carry on this war to the end as a civilized nation, to whom the legacy of a Goethe, a Beethoven, and a Kant, is just as sacred as its own hearths and homes.[8]

Among the scientists described in this book, Nernst, Haber, Ostwald, Max Planck, Paul Ehrlich, Emil Fischer, Philipp Lenard, Wilhelm Wien, and Richard Willstätter all signed what became knows as the "Manifesto of the 93." Ostwald not only signed but, according to author Sven Widmalm, "proclaimed the war to be a good thing since it would lead to Europe's subordination under a higher and more efficient form of civilization, i.e., the German one."[9] Einstein signed a pacifistic countermanifesto, but it received little support and was not released.

Haber's Kaiser Wilhelm Institute Goes to War

Haber turned his Kaiser Wilhelm Institute over to war work. His first efforts were to develop lubricants, antifreezes, and fuels that would allow the German army to operate on the eastern front in the Russian winter. Scientists under his control also began work to develop tear gases that could be fired from artillery shells. Germany had signed the Hague conventions of 1899 and 1907, which dealt with the use of gases in warfare. The 1899 convention stated that "the Contracting Powers agree to abstain from the use of projectiles the sole object of which is the diffusion of asphyxiating or deleterious gases," and the 1907 convention stated that it was especially forbidden "to employ poison or poisoned weapons" and "to employ arms, projectiles, or material calculated to cause unnecessary suffering."[10] The use of tear gas was questionable under these treaties, which were vaguely worded, but the Germans would later argue legalistically that even their use of lethal gases was not covered. They would focus on what exactly was a "poison," on whether an explosive shell filled with a gas had as its "sole object" the diffusion of that gas (since the explosive was also lethal), and whether gas warfare caused "unnecessary suffering" when compared to other methods of being killed. And besides, they would say, the French started it all anyway.

The German effort at developing gases began with an idea by Lieutenant Colonel Max von Bauer, who suggested using shells containing materials that would damage the enemy or make him unable to fight. Falkenhayn, the army chief of staff, instructed Bauer to call a meeting, which Nernst and Carl Duisberg, CEO of the chemical company Bayer, attended. Nernst approved the idea, and someone, possibly Duisberg, suggested using dianisidine chlorosulfate, which caused uncontrollable sneezing and could be readily synthesized from a chemical that was already in stock at Bayer's Leverkusen factory.[11] Several hundred kilograms were loaded into artillery shells, and the first chemical attack took place at Neuve Chapelle on 27 October 1914. It was not a success; the gas dispersed so rapidly that the Allied forces were unaware that the shells had contained anything other than explosives. By the end of the year, the Germans had improved things, using howitzer shells loaded with a somewhat more effective tear gas, xylyl bromide, given the code name *T-Stoff.* Nernst also suggested loading shells with phosgene, a poisonous gas that saw much wider use later.[12]

While working on weapons research at Haber's institute on 17 December 1914, two of Haber's senior colleagues, Gerhard Just and Otto Sackur, mixed two chemicals in a test tube that Just was holding. The explosion blew off Just's hand, but Sackur caught it full in the face. Haber's wife Clara, who was in the next laboratory, ran in and attempted to help Sackur, but he died shortly thereafter. Both Haber and his close friend, Richard Willstätter, the director of the neighboring institute who would also work on chemical weapons, wept uncontrollably at the funeral.[13] The day after the accident, the work continued. Haber viewed a test firing of *T-Stoff* and said that he thought that the small amount of tear gas that could be delivered with artillery shells would be useless, suggesting that trench mortars be used instead, which could fire entire drums of gas a short distance. When he was told the mortars were not available, he came up with what he thought was a better idea—releasing chlorine from gas cylinders and letting the wind blow the gas into the Allied lines.[14]

By the end of 1914, Walther Rathenau and Haber had fallen out, ostensibly over a conflict in management styles.[15] It is possible that Rathenau's decision to dismiss Haber may have been prompted by what he saw as Haber's self-serving dealings with BASF and with Auergesellschaft, the firm run by Leopold Koppel, Haber's patron who had endowed his Kaiser Wilhelm Institute. But Haber's talents were clearly of value to the military, and he was put in charge of the new effort at chemical warfare. Haber's son Ludwig said of him, "In Haber, the [German High Command] found a brilliant mind and an extremely energetic organizer, determined, and possibly unscrupulous."[16]

Haber held only the rank of sergeant in the army reserves, but he held a much higher position in civilian life as director of his Kaiser Wilhelm Institute. That made him a *Geheimrat*—an adviser to the Kaiser. The Kaiser bypassed the normal channels and promoted Haber to the rank of captain, giving him the uniform and military prestige of an officer, which he had been denied a quarter of a century earlier when he had been turned down in his application for a reserve commission during his compulsory military service.

In early April, Haber was ready to give chlorine a try. In the first lethal gas attack at Ypres on 22 April 1915, three future Nobel laureates took part: Haber himself, and James Franck and Otto Hahn as his assistants. After the war, Haber and Nernst talked with Sir Harold Hartley, a British physical chemist who had been involved in the British gas service and had been assigned the task of debriefing the German scientists. Both Haber and Nernst had approached gas warfare in the same way that they had earlier approached other technical problems. They both called the chlorine attack at Ypres an "experiment"—they used the German word *Versuch*—saying that insufficient gas had been used and that the soldiers had lacked imagination.[17]

Release of gas from cylinders required reliable meteorological information on wind speed and direction. If the wind were too strong, the gas would disperse, and if the wind direction shifted, there was a risk of blowback at the German troops, who had not been issued defensive masks. The first attack had to be delayed several times. When the gas was finally released, the French line collapsed, and the German High Command was convinced that more such gas attacks would be a good thing. Chlorine was released four more times that week at Ypres, and Haber's institute immediately began working on developing even more lethal and effective gases. Haber approached military research on poison gases with the same methodical approach he had applied to ammonia synthesis, evaluating gases for suitability using several factors.

There are a great number of poisonous gases, but most of these are not suitable for battlefield use. What is known to chemical warfare experts as "Haber's constant" is a measure of a gas's lethal nature: the product of the weight of the gas and the time required for it to cause death—the lower the Haber constant, the greater the toxicity. The idea was that more gas for less time would work just as well as less gas for more time. Tests were made on cats and the results scaled up to the body weights of humans. Thus, the Germans estimated that chlorine had a Haber constant of 7,500—7,500 mg/m^3

of chlorine gas would kill in one minute, while 750 mg/m^3 would require ten minutes to kill. The gas released at Ypres had a density of about 5,000 mg/m^3. But the Haber constant was not the only factor determining whether a gas was worth using. Hydrogen cyanide (HCN), for example, is more lethal than chlorine—it was estimated to have a Haber constant of 1,000 rather than chlorine's 7,500—and would later be used for executions in gas chambers in the United States for years. But hydrogen cyanide has a drawback as a weapon: it is lighter than air, so it will not form a ground-hugging cloud and will not sink into trenches. Phosgene, $COCl_2$, seemed to be the best choice. The Germans estimated phosgene's Haber constant to be 450—more than sixteen times more lethal than chlorine—and it was almost as dense as chlorine, so it would hug the ground and drop into trenches, shell craters, and foxholes. Its boiling point is a little high—7.5°C—making it unsuitable for undiluted winter use, but mixing it with some chlorine fixed that. Phosgene was more insidious than chlorine. Its odor of new-mown hay was not unpleasant, and the victim might not even be immediately aware that he had been gassed, but a few hours later the lungs would show the same damage as with chlorine, and the victim would drown in his own fluid.

While Haber was experimenting with different gases, Nernst was working on improved gas delivery. He was experimenting with trench mortars, the devices that Haber had first suggested would provide better delivery of dense clouds of gas than would artillery shells.[18] Artillery shells had greater range but a low payload and, because of their high velocity, often embedded themselves in the ground before exploding. The trench mortar would be able to drop an entire drum of gas at low enough terminal velocity for the gas to spread when the drum burst. Nernst, who was known to be careless in the lab, performed some of the research on trench mortars

> at his Berlin laboratory in his usual impatient manner, filling his co-workers with admiration and horror for the off-hand way in which he dealt with new explosives. Usually the preliminary tests were done at the proving grounds at Spandau, ten miles away. However, on one occasion Nernst could not be bothered to go that far and instead packaged the charge into a small disused well in the laboratory court. The bottom appeared to be filled with rubble and it was open on the top. This, Nernst argued, would direct the blast skywards without doing any damage. The firing took place just after noon while [Professor] Rubens gave his daily lecture on elementary physics to an audience of about 300 students. They, as well as their lecturer, were duly startled, first by a terrible bang and immediately afterward by complete darkness. The latter phenomenon turned out to be caused by dense clouds of dust which had been blasted into the lecture room. In his haste, Nernst had omitted to investigate the original purpose of the well, which in fact, was closed at the bottom but opened into a number of ventilation shafts. This system of providing air for the lecture rooms had long

been superseded by a more modern installation and the shafts were filled with dust and dirt that had accumulated in them since Helmholtz's days, forty years earlier.[19]

Haber organized a demonstration of Nernst's trench mortars, using a mixture of phosgene and chlorine, on 25 March 1915.[20] The trench mortars were first used on the battlefield on 30 July and 1 August on the northern front, firing tear gas. Nernst was present that day and investigated the effects on the Allied prisoners who had been captured. As a result of Nernst's service, the Kaiser awarded him the Iron Cross, First Class, and Germany's highest award, *Pour le Merité*. At this point, Nernst seems to have stopped work on poison gases and their delivery.[21] Haber's son Ludwig, who wrote a history of chemical warfare in World War I, believed that Haber pushed Nernst out, seeing him as a potential rival for leadership in the field.[22] Personal rivalry flourished even in the worst of efforts, but in retrospect, Haber did Nernst a great favor if he forced him out of chemical warfare. When Hartley interrogated Nernst after the war, he said that Nernst held a grievance against Haber that his trench mortars had not been used.[23] The army was not enthusiastic about the use of either trench mortars or the cylinder release of gas, because both put gas close to the German lines with the risk of blowback if wind conditions

Figure 4-1. Emma and Walther Nernst during World War I. Kurt M. Mendelssohn, *The World of Walther Nernst: The Rise and Fall of German Science* (London: Macmillan, 1973), 97.

changed. A supporter of Nernst's trench mortars, Major Lothes, was killed at Verdun, and thereafter Nernst could not find support for his weapons.[24]

Haber Takes Charge, with Disastrous Results at Home

J.E. Coates, an English chemist who had been Haber's pupil, said of him,

> The war years were for Haber the greatest period of his life. In them he lived and worked on a scale and for a purpose that satisfied his strong urge towards great dramatic vital things.... To be a great soldier, to obey and be obeyed—that, as his closest friends knew, was a deep-seated ideal.... It must not however be supposed that he exalted and enjoyed war *as such*, but the coming of the war brought out another side of his nature and transformed him into a Prussian officer, autocratic and ruthless in his will to victory.[25]

At the peak, Haber had 2,000 people working for him, of which about 150 had chemistry degrees.[26] His institute's buildings were surrounded by barbed wire, and Haber kept appropriating lab space in neighboring institutes. The organization was top-down. Hartley said of him, "Haber seems to have become rather an autocrat, and consequently much of the research was carried out as prescribed tasks, little freedom being left to individual workers."[27] Hartley's inquiries indicated that Haber was not an effective communicator or manager and was not willing to take criticism, and that as the war went on he became exhausted and despondent.[28] Haber's own life was a Potemkin village—a successful external facade concealing an internal core of family stress, illness, anxiety, and insecurity. As the war began to go badly, the conflict between external and internal became more difficult, and he became less and less able to manage it all.

When Haber transformed his institute into a center for chemical warfare research, his wife, Clara, strongly objected. She had opposed his work on gas warfare from the start, and having witnessed the January laboratory accident that had resulted in Sackur's death can only have strengthened her opposition. While in the laboratory, she had seen the results of animal tests of the different gases. Clara pleaded with Haber to give the work up, but he was adamant on both patriotic and personal grounds. Clara was said to be nervous and agitated generally, and fourteen years of marriage to Haber cannot have helped. She had continued to correspond with Richard Abegg, her friend and former teacher, and her letters to him talk of being trapped by what she saw as "a genius... [with a] sovereign contempt for every rule of normal behavior."[29] When Abegg died in 1910, Clara was deprived of her confidant. She concentrated on the household tasks and on her son, Hermann—things that Haber found uninteresting. Friends described how Haber would joke about Clara's agitation: "When Fritz Haber was asked one evening where [his wife] was after a trip on a steamer on which rain had wet them through, he answered, 'She is at home worrying about which relative might have contracted which sickness and how.' "[30] Clara had had reservations at the time of their marriage, but Haber's enthusiasm had convinced her. She had given up her own career in chemistry for marriage, and although she continued to work in the laboratory occasionally, the Habers were certainly not Marie and Pierre Curie. By 1915 she saw her husband as uninterested in their marriage and directing a massive research project into a subject that she understood well, and that horrified her.

A week after the chlorine attack at Ypres, Haber returned home from the front for a brief leave. On the night of 1–2 May 1915, Clara took her husband's service revolver, went into the garden, and shot herself. Her son, Hermann, heard the shot and found her dying. The next morning, Haber, who was under orders to return to the front, left his twelve-year-old son with his dead mother and in the care of an aunt. Haber's troops were conducting the first chlorine attack on the eastern front at Bolimov,[31] and Haber saw it as his duty to be there.

What led Clara to suicide that night is not entirely clear. That she was unhappy in her marriage and that she was opposed to Haber's work in chemical weapons are both supported by reliable testimony, and the suicide was only days after Haber's first use of gas on the battlefield. Clara's sister Lotte and other family members had also committed suicide,[32] and Clara's unhappiness was so extreme that it might be called mental illness. But there were unsubstantiated rumors, fed by the accounts of servants, that the Habers had had a reception that night, that Clara came upon Fritz in a compromising position with Charlotte Nathan (who would later be his second wife), and that there was a jealous confrontation between Fritz and Clara. Charlotte later claimed that her first meeting with Fritz was not until the spring of 1917,[33] which was almost certainly untrue, as Fritz had regularly visited the dining room and salons of the German Society of 1914, a patriotic businessmen's club of which she was manager. One of Haber's biographers includes a photo of Fritz and Charlotte dated 1916.[34] But whatever its cause, Clara's suicide was not an impulsive act after an argument, as she wrote several letters before killing herself, none of which has survived.

Haber's reaction to his wife's suicide was to throw himself more deeply into his public life. He returned immediately to the front and redoubled his research efforts.

Chemical Warfare as a Game of Chess

After the chlorine attack at Ypres, the Allies immediately began to make plans to retaliate and to develop defensive strategies. Haber's son Ludwig wrote,

"The story of chemical warfare is one of imitation: the Germans usually led and the Allies followed."[35] The British made their first chlorine attack at Loos on 24 September 1915. The attack was not well planned, and the British suffered more than 2,500 gas casualties from their own release, apparently more than the Germans did.[36] The French were the first to use chemical artillery shells on a large scale, and were ineffectively firing tear gas shells by July.[37] Both sides worked on perfecting gas masks and other protection devices, which became increasingly sophisticated as the war progressed. Haber's friend and neighbor Richard Willstätter worked on developing a "three-layer cartridge" gas mask filter, for example, where each layer provided a different chemical function.[38]

Haber saw gas as a psychological weapon—soldiers feared gas more than explosives. Wilfred Owen, the British soldier-poet who died in the war, expressed the horror of gas:

Dulce et Decorum Est

Bent double, like old beggars under sacks,
Knock-kneed, coughing like hags, we cursed through sludge,
Till on the haunting flares we turned our backs,
And towards our distant rest began to trudge.
Men marched asleep. Many had lost their boots,
But limped on, blood-shod. All went lame, all blind;
Drunk with fatigue; deaf even to the hoots
Of gas-shells dropping softly behind.

Gas! *Gas!* Quick, boys!—An ecstasy of fumbling
Fitting the clumsy helmets just in time,
But someone still was yelling out and stumbling
And flound'ring like a man in fire or lime.—
Dim through the misty panes and thick green light,
As under a green sea, I saw him drowning.

In all my dreams before my helpless sight
He plunges at me, guttering, choking, drowning.

If in some smothering dreams, you too could pace
Behind the wagon that we flung him in,
And watch the white eyes writhing in his face,
His hanging face, like a devil's sick of sin,
If you could hear, at every jolt, the blood
Come gargling from the froth-corrupted lungs
Bitten as the cud

Of vile, incurable sores on innocent tongues,—
My friend, you would not tell with such high zest
To children ardent for some desperate glory,
The old Lie: *Dulce et decorum est*
Pro patria mori.[39]

The last Latin lines translate as "It is sweet and fitting to die for one's country."

Haber saw gas as a psychological weapon that could "turn soldiers from a sword in the hand of their leader into a heap of helpless people."[40] The gas mask, especially the heavy British single-box respirator, was one more burden for the soldier to carry into battle. Soldiers in the trenches found themselves constantly sniffing for gas, and a soldier in a gas mask, even if it were functioning, was half-blinded, unable to aim properly or to see peripherally. And when a gas attack occurred, and the concentration of gas became so high that it began to overcome the filter and to be felt in the throat, the desire to rip the mask off and breathe deeply became almost irresistible, and some did succumb to the urge.

According to Haber, gas was a natural stage in the evolution of warfare—just as technological advances in artillery and machine guns had led armies to dig trenches, the stasis of trench warfare led armies to develop gas in an attempt to make trenches uninhabitable. Haber believed that gas was of greatest advantage to the most industrialized nations—the Germans were best at it, the British better than the French, and the Russians hopeless. He saw conventional warfare as a game like checkers, but gas warfare like chess—each new gas required the enemy to respond by developing a new gas mask filter.[41]

The Germans were not the only ones who were developing new gas strategies. In April 1917, the British introduced the Livens projector, a trench mortar. The British used the projector at Arras, dumping hundreds of drums, each eight inches in diameter and carrying 30 to 40 pounds of chemical agent. These could be dropped in a concentrated pattern and had the gas density of a cylinder attack, but provided the added element of surprise and of increased range (1,200 to 1,900 meters). They were the most effective Allied innovation in offensive chemical warfare. The Germans were unable to respond with an equivalent projector until August 1918, three months before the end of the war. The Livens projector was essentially the same trench mortar that Nernst had proposed without much support from Haber and that the German army had refused.

By 1917, Haber had developed a much better gas, dichlorodiethylsulfide, which was to become known as "mustard" because of its slight mustard-like odor. Haber had first looked at it in 1916 and rejected it because he believed its toxicity was too low, and the French had done the same. But mustard gas had other properties: unlike chlorine and phosgene, which had their principal effects upon the lungs, mustard gas was a blistering agent, causing skin burns, blindness, and internal and external bleeding. Soldiers often took four to five weeks to die, putting a further load on the enemy's medical services, and the pain was so bad that most soldiers had to be strapped to their beds. It was a much more terrifying weapon than chlorine. Because mustard gas attacked the skin, soldiers had to cover every inch of the body in a poncho during an attack. And mustard had another advantage—where phosgene and chlorine dissipated quickly, mustard was actually not a gas but a liquid sprayed as an aerosol. It was persistent, poisoning grass, plants, and the very earth for a long time. It could be used to deny territory to the enemy, to support the flanks in an infantry advance, and to cover a retreat. Mustard gas was by far the most deadly agent used in the war.

Haber was reluctant to use mustard—not for moral reasons, but because he knew that in a few months the Allies would be using it, too, and he feared that it would allow them to break the German lines. He passed the buck to General Erich Ludendorff, who was effectively running the war effort as deputy chief of staff to the pliant General Paul von Hindenburg, who had replaced Falkenhayn as chief of staff in 1916, advising him that mustard should not be used unless the general was confident that he could win the war in six months. Of course, Ludendorff ordered its use, and of course, Haber was right about the consequences. After the first German use of mustard in July 1917,

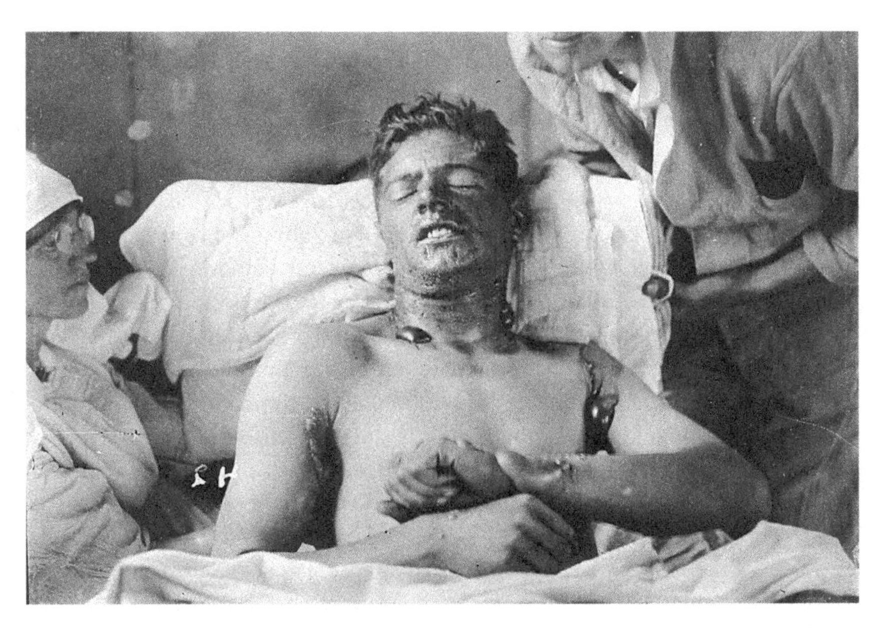

Figure 4-2. Unknown Canadian soldier suffering from mustard gas burns, World War I. Library and Archives Canada, W.L. Kidd collection, C-080027.

the British examined an unexploded shell and had the compound identified within three days.[42] While it would take the Allies until the spring of 1918 to devise an industrial synthesis for mustard gas and be able to use it against the Germans, they then began to do so with a vengeance, and Germany was unable to respond defensively. Had the Germans not surrendered in November 1918, the Allied plans for 1919 included caterpillar-driven armored vehicles that would have navigated the battlefield spraying mustard gas into the German positions.[43]

In July 1916, Haber's patron Koppel, who had supported him as director of his Kaiser Wilhelm Institute, approached the Ministry of War with the idea of a new and separate Kaiser Wilhelm Institute for the Science of Military Technology and another for Applied Chemistry and Biochemistry—that is, chemical warfare. Haber likely was behind the proposal. Having built his empire, he must have wanted to keep it after the war, but at the same time wanted his original Kaiser Wilhelm Institute for Physical Chemistry and Electrochemistry to return to its prewar civilian status. The Ministry of War pledged six million marks for postwar military research. In October 1918, a month before the end of the war, Haber realized that his association with chemical weapons might prove to be a problem after the war—he might be prosecuted as a war criminal—and attempted to back out of the plan for the new institute. But even on 4 November—a week before the armistice—the war ministry was still planning on giving him use of the funds.[44]

Germany near the End, and America Enters the War

In April 1917, Nernst's second son, Gustav, was killed at Verdun. The opening sentence of the preface to Nernst's book *The New Heat Theorem*, written in December 1917, reads, "Nothing is as good as physics to divert the mind from the present time which, in spite of the greatness achieved by our people, is nevertheless to be deplored."[45] (This translation of the German edition is by Nernst's biographer Kurt Mendelssohn, who had worked in Nernst's institute. The authorized postwar English translation of the book removed the jingoistic reference to the German people, reading, "In times of trouble and distress, many of the old Greeks and Romans sought consolation in philosophy, and found it."[46]) Nernst had known that the war was lost since the failure of the Schlieffen plan, but he had soldiered on, losing both of his sons. Nernst's hands were not as bloody as Haber's—he had not been directly involved with any attacks using lethal gases. But he had suggested use of phosgene and had likely left the chemical warfare effort only when pushed out by Haber. Nernst, like Haber, would be on the list of those that Allies would want to charge as war criminals.

The Germans were at near-famine levels by 1917. They were cut off from all overseas imports and had to feed large armies in the field, and the civilians were the ones to suffer most. A new German verb was coined to describe the activities of the civilian foraging parties—*hamstern*, to "hamster."[47] While the British were also struggling, they had the option of supply from overseas, especially from America and their colonies. In an attempt to stop supply to Britain, the German admiralty relied on the submarine. According to the then-accepted laws of warfare, a naval ship was not allowed to attack an unarmed enemy merchant ship without first demanding its surrender. This was, of course, completely impractical for a submarine commander, who had no crew to take over a prize ship, and in any case could not wait while the merchant captain used his new radio to summon help. During the war, the German navy authorized unrestricted submarine warfare several times, allowing the sinking without notice of English merchant ships or of neutral ships carrying war supplies. This caused outrage in the United States, particularly when American nationals were killed, as they were in the sinking of the British liner *Lusitania* in 1915. The British navy had cut the German transoceanic telegraph cable early in the war, and almost the only news from Europe that Americans could receive was from Britain and France, so Americans were especially susceptible to Allied propaganda. The stories of atrocities during the Belgian invasion and of the horrors of the gas attacks swung American opinion from neutrality to the Allied side. President Woodrow Wilson was determined to keep the United States out of the war, but the Preparedness Movement, led by Theodore Roosevelt, advocated the buildup of the military so that the United States would be ready to enter the war. In early 1917, the Germans had restricted their submarine attacks to armed merchantmen because of international pressure. Even with restrictions, the Germans were sinking so many ships that the British situation was becoming desperate, and Admiral John Jellicoe, the commander of the British Grand Fleet, warned that if the current trend continued, Britain would need to sue for peace by the summer of 1917. In early 1917, the Kaiser, acting on Ludendorff's advice, attempted to increase the pressure by again allowing unrestricted warfare. Nernst was present at a 1917 meeting with Ludendorff and the Kaiser where the renewal of unrestricted submarine warfare was adopted. Nernst was in despair at the decision, believing that this would bring America into the war and ensure Germany's defeat,[48] which is what in fact happened.

Langmuir's War

The United States entered the war on 6 April 1917 almost completely unprepared, with only two days' stock of artillery shells and no gas masks. Only the navy had attempted to involve scientists in the planning of the war in any significant way. Josephus Daniels, the Secretary of the Navy, had established the Naval Consulting Board in 1915 in order to try to organize research into the problems the navy would face if the United States were to enter the war. Thomas Edison was recruited to head the board, and eleven scientific and engineering societies were each asked to nominate two members; Willis Whitney, Irving Langmuir's boss, was nominated by the American Chemical Society. Edison was then sixty-eight years old, but he was far from a titular board head. He served for months at sea and came up with ideas for submarine and torpedo detection, for concealing Allied shipping, for avoiding torpedoes and minimizing damage, for secure communication between ships, for blinding enemy submarines, and for an improved intraship telephone, among others.[49]

Despite the time he had spent in Nernst's laboratory in Germany (or perhaps because of it), Langmuir was a strong supporter of the Allies, marching with his wife in a Preparedness parade in 1916.[50] On 11 February 1917, Whitney pulled Langmuir into the war effort. Early in 1917 the Naval Consulting Board had authorized the Submarine Signal Company of Boston to build a research station at Nahant, Massachusetts, where Langmuir would spend most of his time during the war's duration. From Langmuir's diary:

> 11 February 1917: Sunday: Went out [with] Dr. Whitney with Hawkins & Coolidge. W[hitney] told us of the problems put up to Naval Consulting Board by Navy Dept. & asked our help.
> 23 February 1917: With Naval Consulting Board went out to a "tug" to Boston Light Ship to hear demonstration of submarine signaling by Sub Signaling Co. . . .
> 24 February 1917: Several of us go to Newport to see torpedoes made. See three fired.
> 3 March 1917: Meeting Committee [on] special problems [of the] Naval Consulting Board. Discuss submarines & torpedoes, detection and defense.[51]

Langmuir then began working on antisubmarine research, and suggested using stereophonic sound to detect submarine direction. His device worked in this fashion: Two microphones, separated at some distance, were lowered into the water on a boom. The microphones were connected to separate audio channels leading to an observer's two earphones. The boom was then swiveled until the sound of a submarine struck both the observer's ears at the same time, and the direction to the submarine was then known to be perpendicular to the boom. Langmuir began interviewing and testing blind volunteers to see how they used sound to navigate, and he found one man from Boston who could navigate by street noise alone. He could detect buildings, fences, the size of rooms, and any obstacles placed between himself and a speaker, all from transient echoes. Langmuir calculated that in an ordinary room, only

10% to 30% of the sound from a speaker is transmitted directly, and that the rest came from echoes that could be used for navigation, which is what the blind volunteer was doing. Since there was a war on, Langmuir could not pursue this as far as he wished, but he sketched out some ideas for a stereophonic doctor's stethoscope, and he would later work on stereophonic sound with RCA and Leopold Stokowski, conductor of the Philadelphia Orchestra.

America's war was too short for any of Langmuir's work on submarine detection to be put into practice, and before the end of the war he was pulled out of the naval effort and sent to Muscle Shoals, Alabama, to work on getting an American ammonia plant going that would apply the Haber-Bosch process. By the end of the war, the United States was not even close to duplicating what Haber and Bosch had done in that regard. Langmuir was copied on a report on the progress on ammonia production that was written nine days after the armistice: "The apparatus for work at one atmosphere pressure is just about completed; that for 30 atms has not yet been finished. It is planned to try the rare metals, such as zirconium, titanium, and uranium, and their alloys, as catalysts."[52] The American Haber-Bosch plant was mothballed when the armistice was signed.

Lewis's War

When the United States entered the war, the American public had little actual knowledge of chemical warfare as it was being practiced. While the French and British had used the news of the Germans' first gas attacks in their propaganda, they had been vague about the actual effects because they did not want to scare the Americans off from entering the war. By the spring of 1917, the Allies imposed a total news blackout on gas warfare since, in the words of the British assistant secretary of war, they feared that it would result in an "unreasonable dread of gases on the part of the American nation and its soldiers."[53] As the United States prepared for war, chemical warfare received little attention. It was not until February 1917—two months before the United States entered the war—that development of a gas mask was considered. Even if gas masks had been available, no one in the military had sufficient knowledge of the practice of gas warfare to organize a program of defensive training. The United States entered the war on 6 April 1917, and by mid-July 12,000 U.S troops were thirty miles from the front with no defensive gas training and no gas masks. It was not until 17 August 1917 that General John Pershing organized the army's Gas Service, and even then its chief, Amos Fries, held only the rank of lieutenant colonel.

The results were to be expected. The majority of the U.S. troops entered the fight during and after the German spring offensive of 1918. The Germans had a field day with the green U.S. troops, whose casualty rate from gas was extremely high. On 27 May 1918, the American Expeditionary Force finally installed a compulsory program of gas officers in each unit. The gas officers were under the direct control of each unit's commanding officer and were responsible for training the troops in the use of gas masks and shelters, inspection of equipment, and knowledge of enemy tactics and material. One especially important policy was instituted: when in combat, the "all clear" signal, instructing the troops to remove their protective equipment, could only be issued by a gas officer.

This was a fine system in principle, but the gas officers still needed to be found and trained. Many of them were to be drawn from the ranks of American chemists—more than 10% of American chemists served in the Chemical Warfare Service (CWS).[54] Gilbert Lewis, then forty-two years old, enlisted in December 1917, was commissioned as a major, and was sent to France in January 1918 as director of the CWS laboratory in Paris. He reported to Lieutenant Colonel Fries, who sent him to the front to observe the German and American performance during the German spring offensive. Lewis's report impressed Fries enough that he was put in charge of the Defense Division, where he was responsible for organizing the school for training gas officers. As many as 200 gas officers a week went through the school, and American gas casualties dropped dramatically.[55] Lewis was promoted to lieutenant colonel and was awarded the French *Croix de Guerre* and the American Distinguished Service Medal. But his war work was not completely defensive, as the citation for his American medal mentioned that he secured "a better and more effective use of gas, especially mustard gas, against the enemy, thereby rendering services of great value to our Government."[56] Lewis was not, however, responsible for the poisonous gas known as *Lewisite*, which was developed but not used during the war—that was done by Captain Winford Lee Lewis, another American chemist-soldier.

There is a strange story that Lewis was fond of telling from those days. There was a running argument among Allied chemical officers during the war about the use of hydrogen cyanide, which the French favored and everyone else saw as useless. (It was in fact quite lethal, but impractical as a battlefield weapon for other reasons, as discussed earlier in this chapter.) In one of these multinational arguments at which Lewis was present, the English chemists claimed that the French had done their tests on hydrogen cyanide using dogs, and that humans were much less susceptible to the gas than were dogs. Lewis said that he watched as an English chemist, without a mask, attempted to prove the point by entering a gas chamber leading a good-sized dog, undoubtedly French. The chamber was filled with hydrogen cyanide, and a short while later the Englishman came out, dragging the dead dog behind him. Lewis said that he regarded the heroism of the English chemist as one of the most striking examples of devotion to duty that he had ever seen.[57] Most people would likely describe the chemist's behavior as stupidity and cruelty rather than heroism. Twenty-eight years later, Lewis would die in a laboratory filled with hydrogen cyanide.

Figure 4-3. Lewis in World War I uniform. Courtesy of Gil Lewis.

Haber Remarries, and the End of the War

In October 1917, Fritz Haber remarried. His bride was Charlotte Nathan, the woman who was rumored to have been the cause of a jealous confrontation between Clara and Fritz on the night of Clara's suicide. Charlotte, a vivacious woman twenty years younger than Fritz, was from a Jewish family. She pushed for marriage, and he resisted, possibly in consideration for the feelings of his son, Hermann. He eventually agreed to the marriage but insisted that she convert to Christianity so that they could be married in a church. She objected, he was adamant, and she agreed. Fritz wore his officer's uniform to the wedding, complete with spiked helmet. The couple quickly had a daughter, who was to be followed by a son (Ludwig Fritz "Lutz" Haber, the author of a history of chemical warfare) in 1920. Outside of a few brief interludes, it was not a happy marriage. Haber's friends disliked Charlotte, he was as work-obsessed as ever, and she was not as long-suffering as Clara had been. Hermann

Figure 4-4. Hermann, Charlotte, and Fritz Haber at the wedding of Fritz and Charlotte. Archive for the History of the Max Planck Society, Berlin-Dahlem, Germany.

was unhappy and resentful, and Haber had his usual recurring health problems, for which he blamed Charlotte. They would divorce in 1927.

By the time of his marriage in the autumn of 1917, Haber had come to the same conclusion that Nernst had reached three years earlier after the battle of the Marne—the war was lost. Haber became despondent and pessimistic, believing that the German army would be overwhelmed by gas, for which there would be no defense. In the spring of 1918, Ludendorff knew that the Americans were arriving in force and would soon overwhelm the German defenses. He ordered a massive offensive on the western front, employing 500,000 troops that were shifted from the Russian front after Germany negotiated a Russian surrender and armistice. On 21 March 1918, the Germans fired a million shells and broke through the British line in force. The Kaiser declared 24 March to be a national holiday, and many Germans believed that the war would be won. But the Germans took high casualties and outran their supply lines, and the attack bogged down when the hungry German troops stopped to loot shops and farms. On 15 July, Ludendorff launched a final attack, but the French struck back, again on the Marne River. The Germans had suffered a million casualties between March and July, the German people and army were starving, and the Americans were now in the war in force. In September, Ludendorff told the Kaiser that the war was lost. The German General Staff never publicly acknowledged that the army had been beaten in the field, a denial that would give rise to the belief after the war that Germany had been "stabbed in the back" by war profiteers and Jews. On 11 November 1918, an armistice took effect and the gas attacks stopped.

After the war, Haber never apologized or expressed regret about chemical warfare. His son Ludwig says that Haber claimed that

> it was the French who had first used bullets and shells with toxic materials. This was... Haber's opinion as recorded by Hartley in 1921 and subsequently before a subcommittee of the Reichstag.... He argued that gas did not cause undue suffering and therefore was not inhuman. His logic satisfied sensitive politicians: the German MPs felt absolved of all guilt—had not the army merely done its duty by following the Allied example?—and in their report managed to blame it all on the French.... Though Germany was defeated, [Haber] did not change his opinions: other Germans might talk about "errors" or write of self-inflicted "moral damage," but Haber never retracted: he was not aware of having broken Conventions or rules. He had done what was in the best interest of his country.[58]

In the end, gas was not as successful as Haber had hoped. He told Hartley in 1921 that, "looking back, [gas] wasn't such a godsend to the Germans." Only 5% of the war's casualties were from gas, and the Allies had managed to find defensive measures for every new gas or technique the Germans had introduced, although sometimes with a delay long enough to cost many lives.

Gas had not broken the deadlock of trench warfare and had not proved to be a wonder weapon.[59] The decision by all parties to forgo the use of gas in World War II may have been based as much on their experience in World War I as on any new sense of morality. For the Germans, use of gas in World War II would have been inconsistent with their new plan for blitzkrieg, because gas would slow the attacker up. The military on both sides hated gas. It was a horrible way to die, and it did not work very well, either.

After the war, Haber and Nernst were on a 1920 Allied list of war criminals to be put on trial. The German government issued passports to both of them under false names to facilitate their flight if necessary. Haber grew a beard and sent his family to Switzerland, and both Haber and Nernst moved their money to Swiss banks. In the end, the Allies decided not to demand extradition and turned the matter over to the German courts, and neither Haber nor Nernst was prosecuted.

Science after the War

The Allies imposed a draconian peace on the Germans, refusing to lift the blockade of food until the Germans accepted their terms (the Allied starvation of civilians after the armistice could be considered a war crime in itself). On the international scientific scene the Allies were harsh, as well. German scientists were not forgiven for their support of the war, and many scientific meetings and international associations were closed to them. Because of a shortage of cash, they were cut off even from access to the journals from Allied countries. Theodore Richards, the Harvard chemistry professor who had been Gilbert Lewis's dissertation adviser, was especially bitter toward the Germans. Max Born, a German physicist who was one of the leaders in developing the new quantum mechanics, wrote to both Lewis and Langmuir asking for help in finding a position for Paul Epstein, a young physicist (whom he barely knew) who had suffered in the redrawing of national boundaries after the war—the Poles thought Epstein was Russian, the Russians thought he was Polish, and neither country would allow him a research position. Both Lewis and Langmuir agreed to help and sent Born scientific reprints to help ease the shortage of journals in Germany.[60]

The German scientists who had signed the 1914 Manifesto of the 93 were those most scorned. Hendrik Lorentz, the Dutch physicist, had convinced Max Planck during the war of the truth about German atrocities in Belgium. Planck was the only German scientist to retract his signature to the manifesto during the war, although Emil Fischer had attempted to convince the others among the ninety-three to issue another manifesto calling for an end to the war. Haber, whose process had provided the German the munitions, was at the top of the list of scientific pariahs. It came as a shock when, in November 1919,

Haber was awarded the 1918 Nobel Prize in Chemistry for "the synthesis of ammonia from its elements"; it is likely, however, that the Nobel Committee for Chemistry and the Royal Swedish Academy of Sciences were unaware of his role in chemical warfare at the time of the award.[61] He was awarded the prize alone, although both Carl Bosch and Robert Le Rossignol had been instrumental in the development of the process and had their names on the fundamental patent.

The Allied press was outraged, as were Allied scientists. Lord Moulton, who had headed the British organization responsible for poison gas, accused the Germans of waiting to launch the war until their "Haber factories" were ready to make munitions, and claimed that the Germans (specifically Haber) had unleashed chemical warfare on the world.[62] Nobel Prizes had been announced but not awarded during the war, so the Nobel banquet of 1920 was intended to celebrate all prizes from 1914–19. The German Richard Willstätter would receive the 1915 chemistry prize, and two other Germans, Max Planck and the rabid nationalist Johannes Stark (more on him in chapter 6), would receive the 1918 and 1919 physics prizes, respectively. Nobel's testament had tried to exclude nationalism from the awards and to make Sweden as neutral in its scientific as in its political stance, but the Allies viewed the Royal Swedish Academy as pro-German.

Some Allied scientists boycotted the Nobel ceremonies. The Englishmen William Henry Bragg and William Lawrence Bragg were to receive the 1915 physics prize, the American Theodore Richards the chemistry prize for 1914, and the Belgian Jules Bordet the prize for physiology or medicine for 1919. All refused to attend the presentation ceremonies. Richards was particularly anti-German, refusing to forgive anyone who had signed the Manifesto of the 93 and did not apologize.

Haber's Nobel Prize did not come out of the blue. Haber had been nominated for the prize in 1912, 1913, 1915, and 1916, and Bosch had been nominated in 1915 and 1916.[63] In 1916, the committee agreed that the practical utility of Haber's synthesis for ammonia had been demonstrated—it was hard to argue with this, as the Oppau plant was by this time producing 80 metric tons of ammonia a day. But there were two reservations that caused the committee not to recommend Haber for the prize: first, the details of the process had not been disclosed in the scientific literature (for obvious commercial and military reasons), so it was not possible to see whether Haber deserved the prize alone or should share it with Bosch; and second, in view of the use of the process by Germany's war machine, it was difficult to see how it followed Nobel's instructions that the prize be awarded for an achievement that had "conferred the greatest benefit on mankind." The committee advised that Haber should not receive the prize because of "reasons of opportunism, that perhaps argue against a neutral country rewarding an invention that at present to an eminent degree must be said to constitute a weapon in the struggle between the great powers of Europe."[64]

In 1918, Haber was up for consideration again, and Peter Klason, a strongly pro-German member of the chemistry committee, wrote the report on Haber, saying that Haber's invention would be of great benefit to mankind in the production of fertilizer after the war. Olof Hammersten, another pro-German committee member who detested England and hoped for a German victory, nevertheless opposed an award to Haber because the process was still secret. Hammersten also noted that it was politically unwise to award Haber a prize. The committee endorsed Haber by majority vote, but Hammersten registered a dissent with the academy, which supported him in its vote, and the 1918 prize was reserved for consideration the next year. With the war over in 1919, Hammersten removed his dissent, although the Haber-Bosch process was still as secret as ever. Without a formal objection from a committee member, the academy voted in 1919 to award Haber the 1918 prize retroactively.[65]

Svante Arrhenius, who was a member of the Nobel Committee for Physics and a controlling force behind the chemistry committee, attempted to serve as a conciliator between the pro- and anti-German forces within the Royal Swedish Academy. When Haber's prize was under consideration in 1919, Arrhenius opposed the award not on scientific or moral grounds, but on the basis that German tax rates were so high that Haber would receive little of the prize money. Sven Widmalm, who has written on the Nobel awards to Haber and the other Germans, sees Arrhenius as thinking in political terms, not wishing to be explicitly anti-German but thinking that a prize for Haber would be too blatant a demonstration of support for Germany.[66] Perhaps so, but Arrhenius had the power to block awards at least for some time when he wished to do so, as he had done with Paul Ehrlich (chapter 2) and was then doing with Nernst (chapter 6). That Arrhenius did not block Haber's prize indicates that he either tacitly supported the award or at least did not much care. The Swedes said that they had awarded Haber the prize out of a desire to be scientifically neutral, but the Swedish neutrality was as biased scientifically as it had been diplomatically during the war: "strict neutrality" toward the Allies and "benevolent neutrality" toward the Germans.[67]

In his letter accepting the Nobel Prize, Haber showed that he viewed the prize as a Swedish endorsement of German science and as a personal vindication: "Being in a wretched, dark ignorance about the future of my Fatherland, this award appears to me as a tribute to the accomplishments of German science from professional colleagues in Sweden who, as much by their scientific eminence as by their neutral position between the larger countries, have been called upon to exercise a superior and impartial judgment."[68] Haber had no regrets: he believed that he had behaved as a scientist and as a patriotic German, and he saw his Nobel Prize as recognition for himself and for Germany.

5

The Lewis-Langmuir Theory

Lewis, Langmuir, and Harkins

In March 1917, Gilbert Lewis picked up the latest edition of the *Journal of the American Chemical Society.* As he read, he became increasingly angry. William Harkins, a chemist whom he knew well—he had been Harkins's boss at MIT in 1910—had written a paper on surface chemistry. Harkins had a reputation for claiming credit for work that others had done, and it seemed to Lewis that Harkins had done it again, this time to Irving Langmuir.

Lewis and Langmuir had not known each other well when they had spent two days together at an American Chemical Society meeting in New York that past September, but Lewis had been impressed with Langmuir's presentation of his ideas on the chemistry of surfaces—the chemistry of oil films on water, for example. Both Lewis and Harkins had sat in the audience at Langmuir's talk. After Langmuir had finished speaking and the discussion had begun, Harkins had made some odd comments that could be paraphrased as "I am glad to see that your results, Dr. Langmuir, support my theory"—a theory unknown to anyone in the audience. Six months after that meeting, this new paper by Harkins had appeared, and the ideas he expressed seemed to have been lifted directly from Langmuir. Without any prompting, Lewis wrote Langmuir to offer his support:

> I am very sorry to say what I believe, however, in fairness must be said regarding a recent publication by Harkins in the *Journal of the American Chemical Society* (March 1917), especially since I have known Harkins very well for a number of years. When I was in Chicago last fall Harkins talked with me for a good many hours about the various investigations which he had under way, and a considerable part of this time was spent in the discussion of surface tension. I feel perfectly sure that at that time he had no fundamental theory of surface films and was interested primarily in making more exact measurements than other men had made. When you spoke to

me later of your work on surface films and your theory that the molecules in these films possess a definite arrangement and direction with the polar group of each molecule pointing toward the polar liquid on which the film was developed, it seemed to me an absolutely new and vital contribution to our conceptions of intermolecular forces. I was therefore very much surprised at the time of the meetings in New York when, in the discussion of your paper, Harkins stated that he was pleased to observe that you had been led to the same conclusion which he had reached from a different point of view. It seemed to me that he was attempting to borrow some of the credit of your discovery. In his last paper, he seems to be attempting to take it all, and makes a number of statements identical with those you stated as essential to your theory, without anywhere mentioning your name. My suspicion that Harkins has not been honest in giving the sources of his ideas is borne out by several other incidents of a minor but similar character. I feel very strongly that something ought to be done to prevent this sort of piracy, but I confess I do not know what it should be. I have, however, felt it my duty to write to you what I think regarding the matter.[1]

When we left Lewis and Langmuir at the end of chapter 2, they had both established themselves as prominent young physical chemists by 1912, Lewis at MIT and Langmuir at General Electric. Lewis was developing his "new system of thermodynamics," to which his concept of *activity* was central, and Langmuir had begun his studies of surface chemistry, which had arisen from his development of the modern gas-filled incandescent lamp. By the time of the 1916 American Chemical Society meeting, only four years later, they were perhaps the most prominent young American physical chemists.

Lewis at Berkeley

During his time at MIT, 1906–12, Lewis published more than thirty papers on experimental and theoretical thermodynamics and was promoted to full professor. His boss, Arthur Noyes, served as acting president of MIT for two of those years, and Lewis temporarily took Noyes's place as director of MIT's Research Laboratory for Physical Chemistry. With Noyes's help, Lewis developed some of the social and managerial skills that he had lacked during his time at Harvard.

In 1912, Lewis moved from MIT to take the position of chemistry department chairman and College of Chemistry dean at the University of California in Berkeley. While he had been happy at MIT with Noyes, Lewis had his own ideas, and Berkeley offered him a chance to build his own department. The University of California had ambitions to be one of America's leading schools, and its president, Benjamin Wheeler, needed a young research scientist of Lewis's stature to lead the effort in chemistry.

Only a decade earlier, Berkeley's College of Chemistry had been led by Willard Rising (known as "Bewildered" Rising to the students), who had simultaneously served as both dean of the college and analyst to the California State Board of Health. He had supplemented his academic income by attesting to the purity of Royal Baking Powder in advertisements in the local newspapers. Rising had resigned as dean in 1901 and had been replaced by Edmund O'Neill, who had improved the department's professionalism and had assisted Wheeler in recruiting Lewis. When Wheeler approached Lewis, Lewis negotiated a good deal for himself—he was allowed a university-supported personal research assistant each year, was granted control over all faculty hiring, held the positions of both dean and department chairman (which he would continue to hold until his forced retirement from them at age sixty-six in 1941), and received a commitment to substantial increases in budgets for the College of Chemistry, including new facilities, staff, and salary increases all around.

When Lewis arrived in 1912, Berkeley had only five chemistry faculty members—O'Neill, Edward Booth, Walter Blasdale, William Morgan, and Henry Biddle. Booth was elderly and was allowed to continue on the faculty until his retirement in 1917. O'Neill had been involved in the decision to hire Lewis and helped him in running the department, and Blasdale stayed on at Berkeley until his retirement in 1941. Lewis saw Morgan and Biddle as problems, however.

Morgan was not interested in research and had been in charge of teaching freshmen since 1903, with an educational philosophy that did not match Lewis's. Lewis had President Wheeler put Morgan on sabbatical leave for the 1912–13 school year. Morgan took the hint and resigned to go to Reed College the following year, and Lewis replaced him with Joel Hildebrand from the University of Pennsylvania. Hildebrand would become Lewis's close friend and confidant for the next thirty-three years, joining him in the army in France during World War I, managing the freshman courses (at the end of his career, Hildebrand would estimate that he had taught chemistry to 42,000 Berkeley freshmen), and taking the position of department chairman when Lewis retired from administrative positions in 1941.

Biddle was competent professionally but was involved in something to which Lewis had strong objections—industrial consulting. In 1914, Biddle began to do a great deal of consulting work, for which he received substantial payment. Worse yet (according to Lewis), he used his graduate students as unpaid labor for his consulting, and he applied for the patents in his own rather than the university's name.[2] Lewis persuaded Biddle to leave for a full-time industrial position and thereafter forbade all consulting, which he viewed as a distraction from the pursuit of pure science. He removed all industrial chemistry courses, such as petroleum chemistry, from the Berkeley curriculum.

Lewis brought three scientists with him from MIT—William Bray and Richard Tolman as assistant professors, and Merle Randall, who had been his first doctoral student, as his personal research assistant. Lewis would keep Bray as a friend until his death in 1946, only a month before Lewis's own. Tolman would soon leave, first going to Illinois and then in 1922 rejoining Arthur Noyes, who had by then moved to CalTech. Lewis hired George Ernest Gibson in 1913 from the University of Edinburgh, and Gibson and Lewis became close friends and neighbors in their summer houses. In 1919, Wendell Latimer received his Ph.D. as Gibson's student and was invited to join the Berkeley faculty. In the years that would follow, Latimer would assume much of the department's administrative load. After Lewis's retirement from the position in 1941, Latimer would be appointed dean of Berkeley's College of Chemistry.

In his time under Theodore Richards's thumb at Harvard, Lewis had found that his heavy teaching load left him no time to think. When he had moved to Noyes's Research Laboratory for Physical Chemistry at MIT in 1906, there were no undergraduates to be taught, and Lewis had flourished. During his entire thirty-four-year tenure at Berkeley, Lewis arranged things so that he did not teach a single undergraduate or graduate course. But despite his personal aversion to teaching, Lewis imposed his educational philosophy on the department. As far as undergraduates were concerned, Lewis believed that it was important to teach chemical understanding rather than rote chemistry and to assign the strongest faculty members to the best students. Every professor (other than Lewis) was required to run undergraduate laboratory sections. I have spoken with a number of Berkeley alumni who remember Nobel Prize winners such as William Giauque or Melvin Calvin reviewing their freshman qualitative analysis results.

For graduate education, Lewis believed in the importance of research. If a graduate student felt the need for a chemistry or physics course, he or she was welcome to take it, but that was not how the student would be judged. When John Gofman arrived at Berkeley to begin graduate work in 1940, Lewis told him during his initial interview, "Some of the graduate students should take a course or two but they don't bother much with courses; get your research started within the next few weeks."[3] Lewis removed all petty chemistry stockroom controls from graduate students, giving them free run of the department for their research, and encouraged them not to wait to be assigned a project by a faculty member, but to pursue their own research ideas. Harold Urey, Gerhard Rollefson, and Joseph Mayer received their Ph.D.'s under Lewis's nominal direction, but in fact they worked on problems of their own choosing unrelated to Lewis's research, and they published the results under their own names. Unlike many prominent scientists, Lewis was not interested in attaching his name to work in which he had not fully participated.

Lewis regarded the weekly research seminar, where a department member or guest speaker would give a research paper and where discussion was open to all, as the most important event of the week. Room 101 in Gilman Hall was a tiered auditorium, where the students and postdoctoral fellows sat in the bleachers, and the faculty sat at a long table in the front. The seminar began precisely at 4:00 P.M. each Tuesday. Lewis would light the first cigar of the meeting and say, "Shall we begin?" Attendance was compulsory, and anyone—student or faculty member—could be asked to discuss his or her current research without prior notice. Hildebrand recalled that, shortly after he joined the department in 1913, E.Q. Adams, a graduate student, interrupted Lewis during one of the seminars with "No, that isn't so!" Hildebrand was aghast—such behavior would have had a student at the University of Pennsylvania, where Hildebrand had been previously employed, tossed out—but Lewis just turned to the student with interest, saying, "No? Why not?"[4] At another point, when a student interrupted with a comment, Lewis paused, thought for a moment, and responded with "That was an impertinent remark, but it was also a pertinent remark."[5] But if one were to interrupt Lewis successfully, it took a certain amount of confidence. Glenn Seaborg, Lewis's research assistant in 1938–39, said, "Although Lewis dominated the scene through sheer intellectual brilliance, no matter what the topic, anyone was free to ask questions or speak his piece; in the latter instance, prudence suggested that the comment had best not be foolish or ill-informed. If Lewis had any weakness, it was that he did not suffer fools gladly—in fact, his tolerance level here was close to zero."[6] Hildebrand said of the Berkeley chemistry department,

> The members of the department became like the Athenians, who according to the Apostle Paul, "spent their time in nothing else, but eager to tell or to hear about some new thing." Any one who thought he had a bright idea rushed to try it out on a colleague. Groups of two or more could be seen every day in offices, before blackboards, or even in the corridors, arguing vehemently about these "brain storms." It is doubtful whether any paper every emerged for publication that had not run the gauntlet of such criticism. The whole department thus became far greater than the sum of its individual members.[7]

Lewis was often uncomfortable in larger scientific settings, but within the Berkeley chemistry department he felt at ease. He built the department as a support system for himself, hiring his own research assistants and Berkeley's new faculty members from among the best of Berkeley's graduates—from 1914 through 1936, no one other than a Berkeley graduate was hired for a tenured faculty position in Lewis's department. The department became one of the world's greatest centers of physical chemistry, but it also became completely inbred. Lewis filled it with *his* scientists. He removed all qualifying titles—there were no professors of organic or inorganic chemistry, only professors of chemistry—saying that no one would be allowed to stake out a particular area of research. In fact, they were all professors of physical chemistry—physical organic chemistry, physical inorganic chemistry, and so forth—because that was the subject that interested Lewis.

When he moved to Berkeley at age thirty-seven, Lewis married Mary Hinckley Sheldon, the daughter of a professor of philology at Harvard. Three children followed—Richard in 1916, Margery in 1917, and Edward in 1920. For Lewis, his family was always secondary to his work and to the camaraderie of the chemistry department. Although the Lewis family purchased a home in Inverness, a beach community two hours from Berkeley by ferry, Lewis did not keep a permanent home in Berkeley but rented a different Berkeley house each school year. During the summers, the family stayed in Inverness while Lewis lived at the Berkeley faculty club, where he spent his leisure hours playing bridge. Both his sons would become chemists, but Lewis's son Edward told me that he could not remember a single intimate conversation with his father.

Langmuir, Family, and Children

Like Lewis, Langmuir also married in 1912, to Marion Mersereau. The two shared a love of the outdoors, exploring caves and climbing the Adirondack mountains. Langmuir taught her to ski and sail-skate.[8] Langmuir's marriage and family life seem to have been idyllic. He sent his wife flowers once a week throughout their marriage. When they were unable to have children of their own, the Langmuirs decided to adopt. In 1918, they adopted a two-year-old son, Kenneth, followed by a daughter, Barbara, two years later.[9] Throughout his life, Langmuir would be involved not only with his own children, but also with the Boy Scouts and with children's science education and outdoor camps. From Albert Rosenfeld's biography of Langmuir:

> When people came visiting—or when the Langmuirs went visiting—the grownups were often neglected by Langmuir in favor of the children, whom he collected around him with the ease of a Pied Piper. Their parents would find them in the back yard, where Langmuir could show them all sorts of tricks using whatever was at hand—sticks, stones, pieces of string, buckets, hoses, and whatnot. He might tie a stone on the end of a string and say, "Now let's talk about the pendulum." Or he might be whirling a pail of water over his head while the children wondered why the water did not come pouring out of the upside-down bucket.[10]

For children too young for that sort of thing, Langmuir would sit them on his knees and bounce both knees together, chanting "in phase," or bounce both knees alternately, chanting "out of phase."[11] A twelve-year-old neighbor approached Langmuir with a design for an invention he had made, a perpetual motion machine. It was the usual such device—a paddlewheel, driven by falling mercury; the rotation of the paddlewheel would pump the mercury up to a level where its fall could drive the paddlewheel. Langmuir could have told him that it would not work and explained friction and the conservation of energy, but he knew that the boy would learn nothing from this—he would think he was being told that his idea was stupid. Langmuir instead took him into the shop at GE, had the boy's device constructed, and then helped him figure out why it did not work as he had expected.[12]

Langmuir at GE

After his success with the gas-filled incandescent lightbulb, Langmuir continued to work through the 1910s on the practical problems facing GE, but always starting from investigation of the fundamental properties of gases, electrons, and metals. This was the dawn of the age of radio, and GE was developing the vacuum tubes that would be needed to transmit, detect, filter, and amplify radio signals. Vacuum tubes are related to electric lamps—both involve metals, vacuums, filaments, currents, and bulbs—but there is a fundamental difference: in a lamp, the electrons pass through a solid filament, while in a vacuum tube, they are pulled across space from a negatively charged cathode to a positively charged anode. In order to design vacuum tubes, an engineer would need equations to predict how many electrons would be emitted from a vacuum tube's cathode. When the researchers at GE attempted to apply equations from the scientific literature, they found that electrons did not behave as expected.

When metals are heated, they emit electrons into space even if there is no positive charge pulling at them, a phenomenon known as *thermionic emission.* The effect had been observed as early as 1873 by Frederick Guthrie, Thomas Edison, and others. In 1901 Owen Richardson, an English physicist, had derived an equation that described how thermionic emission *should* behave. But by 1913, there were so many discrepancies between Richardson's equations and experimental observation that some began to suspect that no unifying theory was possible, and that everything was due to secondary chemical effects. Langmuir showed that thermionically emitted electrons accumulate around a heated object and form a *space charge,* a cloud of electrons that inhibits further emission of electrons. He was able to work out equations to predict that the magnitude of the space charge between two electrodes is proportional to the 3/2 power of the voltage, irrespective of the shape of the electrodes. While Langmuir derived this equation from first principles, he might have come to the answer more quickly by reading the literature, as it had previously been derived for positive ions by Clement Child;[13] it is now known as the Child-Langmuir equation and correctly predicts the emission of electrons. Langmuir's research was instrumental in the development and introduction of hundreds of new vacuum tubes by GE, just in time for the radio boom.

Langmuir seemed to be unable to avoid making money for GE, even when he was not trying to do so. His earlier discovery that molecular hydrogen dissociated into atomic hydrogen on a hot tungsten filament, made when he had investigated the properties of gases in the electric lamp, ended up in a GE product. Langmuir described how it led to a welding torch that could produce much higher temperatures than an oxyacetylene torch: "Two tungsten rods, as electrodes, are held at a definite angle to one another by easily adjustable clamps, and a jet of hydrogen is directed from a small nozzle along each of these rods near its end. The hydrogen thus bathes the heated parts of the electrodes and forms a gentle blast of gas which passes through the arc between the electrode tips, and blows the atomic hydrogen away from the electrodes so that these are not unduly heated."[14] The hydrogen in the welding torch could be viewed as a transport mechanism for energy: electric energy was pumped into the tungsten electrodes to heat them; at the tungsten surface, the molecular hydrogen dissociated into atoms; and the atoms were then blown some distance away toward whatever object was to be welded, where the hydrogen atoms recombined into molecular hydrogen, releasing a great deal of heat. And in 1916, Langmuir examined a mercury diffusion pump and realized that the vacuum was being degraded by backflush of mercury gas. By adding a cooling chamber to condense the gas, he developed the mercury-condensation pump, which became a laboratory and industrial standard.[15]

But Langmuir's assistant Harold Mott-Smith said that Langmuir did not always share credit fairly, or that he was "*adjustable* about giving credit."[16] Mott-Smith says, for example, that the idea for the atomic hydrogen welding torch was based on an experiment by R.W. Wood, who was a physicist at Johns Hopkins and a consultant to GE, and that Langmuir did not like to read the scientific literature, which sometimes led him to take credit that should have been given to others. Langmuir could also be dogmatic; he refused to accept the idea that a television tube could by constructed using luminescent phosphors, because he believed the tube would be too dim. A phosphor-based tube would eventually become the basis for commercial television. According to Humboldt Leverenz, later research director at RCA, "[Langmuir] told people pointedly that it was obvious that [phosphors] can't be considered for producing a television picture. And he refused to his dying day [1957] to have a television set in his house."[17] RCA, not GE, would take the lead in the development of television.

Surface Chemistry—Oil Films on Water

Within GE, Langmuir had the freedom to point his research wherever he wanted, whether or not it was directly product-related, and he continued to explore the behavior of chemistry in two dimensions—the chemistry of surfaces. He had begun his work on surface chemistry with the discovery that hydrogen formed a layer one molecule thick on the surface of the electric lamp's bulb. He began to see the same sort of behavior elsewhere, as he puzzled over the nature of oil films on water. Why did some oils form films and not others? How far would the film spread, and how thick was the film? He looked at two classes of oils: pure hydrocarbon oils (such as mineral oil) that were *hydrophobic*—they had virtually no affinity for water—and fatty oils that contained both a hydrophobic part and some chemical groups, such as alcohols and acids, that were *hydrophilic* and had an affinity for water. If a drop of completely hydrophobic oil were placed on water, it did not form a film, instead forming an insoluble sphere on top of the water. But a drop of fatty oil, containing a hydrophilic group, behaved differently. If the hydrophobic part of the oil were small, with five carbons or so, the oil dissolved in the water, because the attraction between the hydrophilic group and the water was strong enough to pull it into solution. But if the hydrophobic part of the oil were large, with eighteen carbons or so, the oil would form a very thin film on the surface of the water.

Langmuir was not the first to study the behavior of oil films on water. Around 1770, Benjamin Franklin had poured "not more than a teaspoonful" of what was probably a vegetable oil on the pond at Clapham Common and observed that it "extended itself gradually . . . making all that quarter of the pond, perhaps half an acre, as smooth as a looking glass," saying that it seemed as if "a mutual repulsion between its particles took place."[18] If he had wished to do so, he could have easily calculated the maximum size of the particles (molecules, in modern terminology) from the thickness of the film, which he could have determined from the volume of oil applied and the surface area covered; it works out to about two ten-millionths of a centimeter, given a teaspoonful of oil and a half-acre covered. But Franklin was more interested in the calming nature of the oil on the water (the reduction of surface tension, in modern terminology) and did not make this calculation. Lord Rayleigh repeated Franklin's experiments more than a century later, making the necessary measurements and calculations. Agnes Pockels, a self-educated German scientist, read Rayleigh's paper on the subject in 1890. She had been working on experimental observations of oil on water in her kitchen since 1881 and had developed both experimental apparatus and a theory that she communicated to Rayleigh in a letter. Rayleigh saw her work as more sophisticated than his own, and he arranged to have it translated and published in the journal *Nature* within weeks.[19]

In a long paper on the subject,[20] Langmuir explicitly noted the contributions of both Rayleigh and Pockels, as well as those of contemporary European scientists who were working in the area. Where Langmuir's theory differed from other workers' was in its attention to the specific chemical orientation of the oil molecules. He favorably quoted Lewis's valence theory, which is the subject of a later section of this chapter. He also differentiated between what he called *physical* forces (such as gravitational, electrical, and magnetic forces), which deal with molecules as simple particles (usually spheres) and can be described by equations expressed in terms of distance uniformly in all directions, and *chemical* forces, which deal with interactions involving particular atoms and functional groups within a molecule, where the forces are local and may have a directional component. Within chemical forces, he differentiated between *primary* and *secondary* valence effects; primary effects include what we would now call covalent and ionic bonds, and secondary effects are what we would now call hydrogen bonds—the interactions between the hydrophilic parts of the oil molecules and the water molecules (hydrogen bonds are discussed more fully in chapter 9). Langmuir believed that the forces responsible for the behavior of oil films on water could be described entirely in terms of primary and secondary chemical effects and that the molecules in oil films could not be regarded even approximately as spherical.

Langmuir's explanation of the behavior of oil films was this: in large fatty oils, the hydrophobic group was too large for the molecule to be pulled into water solution. The molecules would arrange themselves side by side; the hydrophobic parts would clump together sticking up out of the water, and the hydrophilic parts would be drawn down into the water—forming, once again, a layer one molecule thick. Langmuir derived equations that described the behavior of these films. He found that the equations were two-dimensional analogues of the normal three-dimensional equations for gases, liquids, and solids. If the molecules in the film were mobile, the film could be thought of as a two-dimensional gas; if they were less mobile, as a two-dimensional liquid; and if they were immobile, as a two-dimensional solid.[21] He could even calculate the molecular dimensions from the dimensions of the film; if the molecule consisted of a long saturated chain of carbon atoms, connected to a hydrophilic carboxylic acid group at one end, the area covered per molecule did not change as the length of the chain was increased, but the film thickness did. Comparing the film's thickness to the carbon–carbon bond length in diamond, Langmuir showed that the length of the hydrophobic carbon chain indicated that the carbon molecules were not stretched out in a straight line, but were kinked like this:

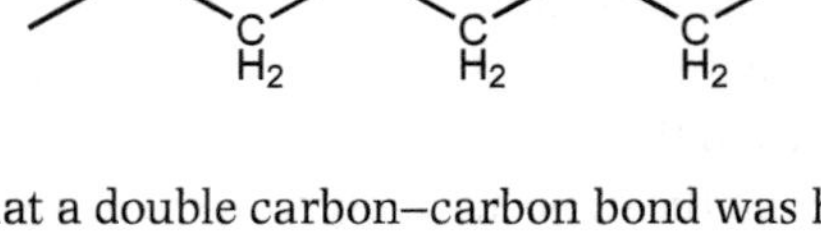

He also found that a double carbon–carbon bond was hydrophilic. When a double bond was substituted for a single bond in one of his fatty oils, the area covered was observed to increase and the film's thickness thus must have decreased (the product of the film's thickness and the area covered equaled the volume, and the specific volume of the oil was virtually unchanged by the substitution of a double bond for a single bond), indicating that the double bond had been pulled down into the water.

Langmuir presented his results in the New York meeting in 1916 (which Harkins and Lewis attended), published a short abstract[22] of his talk in 1916, and published a full exposition in 1917.[23] This was the work that, in the letter that begins this chapter, Lewis called "an absolutely new and vital contribution to our conceptions of intermolecular forces," and that Lewis accused Harkins of trying to appropriate in an act of scientific "piracy."

William Harkins

William Draper Harkins was born in 1873, two years before Lewis. He received his bachelor's degree in chemistry from Stanford in 1900, whereupon he was immediately made professor and chairman of the chemistry department at the University of Montana in Missoula. While at Montana, he continued his graduate work and in 1907 received the first chemistry Ph.D. awarded by Stanford. In 1909, he went to study with Fritz Haber in Karlsruhe, where he developed new and more accurate methods for the measurement of surface tension. The following year, Harkins joined Noyes's Research Laboratory for Physical Chemistry at MIT as a research associate, where he worked under Lewis's direction. After a brief return to Montana, he joined the faculty of the University of Chicago as a lowly instructor in 1912. Six years later he was a full professor at Chicago. Harkins worked on the structure of the atomic nucleus, where he was the first to show the enormous mass loss in the then-hypothetical nuclear fusion of hydrogen to produce helium, positing this fusion as the source of stellar energy. Harkins also showed that atomic nuclei with even-numbered masses were more stable than odd-numbered masses.

His work on surface chemistry was a continuation of his work with Haber on surface tension, and he published two papers on the subject in 1917, the second of which presented a large body of careful experimental data that he had been amassing for years. There was no question that this experimental work had been done by Harkins and that it was well worth publishing. The problem was that he also presented a theory of surface chemistry, which was essentially the same as that which Langmuir had presented at the New York meeting of the American Chemical Society the previous year, which Langmuir had submitted for publication but which was still in press. In the first of Harkins's two papers,[24] which Lewis had read just before writing to Langmuir, Harkins did not even mention Langmuir's work. In the second paper, he made a frantic claim that he deserved priority for his theory:

> A part of the work in this paper was done by Harkins in 1909 ... at the suggestion of Professor Fritz Haber.... Parts of the paper were presented at the Boston

meeting of the Society in 1909, and at the New Orleans meeting. Another part was prepared in abstract for the presentation at the Symposium on Colloids held by the American Chemical Society in September, but was not accepted. An abstract of a paper presented at this Symposium by Irving Langmuir gives somewhat similar views to some of those developed by us, and for this reason it has been thought best to publish at once the data we have collected.... The work of Langmuir and that which we present in this paper, while developed independently from somewhat different starting points, are alike in that we have both developed *the same fundamental idea: that surface tension phenomena in general are dependent upon the orientation and packing of molecules in surface layers, and that the forces involved in this action are related to those involved in isolation and adsorption.*[25]

Harkins already had a reputation for dubious priority claims, as Lewis noted in his letter to Langmuir; in fact, his nickname was "Priority" Harkins.[26] Jacob Bigeleisen, who was at Chicago with Harkins in the 1940s, describes him as "a scientist who several times took the ball to the four-yard line"—almost, but not quite, coming up with an important discovery. Harkins was indeed uncommonly unlucky in his career: Francis Aston published evidence from mass spectroscopy for the existence of chlorine isotopes of mass 35 and 37 in February 1920, while Harkins published similar results in April of the same year. Harkins realized that the Wilson cloud chamber could allow determination of the energy and mass of nuclear reactions and analyzed thousands of alpha-particle tracks in nitrogen and argon, but Patrick Blackett, using equipment identical to Harkins's, was the first to show in 1925 that nitrogen captured an alpha particle and emitted a proton, forming oxygen-17; Harkins confirmed Blackett's results the following year. Harkins was the first to predict the existence of the neutron in 1920 as the union of a proton and electron, but it was not until 1932 that James Chadwick observed the neutron.[27] Bigeleisen remembers Harkins coming to seminars with his laboratory notebooks. Whatever the subject, Harkins might go to the opaque projector and produce a notebook page that showed, at least to his own satisfaction, that he had done the important work on the subject first.[28]

Langmuir's response to Harkins's claims to a theory of surface chemistry was measured. After Harkins's strange comments at his September talk, Langmuir began corresponding with Harkins in a series of letters beginning in January 1917, possibly because he had heard that Harkins was about to rush a theory of surface chemistry into print.[29] The first letters were cordial, asking for chemical samples or information. On 9 February, Langmuir asked Harkins for a copy of the manuscript that he had submitted to the *Journal of the American Chemical Society*. Harkins replied with an agitated letter on 25 February, saying, "You will find several references to your work in the second paper," "I hope you will let me know at once if there is anything in these references which should be changed to give you proper credit," and "I did not send the papers before because I lost a part of the second paper." He ended the letter with a handwritten postscript, "I can get no word as what has happened to my second paper from the editors, so I am sending only a part of the paper until I can find out what has happened to it."[30]

Harkins's excuse, "The dog ate my homework," did not impress Langmuir. He wrote Harkins a long letter on 9 March 1917 detailing the history of his own ideas and reproving Harkins for claiming that he had arrived at the results entirely on his own:

> It is perfectly evident that we have developed these ideas independently and that we should both publish all that we can on the subject, but I think we should both be frank in admitting the value of each other's work, rather than merely emphasizing minute points of difference.... To avoid duplication in the future, we ought to keep in pretty close touch with one another and come to definite agreement as to just what field we will specialize in. There is no objection to our both publishing papers on the same general subject, but one of us should certainly not work on the same subject as the other, only to find that the other has an almost identical paper ready for publication. Therefore will you not let me know what work you have actually in hand and what particular subjects you would like to reserve?[31]

Langmuir then proceeded to tell Harkins the problems on which he was working.

In a long footnote to his 1917 paper on surface chemistry, which had appeared after Harkins's two papers, Langmuir laid out his own priority claim and, in veiled language, gave some clues to Harkins's behavior:

> An account of this work was given...at the New York meeting of the American Chemical Society in September, 1916, and a short abstract was published [later in 1916]....Since this time I have found that Prof. W. D. Harkins has been developing an essentially similar theory.... He has recently published two papers on this subject....Nearly all the data in the second of these papers...had been worked over by me during the summer of 1916 and it was my intention to publish them together with the material now given in the present paper. Dr. Harkness [*sic*], however, kindly sent me advance manuscripts of his papers, and I have therefore been able to avoid duplication of this work. Harkins had expected that his two articles would appear simultaneously, and, as an unfortunate result he failed in his first article to mention my prior publication. In his second paper, he refers to my work, but, by an oversight, fails to refer to the publication of my general results [in 1916]..., although in the later part of his paper, he refers to [it] in a footnote in which he points out that his conclusions...are the same as mine.[32]

Langmuir then reinforced his priority claims by finishing the footnote with a quote of a large section of his previously published abstract from 1916. The fact that the journal's editor allowed a 1,000-word footnote that had no purpose other than documenting a priority claim indicates that the editor believed that

Harkins had done something shady. Although Langmuir let Harkins's claims pass without formal objection, the story of Harkins's behavior must have circulated quickly, particularly given Lewis's support of Langmuir and the way that Langmuir's footnote described Harkins's failure to reference his work.

Lewis and the Chemical Bond

During his first five years at Berkeley, most of Lewis's research focused on putting experimental substance behind his reformulation of thermodynamics in terms of his concept of activity. He determined the free energy of formation of a number of compounds from their constituent elements, and he measured electrode potentials that would enable other researchers to determine still more free energies by using the Nernst equation. But thermodynamics did not seem to be leading to a solution of what Lewis saw as the most important problem in physical chemistry—the failure of strong electrolytes to follow the law of mass action. Thermodynamics had nothing at all to say about the processes that might be occurring *inside* molecules during a chemical reaction, and that was the knowledge that predicting the behavior of strong electrolytes seemed to require.[33] In an effort to understand the forces that held strong electrolytes together and that caused them to conduct electricity in solution, Lewis puzzled over the nature of the chemical bond.

In order to understand Lewis's revolutionary ideas on the chemical bond, we need to first look at how the chemical bond was viewed in 1916. The idea of the bond had been part of chemistry since at least the 1850s, when chemists began drawing complicated chemical formulas (see chapter 1). They wrote bonds as lines between atoms, and wrote chemical reactions with bonds appearing and disappearing. Once J.J. Thomson discovered the negatively charged electron in 1897, most scientists thought that the electron was involved in chemical bonding through the transfer of charge from one atom to another—Jöns Berzelius's dualistic theory of bonding (described in chapter 1) that molecules are held together by attraction between positively charged and negatively charged atoms, but now updated to include the idea of the electron. As an example, the new dualistic theory would explain the formation of sodium chloride as follows: a sodium atom would transfer an electron to a chlorine atom; the transfer of a negatively charged electron would form a positively charged sodium ion and a negatively charged chloride ion, and the positive and negative ions would have an electrical attraction for each other and would form sodium chloride. A notation from that time showed the chemical bond resulting from this electron transfer as an arrow:

$$\mathrm{Na} \rightarrow \mathrm{Cl}$$

This suggested that for those molecules formed from atoms that had similar electron-attracting powers, two separate chemical substances might be found, depending on the direction in which the electron transfer occurred. These hypothetical substances—called "electromeres"—might be able to be separated, just as isomers could be separated. In the case of NCl_3, for example, one electromere might involve transfer of electrons from chlorine to nitrogen and another from nitrogen to chlorine:

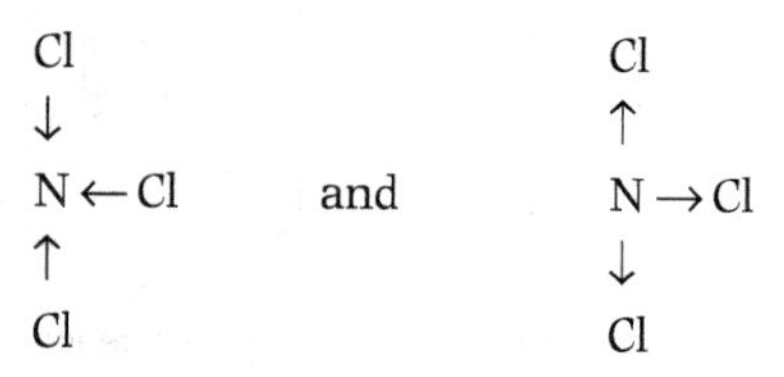

William A. Noyes and Julius Stieglitz, two of America's leading chemists who served as chairmen of the departments of chemistry at the University of Illinois and the University of Chicago, respectively, spent twenty fruitless years trying to separate electromeres.

In 1916, Lewis published "The Atom and the Molecule,"[34] where he broke from the dualistic model with the idea that the chemical bond is due to electron *sharing* rather than to electron transfer, and that what is shared is a *pair* of electrons. But his idea did not appear to him de novo in 1916. He had been mulling the problem over at least since 1902, when he had presented his ideas on the electronic structure of the atom to a Harvard class that he and Theodore Richards were jointly teaching—ideas that Richards had called "twaddle," which may be why Lewis did not publish them at the time. His sketches from 1902 show a cubic model for the atom. As one moves across one of the first two full rows of the periodic table, each atom adds an electron to a corner of the cube. When the first row of the periodic table is filled, the second row begins building a new cubic shell enclosing the filled cube of the first row:*

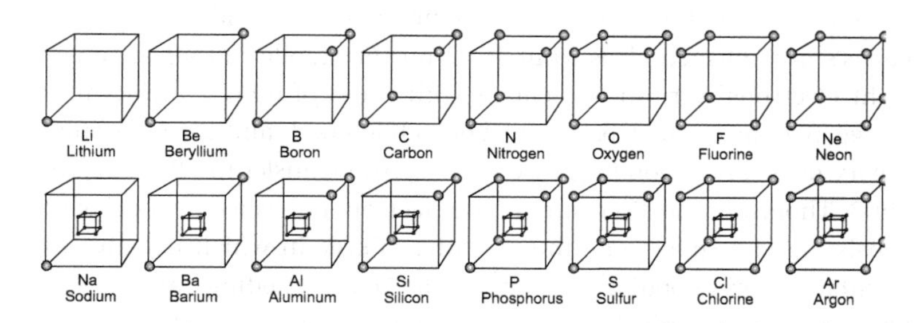

Figure 5-1. The first two rows of the periodic table, showing Lewis's cubic model for the atom. Each new element adds an electron in a corner of the cube. The second row's cube encloses the filled cube of the first row.

* Fig. 5-1 is a simplification of Lewis's 1902 approach. In 1902, he thought that helium had eight electrons and occupied the place of neon.

In the formation of sodium chloride, Lewis's 1902 model would have the sodium atom transfer an electron to the chlorine atom and empty its cube, allowing the chlorine atom to fill its cube:

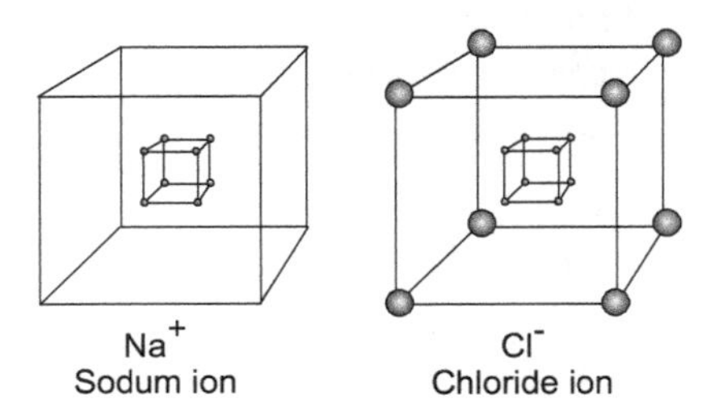

Figure 5-2. Lewis's models for sodium and chloride ions.

This sort of bond is called *polar* or *ionic*, as it is formed between two different types of ions, one positive and one negative. Nothing really new so far—just the Berzelius dualism restated in terms of electrons and cubes.

The inert gases neon and argon (at the far right of fig. 5-1) were discovered in 1898 and 1894, respectively. These inert gases (now called noble gases) were so named because they refused to take part in any chemical reactions, which is why they had taken so long to be discovered—there were no compounds to be found in nature that included them. In Lewis's scheme, the inert gases had their cubes filled with eight electrons. Lewis noted that polar electron transfers seemed to occur in ways that either filled or emptied the outer cube of electrons for each atom, giving them the same electronic structure as an inert gas. In the example of sodium chloride (fig. 5-2), the chloride ion ends up with a filled cube like argon, while the sodium ion ends up with an emptied outer cube and an inner cube like neon.

Lewis was not the only one trying to understand chemical bonding. Lewis's unpublished ideas from 1902 on the stability of an arrangement of eight electrons were in agreement with those of others, including Richard Abegg (whom we met briefly in both chapters 1 and 2), who published a theory of chemical bonding based on electron transfer in 1904. Beginning in 1903, Thomson, the discoverer of the electron, published a model of the atom consisting of a sphere of diffuse positive charge in which negative electrons circulate. It was popularly known as the "plum pudding" model—electrons, like raisins, were embedded in a larger pudding of positive charge—but it was a sophisticated attempt to explain the periodicity of the chemical elements using physical principles; the diffuse positive charge supplied the atom with mechanical stability that a point nucleus with orbiting electrons could not achieve, and the mutual repulsion of the electrons explained why different atoms would have configurations of electrons that would lead to preferential addition or loss of electrons. Thomson's model of the chemical bond was also one of electron transfer, and he represented this transfer by "tubes of force" stretching between two atoms connected by a bond. This model became popular and was endorsed by both William Noyes and Julius Stieglitz in their search for electromeres.

All these theories were more or less a recapitulation of Berzelius's dualistic theories of a century earlier, now recast in terms of the newly discovered electron. Electron transfer explained bonding in *polar* molecules, where the bonding was really between two electrically charged ions, but did not explain bonding in *nonpolar* molecules such as chlorine gas (Cl_2), where two chlorine atoms were bound to each other. In terms of Lewis's theory, why would one chlorine atom, which had seven electrons in its outer cube, give up an electron to another identical chlorine atom? And even if it did, the chlorine atoms would not both have Lewis's and Abegg's magic number of eight outer electrons—one would have eight and the other would have six.

Doodles and sketches of cubic atoms have been preserved from Lewis's time in the Philippines and at MIT, where he was likely trying to determine some rationale for the behavior of strong electrolytes.[35] In 1913 Lewis published a paper in which he puzzled over the problems of polar and nonpolar bonds.[36] He suggested that there might be more than one type of chemical bond—the polar bond, where an electron was transferred from one atom to another, as in sodium chloride; and a nonpolar bond, as in the chlorine molecule.

The important insight came to Lewis in 1915, and his cubic model for the atom was the key to his inspiration. He allowed two cubes to overlap, and if two atoms shared one pair of electrons (a single bond), the two cubes shared an edge. If they shared two pairs of electrons (a double bond), the two cubes shared a face. Using Cl_2 and O_2 as respective examples:

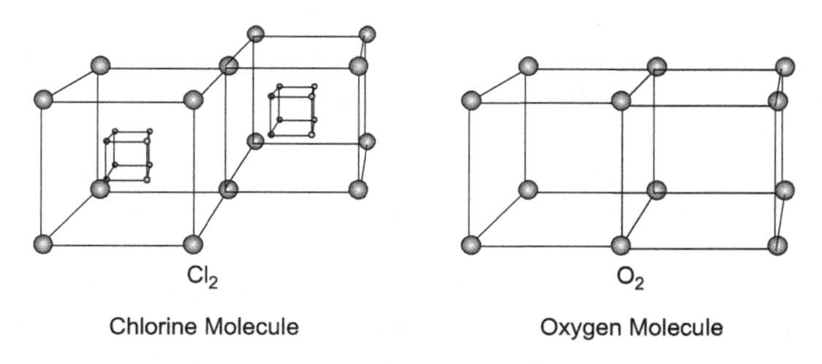

Figure 5-3. Lewis's cubic models for electron pair sharing in chlorine and oxygen molecules—what would become known as the covalent bond.

In each of the molecules shown in fig. 5-3, *both* atoms have eight electrons in their outer cubes, and they have achieved this through sharing electrons. Lewis developed a shorthand notation for this that avoided the necessity of drawing cubes, representing the electrons as dots. Each nonpolar bond was a shared pair of dots between two atoms, and when eight dots surrounded all atoms, the molecule was especially likely to be a stable structure:

:C̤̈l: C̤̈l: Ö::Ö

Cl_2 chlorine molecule O_2 oxygen molecule

Figure 5-4. Lewis dot structures for chlorine and oxygen molecules. This was a shorthand notation for the cubic structures of fig. 5-3.

In 1916 Lewis published "The Atom and the Molecule," where he laid all this out and finally showed his 1902 ideas of the cubic atom.

There was a great deal of extraneous model building and speculation in Lewis's paper—the cubic atomic structure with static electrons stuck at the corners seems quaint today—but the lasting concept from the paper was what later became known as the *covalent bond*, the sharing of an electron pair between two atoms. This ran counter to all physical knowledge of electrical forces: why would two negative electrons attract each other to form a bond? But the great majority of stable chemical compounds contain an even number of electrons, and this suggested to Lewis that the *pairing* of electrons was what was important. He found the chemical evidence for the electron pair so compelling that he argued that at short range repulsive electrical forces must not apply. With the 1923 publication of his book *Valence and the Structure of Atoms and Molecules*, he speculated that magnetic forces might be behind the attraction. Electrons have an electrical charge, but they also behave like tiny magnets. Just as bar magnets attract each other when they are properly aligned, he thought that magnetic pairings of the electrons might be the force behind bond formation,[37] an idea he had only suggested in his 1916 paper.

In 1910, Ernest Rutherford, who had been Thomson's student, showed experimentally that the positively charged nucleus in atoms was extremely small relative to the atomic diameter and that Thomson's model of diffuse charge was hence physically incorrect. Niels Bohr, a Danish physicist who had just received his doctorate, visited both Thomson's and then Rutherford's laboratories in 1911–12 attempting to develop atomic and molecular models that would be consistent with Rutherford's results and would explain radioactivity, chemical periodicity and bonding, and the observed atomic emission spectra.[38] He published three papers in 1913, of which the best known explains light emission by hydrogen atoms using the quantum hypothesis. This accounted for the emission spectra with great precision but did not in itself explain the chemical bond in a way that was useful to chemists. He did discuss the formation of molecules in another 1913 paper using a model that involved sharing of electrons, but he did not identify the electron pair with the chemical bond as a general case. For a molecule with two atoms and many electrons, he theorized that most of the electrons would be in orbits around the two atoms and that "a few of the outer electrons will be arranged... rotating in a ring around the line connecting the nuclei. The latter ring, which keeps the system together, represents the chemical 'bond.'"[39] What separated Lewis's efforts from the efforts of Abegg, Thomson, Bohr, and others was that Lewis saw the electron *pair* as fundamental to the chemical bond. He had changed his mind between his papers of 1913 and 1916—he no longer saw any sharp differentiation between the polar bonds in molecules like sodium chloride and the nonpolar bonds in molecules like chlorine. For Lewis, there was now only one type of bond. The electrons could be pulled (polarized) almost all the way toward one atom, as in sodium chloride, or could be equally shared by both, as in chlorine. He saw most bonds as somewhere between these two extremes, with electron pairs that were partially polarized but still shared.

Bohr's atom had orbiting electrons, while Lewis argued that if the electron were the basis for the chemical bond, the stability of chemical compounds argued for the immobility of the electron—if the electron were mobile, it would be difficult to explain the stability of chemical isomers. Both Bohr's model and Lewis's model ran counter to the laws of physics as they were then understood. An electron in a Bohr orbit undergoes centripetal acceleration, and classical physics predicts that it will consequently radiate energy and collapse into the atom's nucleus. Classical physics also predicts that the static electrons in Lewis's atom should collapse into the nucleus from simple electrical attraction. It would be a mistake to call either Bohr's or Lewis's theory "right" and the other "wrong"—they each explained a subset of the experimental data. Bohr's model could explain the spectrum of the hydrogen atom, while Lewis's model could explain chemical bonding. The physicists and the chemists were talking past each other, while letting off the occasional salvo—the physicist Robert Millikan mocked Lewis's cubic static atom, calling it the "loafer theory" with "electrons sitting around on dry goods boxes at every corner, ready to shake hands with, or hold onto similar loafer electrons in other atoms."[40] It would be another ten years after Lewis's theory before quantum mechanics would be able to explain both chemical and physical phenomena.

Lewis's paper on the chemical bond was at first largely ignored, although Langmuir cited it favorably in his 1917 paper on surface chemistry.[41] Langmuir and Lewis apparently discussed Lewis's ideas over two days at the New York symposium on the structure of matter in the fall of 1916, the same meeting at which Harkins tried to claim that he had come up with the same theory of surface chemistry as had Langmuir. The science historian Robert

Kohler says that "the main reason for the neglect is simply that the problem that Lewis's theory so elegantly solved was not recognized as a problem by most chemists."[42] Lewis's theory was most applicable to organic chemistry, where nonpolar bonding was the general rule, but in 1916 most organic chemists felt no pressing need for a theory of bonding. They were still most concerned with determination of structure, while investigation of organic reaction mechanisms—how the bonds were forming and breaking—was still in its infancy. Most of those who read Lewis's paper did not really understand his idea of the shared electron pair, but saw it as a minor extension to the dualistic orthodoxy.

Langmuir and the Chemical Bond

Lewis went off to serve in World War I shortly after the publication of "The Atom and the Molecule" and did not immediately expand upon his ideas. In 1919, Langmuir reexamined Lewis's paper and began applying Lewis's ideas to his own work. He wrote Lewis an excited letter,[43] telling him that he had some ideas for extension of the atomic model, and began publishing papers in the field. Langmuir fell in love with Lewis's cubic atom, publishing papers with professionally drawn diagrams of connected cubes, but he at first missed the importance of the shared electron pair. Langmuir started examining chemical bonding from his own point of view, eventually concluding in his laboratory notebook of February 1919 that the electron pair was the essence of the chemical bond. When he finally did come to understand the importance of the pair, it came to him in such a flash that he seems to have felt that he deserved credit for its rediscovery.[44]

The idea of the shared electron pair is one of the first things a high school chemistry student learns today, and it is difficult to put ourselves in the shoes of a chemist of 1919. But before Langmuir's "rediscovery" of Lewis's theory in 1919, almost no one seems to have grasped its significance. A pervasive theory—in this case Berzelius's dualistic theory of chemical bonding through electron transfer—can have such a strong hold that it cannot be easily supplanted even by an idea as simple and as intuitive as Lewis's. From Lewis's point of view, Langmuir's attempts to share credit for his ideas on the chemical bond must have seemed like Richards's claims to share credit for his fugacity. By the time Langmuir understood the importance of the electron pair, Lewis's paper had been in print almost three years.

In 1919 and 1920, Langmuir began speaking about what he called his "octet theory"—that molecules formed stable octets, or groups of eight electrons. On 9 April 1919, he read his paper at the American Chemical Society meeting in Buffalo; it aroused such interest that he was asked to repeat it the next day. On 29 April, he read the same paper at the National Academy

Figure 5-5. Irving Langmuir, J.J. Thomson (who discovered the electron), and William Coolidge at the GE research labs in 1923 during Thomson's visit. General Electric Research Laboratory. Courtesy of the AIP Emilio Segrè Visual Archives.

of Sciences in Washington, with 400 chemists from the National Bureau of Standards in the audience. Langmuir did not use Lewis's dot notation for the electron pair but developed mathematical equations to predict the number of electron-pair bonds in a molecule. Lewis's approach was informal and intuitive, while Langmuir's was mathematical and deductive, including eleven postulates. Langmuir took Lewis's insight and turned it into a complex theory.

Langmuir was an inspiring speaker—something Lewis was not. Lewis disliked traveling to scientific meetings, believing that his published work should stand on its own. As Langmuir traveled and spoke, other industrial chemists saw him as a hero—an industrial scientist like them, the inventor of the gas-filled lightbulb, the atomic-hydrogen welding torch, and the mercury-condensation vacuum pump—and they preferred his presentation of the ideas to those of a remote academic in distant California. While Langmuir was careful to give Lewis credit at the start of each talk, that was generally the last time that Lewis was mentioned in the next hour.[45] Langmuir's terminology prevailed—in addition to the "octet theory," he coined "covalent bond," for which Lewis gently chided him, saying, "Sometimes parents show singular infelicity in naming their children, but on the whole they seem to enjoy having the privilege."[46] The new electronic theory soon came to be known as the "Lewis-Langmuir" theory, something that irritated Lewis. William Noyes, chairman of the department of chemistry at the University of Illinois, sent Lewis a manuscript that Noyes had written and asked for his comments. Noyes's manuscript included a structural formula for perchloric acid, which he had unfortunately called a "Lewis-Langmuir" formula. Lewis replied to him in a 1926 letter:

> Your formula for perchloric acid...which you ascribe to Lewis and Langmuir should be called the formula of Lewis, Langmuir and W.A. Noyes, and so on, adding the name of every author who quotes it hereafter. Or perhaps it would be simpler to call it the formula of Lewis, and not mention the names of all the others who have quoted it.[47]

In late 1919, Langmuir and William Noyes exchanged a series of letters. Langmuir espoused his octet theory and Noyes took the older dualistic approach to bonding. The two of them found their exchange so interesting that they agreed to debate in a pair of papers in the *Journal of the American Chemical Society*. When Lewis learned this was to happen he was furious; from Lewis's point of view this was *his* theory, and Langmuir should not be the one responsible for its defense. Arthur Lamb, the journal's newly appointed editor, was caught in the middle. He attempted to mollify both Lewis and Langmuir, but with little success. Both replied to Lamb that priority disputes were distasteful but then proceeded to engage. Lewis wrote Lamb[48] that while Langmuir's "interesting and convincing personality, his admirable methods of presentation, and his opportunity of expounding the new views to many audiences gave the subject an impetus among chemists of all ranks," Langmuir's octet theory was "simply the theory of which I gave a complete though concise exposition in my paper," and that Lewis would himself likely have published some of the concrete examples that Langmuir had published had the war not intervened, although Langmuir was entitled to priority for these examples. Langmuir replied that he "could not be continually giving credit" and that many others had worked on the problem before the two of them. Lamb sent Langmuir a copy of the letter that Lewis had sent to him, and that copy has been preserved in the Langmuir collection in the Library of Congress. On the back of his copy of Lewis's letter, Langmuir made some hand-written notes: "Kossel, Parson, Stark, Thomson, Bohr" (a few of the others who had worked on electronic theory before either Lewis or Langmuir); "I deduce from a more general theory," "It is not for us to decide what part each has played," "Each of us surrounded by those who emphasize our own contributions," and "Lewis's misfortune that he did not publish." Langmuir personally typed an eleven-page single-spaced letter to Lewis in the small hours of the morning, where he argued that he had made significant contributions to the theory that gave him the right to speak on his own.[49]

Even at this late date, Langmuir's letter to Lewis showed that he did not fully understand the significance of Lewis's identification of the electron pair with the chemical bond: in an attempt to minimize Lewis's contribution, Langmuir's letter to Lewis cited Bohr as having come up with the electron pair earlier than Lewis. He based this on Bohr's model for the methane (CH_4) molecule, which had two electrons rotating in orbits between the carbon atom and each of the four hydrogen atoms attached to it. The electron pair in Bohr's methane was accidental, however—there were eight electrons to be parceled out among four bonds, so it had to work out to two electrons per bond. Langmuir's letter to Lewis also cited Johannes Stark as a predecessor for the electron pair. Again, Stark's electron pairs were special cases—sometimes Stark's atoms shared one electron and sometimes two. Langmuir was unwilling to admit that Lewis's electron pair was the essence of his own octet theory. In a briefer letter to Lamb, Langmuir made many of the same arguments that he had made in his eleven-page letter to Lewis.[50] When he was forced to decide, Lamb agreed with Lewis—that the essence of Langmuir's octet theory was in Lewis's 1916 paper "The Atom and the Molecule." The *Journal of the American Chemical Society* declined to publish the debate between Langmuir and William Noyes.

After thinking things over, Langmuir seems to have come to an understanding of the importance of Lewis's electron pair, and he moderated his claims. In a 1920 article in *Nature*, he made his debt to Lewis clear, saying that the idea of the static atom was due to Lewis and reproving another author who had called it "Langmuir's theory." He also corrected the argument that he had made to Lewis about Bohr's priority, saying in print that "Bohr did not, in general, identify a pair of electrons with a valency bond."[51]

Lewis had based his arguments on induction from chemical evidence—he inferred the importance of the electron pair from the fact that the great majority of stable chemical compounds have an even number of electrons. Beyond his suggestion of magnetic forces tying the electrons together, he had not tried to develop a physical model for the electron pair, and he did not see the details of his cubic model for the atom as fundamental. Langmuir saw the cubes as more important and began to try to find some way to unify the electron pair with Bohr's theory, in which it had no special place. Langmuir began corresponding with Max Born,[52] Arnold Sommerfeld,[53] and Bohr[54] on the significance of Alfred Landé's work on the interaction between electrons. Unaware of Lewis's paper (probably because of the war), Born[55] and Landé[56] had investigated a Bohr-like model of an atom with electrons located at the corners of cubes but had eventually discarded it. Langmuir attempted to find a way to revive a Bohr-based cubic model.[57] For a brief moment, Langmuir thought he had succeeded. From a March 1921 letter to his mother:

> Some months ago I promised to write a part of a report on atomic structure which is to be published by the National Research Council. I started to write

it up about two weeks ago and soon found that I began to get new ideas on the subject. These have developed into the most important scientific discoveries that I have yet made. Especially since last Sunday I have been working night and day extending a new theory of atomic structure which is based upon that which I proposed two years ago, but is enormously more valuable in that it takes a quantitative form so that results can be calculated with high accuracy. I am able to prove now with certainty that the electrons of which matter is made are not moving in orbits but remain relatively stationary and I can now calculate exactly the forces that act on each particle. This is a very big step forward. It radically modifies the ideas on the subject that have been most widely accepted during the last few years.[58]

These ideas must have proved unsuccessful, because no subsequent publications appeared.

Lewis had the last word in the literature. In 1923, he published *Valence and the Structure of Atoms and Molecules*, a monograph that laid out his ideas in detail and described the ideas of many of his predecessors, including Bohr, Stark, Abegg, and Thomson. Lewis made it clear that he was the first to have understood the importance of his identification of the bond with the electron pair. In what would prove to be a bad career move, he then left the field of electronic chemical theory behind him just as the physicists' quantum models began to explain chemistry.

By this time, Lewis's dot-pair diagrams were viewed as more useful than Langmuir's equations. The "static atom," which both Lewis and Langmuir had espoused, had come under criticism from physicists. With the 1923 publication of *Valence*, Lewis announced reconciliation of the chemists' and physicists' views: the Bohr orbit that held the electron pair was static relative to the nuclei, while the electrons were mobile within the orbit,[59] a suggestion Langmuir had also made.[60] The physicists did not all applaud. Harold Urey, whom we will meet again in chapter 8, was a graduate student in chemistry at Berkeley at the time and said that "Lewis announced one day in seminar that he believed the Bohr theory was correct.... [Raymond Birge, the chairman of the Berkeley physics department] the next day made a comment in class: 'Yes, after all of us have worked on this, and studied it, and considered it, then Professor Lewis, who has not made any such contributions at all, suddenly announces that he agrees with it, and expects that now that settles the question.' "[61]

Lewis knew that great advances in theoretical chemistry were to come and that his intuitive theory of the chemical bond would soon be supplanted by a mathematical description. Given his intellectual self-confidence, it is unlikely that he felt himself unable to contribute in this area, but he was already drawn to what he saw as an even more challenging physical problem, the dual wave–particle nature of light (discussed in chapter 7). In the preface to his monograph *Valence*, he described his ideas on the chemical bond as part of the "ephemeral literature of science," but it served as a defining reference for most chemists for at least the next sixteen years and was supplanted only when Linus Pauling published *The Nature of the Chemical Bond*[62] in 1939.

This is what Lewis had to say in *Valence* about Langmuir's contributions to the chemical bond:

> In my original paper I contented myself with a brief description of the main results of the theory, intending at a later time to present in a more detailed manner the various facts of chemistry which made necessary these radical departures from the older valence theory. This plan, however, was interrupted by the exigencies of war, and in the meantime the task was performed with far greater success than I could have ever achieved, by Dr. Irving Langmuir in a brilliant series of some twelve articles, and in a large number of lectures given in this country and abroad. It is largely through these papers and addresses that the theory has achieved the wide attention of scientists.
>
> It has been a cause of much satisfaction to me to find that in the course of this series of applications of the new theory, conducted with the greatest acumen, Dr. Langmuir has not been obliged to change the theory which I advanced. Here and there he has been tempted to regard certain rules or tendencies as more universal in their scope than I considered them in my paper, or than I now consider them, but these questions we shall have a later opportunity to discuss. The theory has been designated in some quarters as the Lewis-Langmuir theory, which would imply some sort of collaboration. As a matter of fact Dr. Langmuir's work has been entirely independent, and any additions as he has made to what was stated or implied in my paper should be credited to him alone.[63]

What Lewis meant by this last sentence, of course, was that Langmuir had not added anything of importance, which is not true.* But in the long run, Lewis's intuitive dot structures seemed to be more valuable than Langmuir's equations. Langmuir applied his ideas to the higher rows of the periodic table, but with what was, in retrospect, limited success. His emphasis on the octet did not work well for many higher row atoms, and Lewis noted that it caused him to argue that molecules such as $PtCl_5$ consisted of four covalent bonds and one ionic bond, $[PtCl_4]^+Cl^-$.[64] When Lewis published his monograph *Valence* in 1923, he emphasized the historical credit due Alfred Werner, who had published his theory of coordination chemistry in 1905. Werner's theory

---

* Langmuir's mathematical formulas led him to a realization that compounds with similar electronic structure, such as N_2O and CO_2, should have similar properties, and Langmuir introduced what he called the principle of electroneutrality—that if more than one structure could be drawn with eight electrons around atoms, the preferred structure was one with the least formal charge difference among adjacent atoms. Both these ideas were significant additions to the original Lewis theory.

did not show Langmuir's octet as particularly important in many compounds containing elements from the higher rows of the periodic table. Lewis was likely referring to the octet when he said that Langmuir had "been tempted to regard certain rules or tendencies as more universal in their scope than I considered them in my paper, or than I now consider them."

Kohler has discussed the acceptance of Lewis's and Langmuir's ideas by the chemical community.[65] To summarize Kohler's careful analysis in two sentences, the Americans took up the Lewis-Langmuir theory to some degree, although with some resistance from the advocates of the older dualistic theories of bonding; the Germans ignored it with only a few exceptions; and the British, particularly the organic chemists, adopted it enthusiastically. The idea of the shared electron pair bond became fundamental to the ideas in British physical organic chemistry and physical chemistry.

Lewis's 1916 idea of the shared electron-pair bond was based on chemical thinking; the truth of the idea was so clear to him that he was willing to deny the repulsive forces between electrons at short distances. Although Langmuir "rediscovered" the theory only three years after Lewis, the credit for the idea belongs to Lewis, not Langmuir, just as the popularization of the idea was due to Langmuir, not Lewis.[66] For both Lewis and Langmuir, their work on the chemical bond was secondary to their principal research efforts at the time—thermodynamics for Lewis and surface chemistry for Langmuir. By the late 1920s the physicists' model of the molecule included concepts of chemical bonding, and the shared-electron pair was incorporated in the Pauli exclusion principle. In 1927, Walther Heitler and Fritz London performed the first quantum mechanical calculation of a chemical bond,[67] although it was for the simplest of all molecules—H_2. By 1939, when Pauling published *The Nature of the Chemical Bond*, ordinary chemists were beginning to be familiar with the concepts of quantum chemistry, and much of Lewis's valence model was incorporated into the new formalism.

Arthur Schopenhauer has said, "All truth passes through three stages. First, it is ridiculed. Second, it is violently opposed. Third, it is accepted as being self-evident." Lewis's theory of the covalent bond, based on chemical rather than physical thinking, managed to go from Schopenhauer's first stage to his third in only a decade. But chemists still use Lewis-dot structures as a shorthand notation for explaining structures and reactions. The idea that the electron pair is the essence of the covalent bond, which Lewis was the first to express, lies at the heart of all modern theories of chemistry and biochemistry.

Shifting Loyalties

Although Langmuir engaged in debate and priority arguments with Lewis over the chemical bond, his primary interest was surface chemistry. After

Langmuir's and Harkins's scuffle over their rival theories in 1917, Langmuir continued to elaborate his ideas on the subject, as did Harkins. But Harkins was acquiring a bad reputation.

In 1922, Edwin Bidwell Wilson, an old friend of Lewis's, was secretary of a committee that was running MIT while it was without a president. Wilson asked Lewis for recommendations for hiring a chairman of MIT's chemistry department and mentioned Harkins's name as a possibility, saying, "I think Harkins was talked of and would take the job, but there is something about Harkins that would make me hesitate very long—he seems to be selfish, and unless a man is a real genius he cannot afford to be selfish in an executive position. Moreover, I sometimes think Harkins is a sort of scientific bounder instead of a real scientist."[68] Lewis's reply was straightforward: "Harkins is a sick man, has been ridden by his ambition for years; this, combined with his own illness and that of his family, have greatly impaired his usefulness."[69]

In 1923, Harkins did something that finally provoked Langmuir to make a formal complaint. In a letter to Arthur Lamb, editor of the *Journal of the American Chemical Society*, Harkins objected to a footnote in a paper by Edward L. Griffin that attributed a theory of emulsions to Langmuir. Harkins believed that he alone should have been referenced, claiming that Langmuir had not offered any theory of emulsions, whereas he had. He also objected to a paper that he had heard Joel Hildebrand, Lewis's colleague at Berkeley, had submitted titled "A New Theory of Emulsions," which Harkins thought would be an improper title, since there was already a Harkins theory of emulsions.[70] Lamb wrote to both Langmuir and Hildebrand, asking for their comments.

Hildebrand was irritated, but his response to Lamb was relatively mild: "As a matter of fact, I find that Harkins did suggest that orientation at a liquid interface might cause a natural curve which would determine the type of emulsion. If Harkins would not write so voluminously, some of his nuggets of wisdom might be more easily discovered." Hildebrand also responded that his title was not "A New Theory of Emulsions," but rather "The Theory of Emulsification," which would cover previous work as well. He agreed to add a reference to Harkins's work.[71]

Langmuir, on the other hand, had let Harkins pass before, but this time he pulled out all the stops. He sent Lamb a copy of the 1917 letter from Lewis that began this chapter, saying that while he had previously accepted Harkins's statements that he had developed his theories independently, he no longer believed this to be true: at best, Harkins was ignoring and refusing to reference Langmuir's work, and at worst, Harkins was stealing his ideas:

> I have repeatedly spoken to Harkins about the unfairness of his statements in regard to the work that I have done, particularly in his denial that I have done any work on surface energies, and he has apologized to me most abjectly for his treatment of me in this respect; but he then goes quietly on

repeating the same unfair statements. . . . In view of this persistent attitude on the part of Harkins, I feel that it is nearly time that at least the Jr. [*sic*] American Chemical Society should cease publishing statements of this kind from Harkins; and I think that it is only fair that Harkins should in the future have to base his claims of priority solely upon publications, and he should frankly acknowledge my prior publications on these subjects.[72]

In an attempt to salvage his reputation, Harkins sent Lewis a photostat of the lecture notes that a student had taken during one of Harkins's 1914 lectures. He claimed that a single phrase, "COOH of acid down because both acid and H_2O associated and polar," written by a student ten years earlier, supported his claims for priority over Langmuir. His accompanying letter to Lewis denies that he was even present at the scientific meeting in 1916, where

Figure 5-6. William Harkins. University of Chicago Library.

Langmuir had first read the paper on the surface chemistry of liquids and where Lewis had watched Harkins make comments as to how Langmuir's results supported the Harkins Theory of Surface Chemistry, unknown to anyone other than Harkins:

> Recently Dr. Langmuir has been circulating some letters, one of which I believe was sent to Dr. Hildebrand. Unfortunately as the basis of one of these statements, he cites a letter written by you, but I think he must have misconstrued what you wrote.... [Langmuir] states, in regard to a paper which he read in September, 1916 at the New York meeting of the American Chemical Society..., "Professor Harkins and Dr. G. N. Lewis were present at the reading of this paper." This is entirely untrue so far as I am concerned, because at the time of the reading of this paper, I was in Southern California.[73]

Harkins's reputation had by this time sunk to the point where he was in danger of being drummed out of the scientific community. Julius Stieglitz, then chairman of the department of chemistry at the University of Chicago, wrote Lewis in 1926:

> I am writing to you for your advice and your judgment on a subject of considerable delicacy relating to the standing of our own Professor Harkins and the value of his work.... I had occasion to consult with our new president, Dr. Max Mason. I was quite taken aback when the president told me that he himself was in a quandary about Harkins and had been puzzled about him even before joining the University. The question on his mind, as I understand, is whether Harkins is an original thinker and investigator of high standing or whether he elaborates on the ideas of others and does a good deal of self-advertising in connection with his own work without making sufficiently clear his indebtedness to others in a given field. Of course, I have known for many years of Harkins's personal weakness in talking too much about his own work, in being self-centered in an exaggerated degree and in clashing continually with other workers about priority, etc., etc. I have often advised him urgently to overcome this failing.... I feel quite certain that some of all this had come to President Mason's attention at Wisconsin and that it is the root of the doubts assailing him now.[74]

Lewis replied to Stieglitz:

> I am very glad to have an opportunity of saying a few words on behalf of Professor Harkins. I first became acquainted with him when he came to Boston from Germany to work under my direction.... When I came to California Harkness [*sic*] applied for a position here and he was on my preferential list.... At that time he had shown very little of the unfortunate tendency to which you refer, and which I have been inclined to attribute to ill-health.

In scientific achievements I rate him very high among chemists of his age in this country....

The dispute between him and Langmuir regarding relative priority was even less reasonable than it might otherwise have been since they had both been considerably antedated in several of the important points by English and European investigators....

If we should place the physical and inorganic chemists of this country in two lists ranked according to their success in theoretical work on one hand and in their experimental work on the other, I should think that Harkins's name would appear among the first half dozen in each list.[75]

In his letter to Stieglitz, Lewis made no mention of what he had earlier called Harkins's "piracy" and Langmuir's "absolutely new and vital contribution to our conceptions of intermolecular forces." He now saw Langmuir's surface chemistry as "antedated in several of the important points by English and European investigators." Between 1917 and 1926, Lewis's opinions of both Harkins and Langmuir had changed. Lewis stood by Harkins—whatever Harkins's faults in stealing from others, he had not stolen from Lewis, as Lewis thought Langmuir had.

Both Lewis and Langmuir tried to stake out an area of scientific turf—electronic theory in Lewis's case and surface chemistry in Langmuir's—and to defend it against what they saw as the intrusions of others. In both cases, the intruders—Langmuir in electronic theory and Harkins in surface chemistry—made some unwarranted claims for rediscovering what had already been discovered, but both did make significant contributions. But Langmuir had a magnetic personality and acted openly and confidently, while Harkins was unpleasant and skulked and wheedled; that seems to have made the difference in how they and their work were perceived.

6

Science and the Nazis

Nernst and Haber

In the spring of 1933, Walther Nernst, then cochairman of the University of Berlin's department of physics, arrived at work and went looking for his colleague, Peter Pringsheim, only to be told that Pringsheim had been banned from the laboratory as a Jew. Enraged, Nernst stormed into the office of Artur Wehnelt, his cochairman. Wehnelt had been given the cochairmanship as a sinecure for past service and had always deferred to Nernst. But today he seemed puffed up, and he lectured Nernst on the importance of cleansing the laboratory of Jews and on the requirements of the new civil service law of 7 April, which clearly barred the employment of non-Aryans such as Pringsheim. Nernst was in no mood to listen. He left, hailed a taxi, and went immediately to see his old colleague and sometimes rival, Fritz Haber.[1]

The two would have a lot to discuss. Fully a quarter of the staff of Haber's Kaiser Wilhelm Institute of Physical Chemistry was Jewish. Haber knew all about the new law and the firings; he was in fact considering resigning, but he and Nernst were most immediately concerned with protecting their employees. Under the new law, it did not matter whether an individual had converted and been baptized as a Christian, as Haber had done. The law defined a "non-Aryan" as anyone with at least one Jewish (nonbaptized) grandparent, but there were exemptions for those who had been frontline soldiers. Haber himself was exempt under this rule, as was Franz Simon in Nernst's institute. Perhaps a few could stay under the exemption—but most of the Jewish scientists would have to go, and Haber and Nernst would be responsible for dismissing them.

There is always a temptation to look ahead when viewing the past. Seventy-five years after 1933, we are able to see what was then the future—World War II and the Holocaust. While we may wish to shout warnings at Nernst and Haber as they concern themselves throughout the 1920s with science, business, and their personal affairs, we cannot. But we can watch them as they live and work in the Weimar Republic, the weak democratic German government that had a brief life after World War I. We will return to 1933, but we will first look at the fifteen years between the end of World War I and the Nazi rise to power.

The Weimar Republic

Germany had been governed by a military dictatorship during the final two years of World War I, when generals Paul von Hindenburg and Erich Ludendorff were running the country. In late 1918, the German High Command decided that surrender was inevitable and that it was better to end the war before the Allies broke through and German territory was occupied. Only a few months earlier, German success had seemed possible. Russia had surrendered to Germany in March, and the Germans had moved forty-three divisions to the western front and broken thorough the Allied lines in a dramatic spring offensive. But by August the German offensive had stalled, the Allied blockade was starving Germany, and American troops were arriving in great numbers. Germany was out of men, food, and supplies. In their peace terms, the Allies had demanded an end to the German monarchy, and the Weimar Republic was born in 1919.

Although the laws that provided Jews equal civil rights with other Germans remained on the books, anti-Semitism was a disruptive force from the Weimar Republic's beginning, and neither right-wing nor left-wing extremists had much regard for the law in any case. Wolfgang Kapp, a right-wing journalist backed by ex-military *Freikorps* (paramilitary militias), took control of Berlin for four days in 1920, forcing the government to flee to Stuttgart. Communists attempted revolutions in Saxony and Hanover in 1921 and seized control of the Ruhr region, the heart of Germany's industrial base, only to be put down by the army and the *Freikorps*.

When the Germans were unable to pay the impossible sums that the Allies had demanded as war reparations, the French occupied the Ruhr in 1923, leading to strikes and resistance that further damaged the economy. Uncontrolled inflation in 1923 made the German paper currency worthless. A new face moved to prominence in German politics that year—Adolf Hitler, a frontline corporal who had been decorated in World War I and was now leader of the National Socialist Party. He and General Ludendorff attempted to take control of Bavaria in the 1923 Munich Beer Hall Putsch, for which Hitler served only eight months in prison. Hitler spent his jail time writing *Mein Kampf*, where he clearly described his intentions to eliminate Jews from German society (as well as his contempt for natural science education).[2] After Hitler's release from prison, the Nazi party decided to focus on achieving power through the vote rather than through a coup, although it continued street violence against parties on the left.

The German General Staff was unwilling to admit its own role in Germany's World War I defeat, so someone else had to be made responsible. The myth of Germany being "stabbed in the back" by Jewish bankers and war profiteers appealed to many. Germany's long history of anti-Semitism was expressed in street prejudice but also in pseudo-intellectual language; Houston Stewart Chamberlain, an Englishman who was composer Richard Wagner's son-in-law, had written an 1898 anti-Semitic tract, *The Foundations of the 19th Century*, that was viewed by many to be a profound theoretical statement on the subject. Ludendorff, in particular, was an adherent of Chamberlain's ideas and rejected Christianity on the grounds that it had been founded by a Jew. Although Jews amounted to less than 1% of Germany's pre-World War II population, there seemed to be an almost universal belief in the existence and importance of a "Jewish problem," with different groups espousing solutions that ranged from assimilation to expulsion. But prior to World War II, ordinary Germans were not discussing what was to be the Nazis' "final solution" to the Jewish problem—mass extermination. Before World War I, Germany had been far from the worst case of central European anti-Semitism; Jewish scientists such as Edward Teller, Leo Szilard, John von Neumann, and Eugen Wigner had come to Germany from Hungary because they saw Germany as more welcoming.[3]

But after World War I, *Organisation Consul*, one of the paramilitary *Freikorps*, took on the task of killing prominent German Jews. After a number of lesser successes, *Organisation Consul* assassinated Walther Rathenau, then Germany's foreign minister, in a 1922 broad-daylight attack involving machine guns and hand grenades. Rathenau was a Jewish industrialist who had been chairman of the giant electrical corporation AEG and who had managed to keep Germany's economy running and able to supply the army during the war (see chapter 4). Rathenau was a particularly important assassination target—he had been a close witness to the actions of the military dictatorship during the war, and he could perhaps refute the "stab-in-the-back" claims about Jews better than anyone else.

Rathenau and Nernst had been friends and had served as advisers to the Kaiser. Despite the personal risk of doing so, Nernst, as rector of the University of Berlin, gave a eulogy for Rathenau at the opening of the university's 1922 summer term. Both Nernst and Haber had been supporters of the German monarchy, but after the Kaiser's abdication they supported the Weimar state as the only possible alternative to revolution from the left or right. Both would suffer the effects of anti-Semitism. Although Haber had converted to Christianity, the anti-Semites still considered him a Jew, and the Nazis would call Nernst a "white Jew"—someone who, as they would put it, had Jewish friends, intermingled his family with Jews (two of Nernst's daughters married Jews), and thought like a Jew.

Science in the Weimar Republic

The first two decades of the 19th century saw the emergence of two revolutionary physical theories—relativity and the quantum theory—and the center for both was Germany. In 1900, Max Planck derived an equation that matched the observed emission of radiation from a heated "black body" by assuming that the energy of electromagnetic oscillators was limited in frequency to discrete values or quanta—the assumption that became known as the "quantum hypothesis." Einstein used Planck's quantum hypothesis in 1905 to explain the photoelectric effect—the emission of electrons by a body that is irradiated by light—and in 1907 to derive Nernst's heat theorem, which Nernst had first stated as an unproved assumption. Niels Bohr applied Planck's quantum hypothesis to the electron in the hydrogen atom in 1913, coming up with a formula that precisely matched the observed frequencies of one of the spectral series of the hydrogen atom. Through the 1910s and early 1920s, Arnold Sommerfeld extended Bohr's theory to atoms with more than one electron. All of this is usually termed the "old" quantum theory; the quantum hypothesis was applied individually to particular problems. Starting in the mid-1920s, the "new" quantum theory began to emerge, organizing the subject within an overall theoretical framework. This effort would eventually result in a theory that would unite Bohr's and Gilbert Lewis's theories. Göttingen was its center, where Max Born, Werner Heisenberg, Wolfgang Pauli, Erich Hückel, Enrico Fermi, Fritz London, Friedrich Hund, Walter Heitler, Vladimir Fock, Eugen Wigner, Gerhard Herzberg, Maria Göppert-Mayer, Paul Ehrenfest, J. Robert Oppenheimer, Max Delbrück, Edward Teller, and others were all either permanently or temporarily in Göttingen's Institute of Theoretical Physics, and James Franck directed its Institute of Experimental Physics after 1920. David Hilbert, Felix Klein, Carl Runge, Hermann Weyl, and Richard Courant were in Göttingen in mathematics. Although the Americans may have taken the lead in physical chemistry after the war, Berlin was the place to be within German physical chemistry: there was Nernst's institute at the University of Berlin, where Franz Simon and Arnold Eucken worked. When Nernst became rector of the university in 1922, Max Bodenstein succeeded him as the institute's head. And there was Haber's even larger Kaiser Wilhelm Institute of Physical Chemistry in suburban Dahlem, where James Franck, Gustav Hertz, Herbert Freundlich, Michael Polanyi, Karl Bonhoeffer, and Paul Harteck all spent time.[4]

In 1913, Nernst and Planck had managed to convince Einstein to move from Zurich to Berlin, where he was offered a research position in the Prussian Academy of Sciences, a chair (without the distractions of teaching duties) at the University of Berlin, and the directorship of the planned Kaiser Wilhelm Institute of Physics. Planck and Einstein shared a love of music: Planck played the piano and Einstein the violin. Despite Nernst's and Haber's belligerence in World War I and Einstein's own pacifism, Einstein was friendly with both of them. James Franck recounted one episode between Nernst and Einstein: Nernst was known for defending anything he had published, right or wrong, bluffing his way through until he was absolutely forced to recant, at which point he would simply drop the subject entirely.[5] At one point Nernst was preparing to present a paper that depended upon results Einstein had derived only a short time earlier. When Einstein told him, "I regret very much that I misled you, but what I have published is wrong. It isn't so," Nernst told him, "Einstein, you can't do that. You have published that. You have to stick to that." Einstein replied, "Shall I start an argument with the Lord that he has

Figure 6-1. Haber and Einstein, courtesy of the AIP Emilio Segrè Visual Archives.

not made the world the way I thought he should have done four months ago?"[6] And although Einstein mocked Haber's desire to be accepted by German society, he clearly felt a real affection for him. Haber became an intermediary in Einstein's divorce from his first wife, Mileva. On 29 July 1913, Haber stood with a weeping Einstein on the train platform as his wife and children left for Switzerland and then took him home to spend the night at his house.[7]

Nernst's Heat Theorem Becomes the Third Law

When we left Nernst and his heat theorem in chapter 3, he was attempting to validate it by applying it to published data, including Haber's work on the synthesis of ammonia from its elements—the fixation of nitrogen. Theodore Richards had convinced himself that he, not Nernst, had made the important discovery. Nernst's heat theorem would eventually become known as the third law of thermodynamics, and Nernst would receive the Nobel Prize for his achievement—but not before fifteen years of opposition, led by Nernst's onetime friend Svante Arrhenius.

The heat theorem offered the promise of obtaining free energies, and thus the ability to predict chemical equilibria, using only measurements of heat. If the theorem were validated, it would be enormously valuable, and the validation seemed to be going well. In the introduction to his book *Thermodynamics*, Planck wrote, "A real extension of fundamental importance is the heat theorem, which was introduced by W. Nernst in 1906. Should this theorem, as at present appears likely, be found to hold good in all directions, then Thermodynamics will be enriched by a principle whose range, not only from the practical, but also from the theoretical point of view, cannot as yet be foreseen" (third edition, 1910), and "[the] theorem in its extended form has in the interval received abundant confirmation and may now be regarded as well established" (fifth edition, 1917).[8]

Einstein showed in 1907 that Planck's quantum theory led to the heat theorem when applied to crystals—that the specific heat should be zero at absolute zero temperature and then at very low temperatures increase exponentially with the temperature, which would make the heat and free energy curves asymptotic, as Nernst had asserted. In order to establish the general validity of his theorem, Nernst became interested in the new quantum mechanics and was a regular attendee at the Berlin physics colloquium throughout the 1920s.[9] The key question regarding the heat theorem was where else it was applicable beyond crystals. Nernst had claimed that it applied to all solids and liquids, but he was finding significant resistance on this point.

The heat theorem could be expressed in different ways. Nernst had simply said that the free energy and the heat of a reaction, when expressed as a function of temperature, become asymptotic as they approach absolute zero. The thermodynamic function *entropy* can be interpreted as the measure of disorder in a system, and many scientists had chosen to reexpress the heat theorem in terms of entropy. In this form, the entropy of any substance to which the heat theorem applied had to approach zero as the temperature approached absolute zero—the disorder would disappear completely. Nernst himself preferred not to use the concept of entropy, but he certainly understood it. The two formulations—one in terms of entropy, and one in terms of the behavior of free energy and heat of reaction near absolute zero, which was Nernst's preference—are equivalent.

Planck suggested that Nernst had overstated his case in implying that the entropy would go to zero at absolute zero temperature for solids or liquids that were mixtures of more than one component.[10] When a substance is pure, each atom is surrounded by other identical atoms. When two substances are mixed, there is a fundamental disorder at the microscopic level. For any mixture, an atom of one substance will generally be surrounded by a random collection of atoms of both substances, and this disorder will not disappear even as the temperature approaches absolute zero. Lewis, who was still nursing whatever grudge he had against Nernst from 1901, showed experimentally that Planck was likely correct—according to Planck and Lewis, the third law applied to pure crystals, but not to solutions or glasses composed of more than one chemical component.[11] Lewis's results could not have helped Nernst's candidacy for the Nobel Prize.

Nor could Richards's continuing complaints that Nernst had stolen his work. Richards had come to believe that the essence of the third law could be found in his own 1902 paper, where his graphs were suggestive of what Nernst had proposed as a universal principle. Although Nernst later would say, "I cannot admit that the thermodynamics in [Richards's paper] had any influence on my deliberations owing to its vagueness and incompleteness,"[12] he had almost certainly seen Richards's data before arriving at his heat theorem. In a letter to Arrhenius, Richards later complained that during Nernst's 1904 visit to America, Nernst had learned of Richards's 1902 paper and had gotten his ideas there.[13] Possibly so, but Nernst might just as well have stayed home and read it in German, the language in which Richards had also published it in Wilhelm Ostwald's journal, *Zeitschrift für physikalische Chemie*.[14] Richards made repeated claims in print and in letters from 1912 through 1921 that he deserved priority for the third law, writing, "[T]he essential ideas involved are indubitably outlined [in my 1902 paper],"[15] that he feared that Nernst might win the Nobel Prize for "this piece of plagiarism,"[16] and that "no new fundamental principle has been certainly added to the main idea advanced in 1902, which may be said to be the basis of the whole subject."[17] In a published 1914 address to the American Chemical Society, Richards said that his ideas "were afterwards adopted unchanged by Nernst in the development of the [heat theorem] usually named after him."[18] This prompted Nernst to attack Richards in the 1917 edition of his book *The New Heat Theorem*, where he caustically reviewed the fumbling mathematics and thermodynamics in Richards's paper.[19]

Arrhenius and Nernst, once close friends, had fallen out beginning around 1900 in a series of bitter personal and scientific disputes over Nernst's extensions of Arrhenius's dissociation theory, Nernst's support of Paul Ehrlich in his disagreements with Arrhenius on immunology, and Arrhenius's role in the administration of the Nobel estate and prizes (see chapter 1). Arrhenius managed to arrange to be the sole referee of Nernst's work for both the Nobel physics and chemistry committees, writing *six* different confidential negative reports over the years that had the effect of blocking a prize for Nernst.[20] Arrhenius's concerns went beyond the scientific; he said that he acted on ethical and moral grounds because of Nernst's investments in student taverns and his involvement in libel suits.[21] Arrhenius and Richards were friends, and Arrhenius had backed Richards over Nernst for the 1914 Nobel Prize in Chemistry for his work in determining atomic weights, which was announced in 1919 after the end of World War I.[22]

Wilhelm Palmaer, a Swedish electrochemist who was secretary to the Nobel chemistry and physics committees, was active in supporting Nernst's Nobel candidacy. In 1920, he led a rebellion on the floor at the Royal Swedish Academy of Sciences, submitting a thirteen-page report on behalf of Nernst, opposing the Nobel Committee for Chemistry's recommendation that the chemistry prize be reserved for that year. Haber, who had once been Nernst's rival, also supported the award to Nernst, who was also the target of Allied war crimes tribunals, and who was also a representative of German science.[23] By 1921, the pressure in favor of a Nobel Prize for Nernst was becoming too much for even Arrhenius to resist. The committee finally asked for a report on Nernst from someone other than Arrhenius; Hans von Euler-Chelpin, a German-born physical chemist, methodically refuted Arrhenius's earlier reports.[24] In May 1921, Arrhenius wrote to Richards that "our academy will have to decide if Nernst's theorem shall be Nobel-crowned. Most chemists here are strongly in favor of it."[25] Arrhenius followed this with a letter to Richards in October, reporting, "It seems very probable that Nernst will get the Nobel Prize this year. There has been a strong agitation in his favor."[26] Arrhenius responded to Euler-Chelpin's report with an unsolicited seventh negative report of his own, but was unable to convince the committee. In 1921, Nernst was finally awarded the Nobel Prize in Chemistry retroactively for the year 1920.[27]

Nernst and Haber, Nobel Laureates

Both Nernst and Haber had become wealthy, Haber from his ammonia royalties and Nernst from his lamp, and both managed to avoid financial ruin in Germany's hyperinflation in 1923. Haber had in 1919 received the Nobel Prize for 1918. With Nernst's award in 1921, he and Haber were perhaps the world's best-known physical chemists, but neither of them would thereafter match their earlier scientific successes.

After his Nobel award, Nernst left traditional physical chemistry and branched out in several directions, none of which proved very productive. He speculated about cosmic radiation, which suggested to him that nuclear energy might be the power source of the stars and that the "heat death" of the universe, which was predicted by the second law of thermodynamics, was much further off than generally believed. He also became interested in the problem of "gas degeneracy," or how the energy of an ideal gas could be quantized. In 1922 Nernst designed an electric piano, which was developed commercially in collaboration with the electrical company Siemens and the piano manufacturer Bechstein. Nernst attempted to negotiate a cash payout similar to the one he had gotten for his electric lamp, but to no avail—Siemens and Bechstein would offer only royalties.[28] The piano was not brought to market until 1931 under the name Neo-Bechstein Flügel. It was a modified grand, with electrical pickups that made it sound more like a modern electric guitar than a piano. About 15–20 of the pianos were built, of which two are still functioning in German museums, one of which is still being played in performances.[29]

Nernst had invested in a country estate, but he still enjoyed the Berlin nightlife. For many Germans, Berlin in the early 1920s meant poverty, but for those who had hard cash it was exciting, a city of nightclubs and gambling joints full of unsavory characters. Nernst and his wife, Emma, enjoyed Berlin's casinos, and Nernst's daughter recalled that she had received intensive coaching in baccarat at age eight.[30] Nernst enjoyed hosting scientific gatherings, but he offered better food and drinks to the professors than to the students and low-ranking researchers; he would walk around with two cigar cases in his coat pockets, offering good cigars to the prominent and cheaper cigars to the hoi polloi.[31]

With his Nobel Prize, Nernst seemed to become even more self-important and difficult than before, especially in his weekly research seminar. He was known to dismiss questions that he thought uninformed with "I will not permit that question within my seminar."[32] And at some point each semester, a flunky would arrive with a polished mahogany case, the size of a baby's coffin, from which Nernst would extract and display the various awards he had received from the Kaiser and from international scientific societies.[33] In 1921, he was offered the German ambassadorship to the United States, which he declined, probably wisely. It seems unlikely that he would have had the tact or patience to put up with the careful dance of the diplomats, although he would certainly have liked the title. In 1922, he accepted the presidency of the national physical laboratory, the Reichsanstalt, but found it frustrating, full of mediocre people and red tape. When he lost his temper, one of his division heads said, "Mr. President, you are behaving like a prima donna." Nernst replied, "I *am* a prima donna."[34] After resigning from the Reichsanstalt, he continued to use the title *Herr Präsident* and found a way to inform anyone who was unaware of how he liked to be addressed: he would call his home and tell the maid who answered that *Herr Präsident* was speaking, and would she please tell *Frau Präsident* that *Herr Präsident* might be a few minutes late for lunch.[35] After his term as president of the Reichsanstalt, Nernst accepted the post of chairman of the University of Berlin's physics department and returned to academic life in 1924 at sixty years of age.

Even with the fall of the German monarchy, Haber was still the German patriot and continued covert research into poison gas despite the complete ban specified by the Versailles Treaty that had ended World War I. At the end of the war, Germany had large stocks of chemical weapons, many of these in railcars, which were to be destroyed under the terms of the treaty. Haber suggested that Hugo Stoltzenberg, who had helped construct the mustard gas works during the war, take charge of this effort. (Stoltzenberg's son, Dietrich Stoltzenberg, has written a biography of Haber.) The poison gas was supposed to be either incinerated or converted to other substances under the supervision of the Allies. But Haber and Stoltzenberg made sure that the German army was informed of exactly what was going on and managed to preserve knowledge of the processes and as much of the feedstocks and equipment as possible. Haber was recognized as the world's greatest expert on gas warfare, and he began getting inquiries from other countries. He passed these on to Stoltzenberg, who in 1922 negotiated a contract with the Spanish military to help build a factory for the manufacture of mustard gas, which was used against Moroccan rebels. Haber then passed another tip to Stoltzenberg—the Soviet Union had suggested a cooperative arrangement, in which factories for the manufacture of poison gas would be set up on Soviet territory, supplying both Germany and the Soviet Union. When the political climate changed and the Soviet-German joint project collapsed in 1926, Stoltzenberg went bankrupt. With his front man out of business, Haber was at least temporarily out of the poison-gas picture. He complained in 1929 of being out of the loop, and the German army responded by directing that he be used as much as possible in matters relating to chemical weapons.[36]

When the German currency collapsed in 1923, Haber thought he might have a solution to the nation's bankruptcy by extracting gold from seawater. In 1903, Arrhenius had assumed a concentration of six milligrams of gold per metric ton of seawater and had calculated that eight million tons of gold were thus available. If German science could develop a method of extraction and could keep it secret from the rest of the world, it could easily pay off its war reparations debt and be able to back its currency. Beginning in 1920, Haber organized a project to investigate gold extraction, which was headed by Johannes Jänicke, a colleague of his who later unsuccessfully attempted to write a biography of Haber. Haber kept the gold project secret from the Allied chemical weapons inspectors who were in and out of his laboratory, and even from his institute's staff. The project's organization was poor: rather than first investigate the actual gold content of seawater, Haber and his colleagues accepted Arrhenius's guess and made up standard solutions containing gold chloride at a concentration of five milligrams per metric ton. They then spent three years perfecting ways to extract the gold before testing actual seawater in 1923. Rather than just taking samples from the North Sea, Haber approached Degussa and Metallgesellschaft, two companies with whom he had close contacts, for a seagoing laboratory to test the gold concentration at different locations around the world. They equipped the *Hansa*, a German-owned passenger ship, and everyone, including Haber, set sail for New York with everything hush-hush. Haber and his group appeared on the crew list as paymasters and dishwashers, clearly enjoying the secrecy. When passengers noticed them taking test samples and asked questions, they spread the rumor that they were seeking a new energy source.

On this voyage and on others, Haber's team collected and analyzed thousands of samples from the South Atlantic, the Pacific, the Antarctic, and the Arctic. All this led nowhere. The average gold content was only a little more than a thousandth of Arrhenius's estimate of six milligrams per metric ton, and even the highest concentration was less than a hundredth of Arrhenius's value—far too low to make extraction economically feasible.[37]

Haber was known as a big spender, staying at the best hotels and eating at the best restaurants. He was welcoming and always ready to pick up the check but a bit overbearing. This is how one of his biographers, Daniel Charles, describes him:

> Within his institute, Haber acted as patriarch and patron, wandering the halls and monitoring the work of young researchers. "Haber would come into my lab and question me: 'What have you done today, Dr. Alyea? Have you done so and so?'" remembered Hubert Alyea, who later taught at Princeton. "And then, no matter what I related, he would nod his head and say gravely, '*Aber! Aber! Aber!*' ['However! However! However!']." These chance meetings often turned into long lectures, fascinating in the morning but much less so at the end of the day, when they prevented researchers from leaving. Researchers who had plans for the evening were known to escape through ground-floor windows when they saw the "old man" wandering meditatively through the garden in the direction of their laboratory.
>
> Haber's generosity was legendary. He seemed to regard anyone who passed through his institute as a personal dependent, and devoted himself to the task of helping them professionally. As one friend put it, "A few more or less lame ducklings always followed in his wake, attaching themselves to

> him until he managed to find a place for them. His good nature was limitless when it came to such things." ...
>
> The generosity wasn't entirely altruistic. Haber enjoyed passing out gifts and favors in part because it showed the world that he *could*. It demonstrated his own wealth and influence. As Lise Meitner once put it, Haber wanted to be "both your best friend and God at the same time."[38]

In 1927, Haber's second marriage ended in divorce. His wife, Charlotte, was disliked by most of his friends. She understandably had a difficult time with Hermann, Haber's son from his first marriage to Clara, and Haber was not the most attentive of husbands. He had negotiated a lump-sum payout in 1924 of his contract with BASF for royalties on the production of ammonia, so he no longer had a large income, and the divorce settlement with Charlotte put a financial strain on him. The stock market crash of 1929 only made things worse. Haber's health, as always, was not good, and he seemed to be prematurely aging.

By the mid-1920s, anti-Semitism was on the rise and out in the open in German universities. In 1924, Richard Willstätter, Haber's closest friend and a Nobel laureate, resigned as professor of chemistry at Munich over a matter of principle. Three senior appointments at Munich came open that year. The leading candidate for each position was Jewish—and each was rejected for his religion. It was a sign of what was to come. After his resignation, Willstätter declined other positions both in Germany and elsewhere but continued his research, directing his assistants by telephone. He was to remain in Germany until 1938, when the Nazis ordered him out of the country.

In February 1933, on the eve of the Nazi era, Haber wrote to his friend Willstätter, "I battle with diminishing energy against my four enemies: sleeplessness, the financial claims of my ex-wife, worry about the future, and the feeling I've made serious mistakes in my life."[39] Given Haber's disastrous life story up to 1933, it is unclear what he meant by "serious mistakes," but he might have included his trust in the German nation. By this time, Hitler had been elected chancellor, and the Nazis were the largest single German party, although they did not command a majority. After the Reichstag fire of 27 February 1933, Hitler received a grant of absolute power by a majority vote of the Reichstag on 23 March. Nazi storm troopers intimidated the delegates from rival parties, chanting "Full power—or else!" Even the *Deutsche Staatspartei*, which Haber had supported, voted to hand Hitler dictatorial authority.

Nazi Science

When Hitler assumed power, many German scientists, both Jews and non-Jews, counseled patience. They hoped that the anti-Semitic demands of the

Nazis had been only crude appeals to the masses in order to gather votes, and that now that Hitler had achieved the power he had sought, anti-Semitism would go on the back burner. While Jews were less than 1% of Germany's populations, they represented about a quarter of the German academic physicists and chemists,[40] and many doubted that Hitler really meant to destroy Germany's international leadership in science—reasonable men like Max Planck and Carl Bosch were still in leadership positions within German science, and surely they could explain to Hitler how important science was to Germany, and how important Jewish scientists were to German science. Two weeks later, the Nazis passed the civil service law requiring the firing of all non-Aryans from government employment, including all university professors. Some of the Kaiser Wilhelm institutes were semiprivate institutions, and some of their staff members were exempt, but Haber's institute was under direct state control, and all its employees were subject to the law's requirements.[41] While many had hoped that the Nazis would moderate their anti-Semitism once in power, Einstein had no illusions; he seems to have understood that the Nazi attitude was, as expressed by the Auschwitz survivor and writer Jean Améry: "Every Jew is a dead man on vacation."[42] Einstein had fled Germany for Switzerland during Hitler's 1923 putsch attempt and had returned only when the coup was suppressed.[43] When Hitler took power in 1933, Einstein was out of the country and immediately announced that he would not return. Soon thereafter, he renounced his German citizenship.

There was an anti-Semitic movement already in place within German science, led by two Nobel-Prize–winning physicists, Philipp Lenard and Johannes Stark, although the movement had been viewed with distaste by most German scientists during the Weimar Republic.[44] Both Lenard and Stark had been rejected for the Berlin physics professorship to which Nernst was appointed in 1924 because "their unusual views in theoretical physics would harm this school at Berlin."[45] After receiving the Nobel Prize in 1905, Lenard was asked in each subsequent year for Nobel nominations. His response was that the only good recent papers were his, and as he already had a prize, there was no sense in proposing anyone else.[46] Authors Jean Medawar and David Pyke had this to say about Lenard:

> He was a man full of resentment, despite having achieved success beyond the reach of most scientists. He had won the Nobel Prize in 1905, when only 43, for work with cathode rays. But this did not soften him....He became infected by the racial theories of Houston Stewart Chamberlain and *völkisch* ideas which extolled German or Aryan or Nordic superiority over other races.
>
> Lenard's emotions spilled over into his science. His hatred of Einstein, the Jew, worsened when in [1922 for] 1921 Einstein was awarded the Nobel Prize. Lenard actually wrote a letter of protest to the Nobel committee and gave it to the press....When Rathenau was assassinated he

would not respect the national day of mourning, refusing to fly the flag of his Heidelberg physics institute at half mast. This enraged the students, who organized a march on the institute which ended with Lenard being jostled at a meeting and taken into protective custody for a few hours. His next public move towards the right and fierce anti-Semitism came when Hitler was imprisoned after his failed putsch in Munich in 1923. Lenard publicly declared his support for Nazism, together with another Nobel Prize–winning physicist, Johannes Stark. Hitler and his comrades, he said, "appear to us as gifts of God."[47]

Lenard's own Aryan heritage was suspect among some anti-Semites, and there were reports as early as 1916 that his Hungarian ancestors were Jewish; his father was a wine-merchant whose partner was Jewish. In 1939 the Nazis began an investigation into his racial pedigree. The report has not been preserved.[48] Lenard had been a student of the Jewish physicist Heinrich Hertz, whom he had idolized in his youth. Lenard had in fact written a biography of Hertz, which he was forever trying to explain away, saying that only later did he realize that his difficulties in following Hertz's physics were due to the fact that Hertz's thoughts were those of a member of an alien race.[49] Lenard resigned from the German Physical Society when it published a paper in English without a German translation, and he thereafter kept a sign over his office door saying "No entry to Jews or members of the German Physical Society."[50]

Johannes Stark, who was awarded the Nobel Prize in Physics in 1919 for his discovery of the splitting of spectral lines in an electric field, had collaborated with Einstein earlier in his career but then became increasingly isolated and frustrated, involving himself in arguments about priority. Franck described Stark as "in every respect a pain in the neck."[51] Stark took it upon himself to defend the Nazi dismissals of Jewish scientists in a letter to *Nature*, a leading British scientific journal: "Measures brought in by the National Socialist Government, which have affected Jewish scientists and scholars, are due only to the attempt to curtail the unjustifiable great influence exercised by the Jews. In Germany there were hospitals and scientific institutions in which the Jews had created a monopoly for themselves and in which they had taken possession of almost all academic posts."[52]

There was an entire philosophical movement generated around the idea of "German science" as opposed to "Jewish science" (and, for that matter, "English science," "French science," etc.). Ernst Krieck, the rector of the University of Frankfurt, said, "What is the purpose of university education? It is not objective science, which was formerly the purpose of university training, but the heroic science of the soldier, the militant and fighting science," and "We do not know science, but only that science which is valid for us Nazis."[53] In his dedication speech at the Heidelberg physics institute in 1935, Lenard quoted Eric Jänsch to say that Jewish science "is mathematical, full of non-material spirituality, purely intellectual, idealistic, projectionist"; English science is a "conglomeration of unrelated theories"; and German science is "vitalistic, organistic, absorptive, and empirical."[54]

But the man the Nazis selected as Minister for Science and Education was neither Lenard nor Stark, but Bernhard Rust. Rust was a former elementary schoolteacher who had been accused of molesting a young girl in 1930, leading to his discharge from his teaching position for "mental instability." His only claim to competence in scientific affairs was having been an early member of the Nazi Party.[55] Rust and the Nazis had little sense of tact in dealing with scientists:

> A mandatory meeting was called for all professors and lecturers at Berlin University.... It turned out they were to be addressed by the Führer's close friend, Julius Streicher [the editor of the anti-Semitic newspaper *Die Stürmer*, executed after the war as a war criminal]. Strutting up and down the rostrum in jack boots and brandishing the horse whip without which he was never seen, he informed the professors... that all their brains put together... did not amount to one-thousandth of the brain of the Führer.... It was Hitler's own effort to win over the professors to the swastika banner.[56]

Some non-Jewish scientists refused to cooperate with the Nazis; Max von Laue was perhaps the leading example. Laue had been awarded the Nobel Prize in Physics for 1914 for his discovery of X-ray diffraction by crystals. He was a member of the Prussian landed nobility, the *Junkers*. When the Nazis came to power, Laue always left his house carrying a parcel in each hand so that he would not be forced to give the Hitler salute. Nernst's biographer Kurt Mendelssohn says of Laue:

> Laue was the unique case of a member of the Junker class to become an outstanding scientist. Even so, throughout his life he looked and spoke like a Prussian general. His lectures and his discussion remarks consisted of a series of short barks, which issued not very distinctly from underneath a moustache. He himself was acutely aware of this failing and went so far as to undergo a course of speech therapy which, however, proved quite worthless.... At one of the university festivities Nernst met a retired general whom somebody had brought along as a guest. Learning that Nernst was a physicist, the old man recalled that many years ago at the famous officers' training school at Lichterfelde they had a cadet who had proved utterly useless at military life. This was particularly sad since he had come from a good old Prussian family, and the fellow had then decided to study physics. However, the general felt that Nernst probably would never have heard of him, a certain Herr von Laue. Nernst told the general that not only did he know Laue but that Laue had received the Nobel prize. At this the old man's eyes lit up and he said how glad he was for the family that the cadet whom they had had to send away had after all proved not completely useless.[57]

When the Nazis took power and passed the 1933 civil service law barring the Jews from state employment, many looked to Max Planck, dean of German physics and president of the Kaiser Wilhelm Gesellschaft (the parent organization of all Kaiser Wilhelm institutes), to try to fix things with Hitler. Over the twelve-year span of the Third Reich, Planck attempted to intervene as best he could, but he paid an enormous personal cost in doing so and was almost completely unsuccessful. John Heilbron, a biographer of Planck, has written:

> The policy of old men like... Planck became one of salvage, an effort to protect science without offending a regime... they did not understand. They complied openly in small things and did not protest publicly against great injustices, they accommodated to insure that lesser men did not take their places, and they strove to persuade younger colleagues to steer a similar course.... Magda Planck wrote... about her husband's actions: "He has often wished that he could withdraw from official matters and great responsibility." But he could not: "Now everyone counts on his help." The political circumstances demanded his entire, diminishing strength.[58]

Planck had been one of the signers of the Manifesto of 93 at the beginning of World War I, which had caused a break between German and non-German scientists. Based on his experience with that manifesto, he had earlier said that he opposed public declarations, saying that they would not be effective: "Usually, I've seen exactly the contrary. Inconsistency inevitably occurs, the declaration is misunderstood, falsely interpreted, and used to slander its signers."[59] He attempted to work quietly behind the scenes to delay expulsions and to preserve German science.

Once the Nazis took power, the swastika was prominently displayed in lecture halls, and all meetings and seminars began with the Hitler salute. One observer described Planck's conflicts over all of this: "Planck stood on the rostrum and lifted his hand half high, and let it sink again. He did it a second time. Then finally the hand came up, and he said, "Heil Hitler."[60] Soon after the new civil service law was passed, Planck obtained an appointment with Hitler to discuss the prospects for science, and Hitler told Planck not to worry because the Nazis had nothing against Jews, and that they in fact planned to protect them. The enemy was not the Jews, but the communists—but unfortunately all the Jews were communists. Planck could find no response to this. After hearing of Planck's interview and Hitler's remarks, Werner Heisenberg saw this as good news, writing in a letter to Max Born, who was then at Göttingen, "Planck has spoken... with the head of the regime and has received the assurance that the government will do nothing beyond the new civil service law that could hurt our science. Since on the other hand only the very least are affected by the law [because of the exemptions for veterans]—you and Franck certainly not, not Courant—the political revolution could take place without any damage to Göttingen physics."[61]

Franck, who had earlier worked in Haber's institute, was now at Göttingen and knew better. He could have stayed—he had been awarded both classes of the Iron Cross in World War I, which gave him an exemption as a frontline soldier. He resigned his position almost immediately and stayed for a short time, holding seminars at his home. In September 1933, Bohr invited him to Copenhagen, and Franck left that November. As he left Göttingen, his friends stood in silence, overflowing the train platform. Most of his colleagues left, as well: of the twenty-six in mathematics and physics at Göttingen before the Nazis, only eleven remained.[62] A government minister later supposedly asked the Göttingen mathematician David Hilbert, "And how is mathematics in Göttingen now that it is free of Jews?" Hilbert replied, "Mathematics in Göttingen? There is really none any more."[63]

Franz Simon, who was Nernst's student and successor in low-temperature work, chose to leave, as well, for Cambridge in his case. (He later became active in the English atomic bomb program and was knighted as Sir Francis for his service.) He had been wounded twice in World War I and had received the Iron Cross First Class. As he was leaving for exile, he was interrogated at the frontier by a German customs agent and told to empty his pockets. He pulled out his Iron Cross, bearing the inscription "The Fatherland will always be grateful," tossed it on the table, and left for good.

About a quarter of Haber's subordinates were Jewish. Nazi leaders ordered the Kaiser Wilhelm Gesellschaft to dismiss scientists at Haber's institute immediately, rather than waiting until September as the law stipulated. The directors wanted Haber to dismiss low-ranking Jews. Haber felt that these were the ones who needed the most protection, and instead dismissed the two most prominent Jewish scientists, Herbert Freundlich and Michael Polanyi, who he knew could get positions elsewhere. When he was forced to fill out a questionnaire on his own racial ancestry, Haber scrawled diagonally across the page: "My parents and grandparents and both women to whom I've been married as well as their ancestors were all non-Aryan as defined by the law." Although he could have stayed because of his wartime exemption, on 30 April 1933, Haber wrote his resignation letter: "My tradition requires that when choosing coworkers for a scientific post, I consider only the professional and personal characteristics of the applicant, without regard for their racial makeup." Haber made his resignation effective five months later in order to buy some time to help his subordinates find positions. Both Planck and Bosch tried to convince Haber to stay, but he was not to be swayed. Planck tried to appeal on Haber's behalf to Rust, who said, "I'm finished with the Jew Haber," and to Hitler himself, who flew into such a rage that Planck had to leave the room.[64] In the same interview, Bosch told Hitler that the expulsion of Jewish scientists would have serious repercussions for German science; Hitler replied, "Then we will do without physics and chemistry for the next hundred years."[65]

Planck and Haber unsuccessfully tried to suggest their choices for Haber's successor—their first choice was Franck, who was a Jew and was leaving in any case, and they suggested Karl Bonhoeffer and Laue as backups. Bonhoeffer thought that the Nazis were unlikely to take Planck's and Haber's recommendation and accepted another position, and Laue was considered by the Nazis to be unreliable. Haber's attempts to delay the firings of his Jewish colleagues and to find them places abroad were largely unsuccessful, and he was forced to dismiss them himself.[66]

Einstein, now living abroad, had some sympathy for Haber, but it was mixed with an attitude of "What took you so long?" Einstein wrote Haber, "I can imagine your inner conflicts. It is somewhat like having to abandon a theory on which you have worked for your whole life. It's not the same for me because I never believed in it in the least."[67] Haber began looking for a place for himself abroad and even considered going to Hebrew University in Jerusalem after meeting with Chaim Weizmann, a Zionist leader encouraging Jewish emigration to Palestine. This surprised Einstein, who had long been a Zionist: "I'm especially glad that your love for the blond beast has cooled off a bit. Who would have thought that my dear Haber would appear before me as an advocate of the Jewish—and even Palestine's—cause." Einstein continued, saying that German scientists were "men who lie on their bellies before common criminals, and even sympathize with those criminals to a certain extent. They couldn't disappoint me because I never had any sympathy or respect for them—apart from a few fine persons (Planck 60% noble and Laue 100%)."[68] Two English scientists who had been active in Britain's chemical warfare effort during the war, Sir Harold Hartley and Frederick Donnan, arranged a nonteaching position for Haber at Cambridge. With his acceptance of the position, Haber finally gave up on the idea of being the good German, writing, "My most important goals in life are that I *not die as a German citizen* and that I not bequeath to my children and grandchildren the civil rights of second-class citizenship, as German law so demands."[69]

Haber left Berlin on 3 August 1933, but he was too sick to take the position at Cambridge. He spent the final five months of his life shuffling among sanitoria and hotel rooms, dying in Basel in January 1934. Laue wrote an obituary, calling him "the man who won bread out of air" because of his success with the fixation of nitrogen.[70] Planck was concerned about the possible Nazi reaction to a memorial service and delayed it for a year, scheduling it for the first anniversary of Haber's death. Bernhard Rust barred the attendance of all those under his jurisdiction. Lise Meitner, Richard Willstätter, and Carl Bosch, who were not government employees, showed up; Laue sat unhappily in his office a few hundred feet away.[71] "The service . . . took place with dignity, and in a well-filled auditorium; it offered the attendees—foreign dignitaries, nongovernmental employees of the Society, wives of Rust's civil servants, representatives of the military and big business—the enjoyment of a successful, if transitory, expression of defiance."[72]

After Haber's death, Einstein wrote to Haber's son Hermann:

> At the end, [Haber] was forced to experience all the bitterness of being abandoned by the people of his circle, a circle that mattered very much to him, even though he recognized its dubious acts of violence. I remember a conversation with him; it must have been about three years ago after a meeting of the [Prussian] Academy of Sciences. He was quite incensed about the way he'd been shabbily treated during a vote, and to recover he went with me to the *Schlosscafé* on *Unter den Linden*. I said to him, a bit drolly, "Console yourself with me—your moral standing is truly enviable, and here I am happy and cheerful!" And this is what he said: "Yes, all of society never mattered to *you*." It was the tragedy of the German Jew; the tragedy of unrequited love.[73]

Nernst and the Third Reich

Nernst was in no doubt about what he would do when the Nazis told him to fire the Jewish scientists at his institute. He was then sixty-nine years of age. He refused to have anything to do with the new regime and retired to his country estate. According to his biographer Mendelssohn:

> The ascent of Hitler coincided with the end of Nernst's official duties. Having reached retiring age, he gave up the Berlin professorship in 1933, a few months after shaking Haber's hand.... He was informed that in the future his presence on the governing body of the Kaiser Wilhelm Institute was no longer required.... A few years later he was to learn that this slight that he had received from the Nazis was considered a mark of distinction by his friends abroad.[74]

Based on his behavior and statements, the Nazis began to suspect (and perhaps hope) that Nernst himself might have Jewish ancestors. A graduate student, Otto Richter, was working in 1935 on a dissertation on the racial genealogies of Nobel laureates, and he sent detailed questionnaires to the subjects. Nernst did not reply, and Richter complained to the dean of philosophy, Professor Ludwig Bieberbach. Bieberbach, who was the leader of an anti-Semitic movement expounding German mathematics in the same way that Stark and Lenard led the German science movement, took charge, sending the questionnaire to Nernst himself after receiving permission from Bernhard Rust to force Nernst to answer Richter's questions. Unfortunately for the Nazis, Nernst's pedigree turned out to be completely Aryan. (One of Nernst's ancestors, Lieutenant Hermann Nernst, had reportedly been the courier who had brought the news of Blücher's victory at Waterloo to Berlin.)[75]

Nernst, living at his country estate in Zibelle, tried to maintain his scientific contacts, but it was difficult. He no longer felt welcome at academy sessions in Berlin after he had caused a stir by refusing to stand for the singing of the *Horst Wessel Lied*, a Nazi anthem. Frederick Lindemann (later Lord Cherwell), Nernst's English former student who had been working to find positions in England and elsewhere for discharged Jewish scientists, visited him. In 1935 Lindemann requested an audience with Hitler, which was not only denied, but he was told to leave Germany immediately. He spent his last evening in Germany with the Nernsts.[76]

When World War II began, Nernst's health was failing. Nernst's two sons had been killed in World War I, leaving him with three daughters. Two of his daughters had married Jews and were forced to emigrate, Hilde to London and Angela to Brazil. Nernst's Swedish friend Wilhelm Palmaer, who had fought for Nernst's Nobel Prize, was a resident of a neutral country and agreed to help by forwarding mail to them.[77] Nernst's daughter Edith stayed in Germany, visiting every three weeks despite the Allied bombing of the railways. Nernst died at home in bed in 1941. Shortly before his death, he ordered all his papers burned, possibly to protect his correspondents from Nazi snooping. His last words, as he drifted in and out of consciousness, were recorded by his wife and were typical of him: "I have already been in Heaven. It is quite nice there, but I told them they could have it even better."[78]

Nernst had developed a reputation within America as an anti-Nazi. His *New York Times* obituary, printed just days before the American entry into World War II, said:

> [T]o the Nazi regime, men of originality, boldness of imagination and intellectual courage are anathema. It is not likely that out of German universities, as they are now conducted, so towering a figure will emerge. He typifies a period when the German scientist was free to think and say what he pleased, when the *Gelehrter* was honored and not regarded, as now, as an annoyance, from whom nothing socially useful can be expected.[79]

Those Americans who knew him best—his students—were less kind. Nernst had earned the nickname "Chronos" among his pupils, because, like the Greek god who ate his sons, he ate his students.[80] Two of his students, Irving Langmuir and Robert Millikan, the discoverer of the mass of the electron, had won Nobel Prizes. Langmuir refused to contribute to a scientific obituary for Nernst and intimated that Nernst's other American students, including Millikan, would feel the same way.[81] Millikan in fact did write an obituary of Nernst in which he criticized his "Prussian" automobile driving, remarked gratuitously that "[Nernst] was a little fellow with a fish-like mouth," and recounted a story of how Nernst had in 1912 spitefully removed all references to Jean Perrin from a chapter in his book *Theoretical Chemistry* after Perrin had taken some of Nernst's speaking time at a meeting. Millikan did say that Nernst's later work "represented, so I always thought, very bad judgment, but *the third law of thermodynamics is enough to give him a seat among the immortals*."[82] It is likely that no one even thought to ask Lewis for his comments on Nernst. Jacob Bigeleisen, Lewis's graduate student at the time, said that Lewis's dislike of Nernst was common knowledge at Berkeley, although Lewis did not express it directly. The morning that Nernst's obituary appeared, Bigeleisen left it lying on top of a stack of spectra that he knew that Lewis would review and watched from across the room to see Lewis's reaction. Lewis flicked it aside and proceeded to review the spectra.[83] Nernst's lack of American support is not surprising. Anyone who had been through his laboratory still bore the marks, and Nernst had after all been involved in chemical warfare. But Einstein, who was then living in America and who valued above all Nernst's opposition to the Nazis' anti-Semitism, stepped forward to describe Nernst, warts and all:

> Although sometimes good-naturedly smiling at his childlike vanity and self-complacency, we all had for him not only a sincere admiration, but also a personal affection. So long as his egocentric weakness did not enter the picture, he exhibited an objectivity very rarely found, an infallible sense for the essential, and a genuine passion for knowledge of the deep interrelations of nature. But for such a passion his singularly creative productivity and his important influence on the scientific life of the first third of this century would not have been possible.... Nernst was not a one-sided scholar. His sound common sense engaged successfully in all fields of practical life, and every conversation with him brought something interesting to light. What distinguished him from almost all his fellow countrymen was his remarkable freedom from prejudices. He judged things and people almost exclusively by their direct success, not by a social or ethical ideal.... He was an original personality; I have never met anyone who resembled him in any essential way.[84]

The Exodus

In all, about a quarter of the academic physicists and chemists left Germany, most because they were Jews.[85] While Haber, Einstein, and Sigmund Freud got most of the international press as famous émigrés, they were older, and their best work was already behind them. The vast majority of those who left were younger scientists or students, some of whom appear in this book—James Franck, Franz Simon, Max Bergmann, Max Delbrück, Max Perutz.

The behavior of non-Jewish German scientists was not much different than the behavior of the entire German nation when faced with anti-Semitism. Some resisted, some tried to pretend it was not happening, some profited, and some actively collaborated. While the expulsion of Jewish scientists was not a crime at the same level as the later extermination of Jews in concentration camps, there are similarities. Primo Levi has written of the *Lager* (the concentration camps) and the Holocaust, "It could not be reduced to two blocs of victims and persecutors. Anyone who today reads (or writes) the history of the Lager reveals the tendency, indeed the need, to separate evil from good, to emulate Christ's gesture on Judgment Day: here the righteous, over there the reprobates."[86]

At the extremes, it is easy to separate evil and good. In the case of Lenard and Stark, we see only a vindictive and vicious racism. For others, like Laue, who refused to compromise, there is nothing to criticize. Some who refused to cooperate with the Nazis were punished for it: when the pharmacologist Otto Krayer refused to accept a full professorship formerly held by a Jew, he was immediately dismissed and forced to emigrate. At international meetings, the non-Jewish German scientists had to make choices as to how to behave when they met their old colleagues. Many avoided the émigrés, while others talked to them but pretended nothing had happened. Hans Weber, Max Vollmer, Heinrich Wieland, and Adolf Windaus (the last two Nobel laureates), on the other hand, went out of their way to associate and be seen with the émigrés.[87] They must have felt the shame that Levi says a just man feels "when confronted by a crime committed by another,... he feels remorse because of its existence, because of its having been irrevocably introduced into the world of living things, and because his will has proven nonexistent or feeble and was incapable of putting up a good defense."[88]

Some were initially sympathetic to Jewish scientists but withdrew when they saw that they would be unable to help. When Haber resigned his position, Bosch wrote a letter expressing his support and attempted to intervene with Hitler but eventually gave up and stopped answering Haber's letters.[89]

Some became defensive in the face of foreign criticism. Otto Hahn, friend and boss of the Austrian-Jewish Lise Meitner, reflexively defended Germany in an interview. He was on sabbatical leave in America when the civil service law was passed and said that the Nazis' persecution of the Jews was a secondary result of their anticommunism and that Hitler was so ascetic that he "lived almost like a saint." The interview was published under the headline "He Defends Hitler, Denies Man 'Who Lives Like a Saint' Is Guilty of the Atrocities Charged."[90] Some of those who supported the expulsions thought they were taking a balanced view: Karl Freudenberg, an organic chemist at Heidelberg, wrote to George Barger at Edinburgh, who had criticized the dismissals of Jewish scientists:

> There are orders which you simply have to comply with. It is my firm conviction that a cure of the body of the German people was necessary,

> something which probably only very few will deny. The way it has been carried out cannot be subject to lengthy considerations in this country, simply because there are orders, and it does not matter at all what the viewpoint of an individual is. I understand as well that opinions differ in foreign countries whose attitude is that of an observer. It is not to be seen as a rejection of our new German state if we show a grateful appreciation of expressions of help for the individual, which can be found particularly in England.... We have a united central government of great power, and an absolute pure man at its head, who is simplifying the administration with great energy.[91]

Freudenberg was no rabid anti-Semite; he had had close relations with the Jewish physicist Max Bergmann, wrote a positive review in 1940 of a text by the Jewish scientist Hermann Mark, and his daughter was married to a Jew. But if the Jews had to be expelled to purify Germany, if those were the orders, then for Freudenberg that was a sacrifice that must be made.

Some were happy to take advantage of the many open positions. Haber's student Paul Harteck, who had a temporary postdoctoral appointment in Cambridge when the expulsions began, wrote to Bonhoeffer on 16 April 1933, "When you come to Berlin..., don't let the former members of [Haber's institute] talk you into anything. They tend to say the decent Aryans, too, ought to sympathize." On 5 May he wrote: "In London the Jews, the half and quarter Jews of Germany are gathering. If these people have ever had a liking for Germany, it can only have been a very superficial one, because now you really don't notice anything of it." Then, to his surprise, Harteck was fired from his position in Haber's old institute by Haber's successor, who wanted to bring his own staff in. Things, however, did work out for Harteck. Bonhoeffer told him on 22 June 1933, "There are many openings occurring simultaneously, you should actually get one of them." Harteck succeeded Otto Stern as full professor of physical chemistry at Hamburg.

Few of the Jewish exiles returned after the war; Max Born was the most prominent of those who did. When the war ended, the Allied occupiers began a de-Nazification program to remove professors who had actively collaborated. At first, even Laue offered excuses for Germany's moral collapse, saying other countries were no different, and called the de-Nazification efforts by the Allies "Hitler methods." Laue did, however, later criticize German dishonesty as practiced after 1945. Klaus Clusius, a physical chemist at Munich, had the nerve to blame the émigrés for not helping their German colleagues in their postwar difficulties.

The postwar attitude of many non-Jewish German scientists seemed to be that all had suffered, both Jews and non-Jews, and it was now time to move on. Franz Simon said after the war, "In my opinion German scientists as a group lost their honor in 1933 and did nothing to get it back. I admit that you cannot say that everybody should have risked his position or life, but such risks were no longer necessary after the war. The least you could expect after all that happened was that German scientists, as a group, would state publicly and clearly that they regretted what had happened. I did not notice anything of the kind."[92]

7

Nobel Prizes

Lewis and Langmuir

After a long wait, the letter from Einstein finally arrived in the summer of 1926. Gilbert Lewis deciphered the neat German script, translating it to see what Einstein thought of his theories on the nature of radiation and time. It started out on a positive note: "Many thanks for your letter and your papers, which I read with great pleasure. The ideas you suggest in 'The Nature of Light' are the same as I have agonizingly turned over in my own mind without coming to a conclusion. . . . I am completely convinced of the validity of your 'law of entire equilibrium.'" But then came the bad news: "But I cannot agree with your opinion that this law proves the incorrectness of my derivation of Planck's law. . . . [I]t is actually impossible to omit the stimulated emission, and it is therefore in no way an ad hoc assumption. Otherwise the theory would not have satisfied me. I hope this convinces you. Please let me know what you think about it. Meanwhile, I remain, with best wishes, your Albert Einstein."[1] Lewis had been working on his theory of light for three years. With Einstein's letter, Lewis saw his hopes vanish.

Lewis spent the years 1923–32 attempting to make his mark in theoretical physics. But he had allowed himself to become isolated within Berkeley, where the scientists with whom he communicated were those who thought the way he did, many of whom he had trained and selected. He was always confident in his own abilities to contribute, no matter what the subject. Michael Kasha, his last graduate student, described Lewis's approach to entering a new field:

> Lewis's approach to science was to find an area that interested him and then to begin to think about it. He told me that what you don't do is read all the literature on the subject. After you read all that, you'll think it's all done. What you do is read a few papers, maybe listen to a talk, and then begin to work. It turns out that even if something had been done, you'd do it a different way.[2]

This approach had worked well for Lewis within chemistry, but caused him problems as he entered theoretical physics. He failed to read the important papers that would have shown him that he was on the wrong path.

The 1920s began well enough for Lewis. He was ready to consolidate his first twenty-five years of work by publishing books on both thermodynamics and the chemical bond. His goal was to finish what he had begun and then move on to bigger challenges. But as he left, he showed that he still had a great deal to offer in both fields: Before leaving thermodynamics, he proposed a practical solution to the anomalous behavior of strong electrolytes, the problem that had puzzled physical chemists since Svante Arrhenius's dissociation theory in the 1880s. And as he finished his work on the chemical bond, he showed that he was aware of problems with Niels Bohr's model for the atom—problems that would eventually motivate the new quantum mechanics in Germany.

Lewis Leaves Thermodynamics and the Chemical Bond

In 1923, Lewis finished writing *Thermodynamics and the Free Energy of Chemical Substances*, the work that had consumed him since his days as a graduate student a quarter of a century earlier. Merle Randall, Lewis's first doctoral student and now on the faculty at Berkeley, was named as a coauthor, but the book was Lewis's—according to chemist William Jolly, "Lewis would dictate, and Randall wrote it down."[3] The book had a great effect on the practice of chemistry; it changed the way most chemists and chemical engineers thought about planning and analyzing chemical reactions. Thermodynamics, which had been an abstruse field, suddenly became accessible. Lewis's book included tables of the free energies and entropies of a number of substances. These allowed prediction of the equilibrium points of many chemical reactions, and Lewis gave detailed methods for measuring the activity of substances in other reactions, which would allow chemists to determine still more free energies. He intended the book not as a textbook, but rather as a working tool for research chemists:

> A textbook is a sort of table d'hôte to which any one may sit down and satisfy his hunger for information, with no thought of the complex agricultural processes which converted these raw materials into food stuffs, nor of the arts of cookery responsible for the well-prepared meal which is set before him. It has not been our desire to offer such a repast to the reader. Our book is designed rather as an introduction to research, and as a guide to anyone who wishes to use thermodynamics in productive work.[4]

While Lewis might not have intended it to be used as a textbook, *Thermodynamics* immediately became the standard university text on the subject in America, continuing as such until long after Lewis's death, when it went through two revisions by other Berkeley faculty members, the last in 1995.[5]

Shortly before completing *Thermodynamics*, Lewis provided an experimental solution to the behavior of strong electrolytes. Dissociation of weak electrolytes (such as acetic acid or ammonia) in solution follows the mass action law as Arrhenius's theory would predict, but the equilibrium "constant" of strong electrolytes is far from a constant—it can vary by many orders of magnitude as the concentration changes.

For solutions of a single strong electrolyte, one could use the thermodynamic methods Lewis had already developed to find its activity. But for solutions containing several electrolytes, perhaps sharing common ions, determining the activity of a particular substance is not so simple. Lewis began working on this problem in 1906 at MIT and continued for the next fifteen years, and a great many other physical chemists around the world were working on it, as well. There is no thermodynamic solution to the problem—it involves the mutual interaction of the ions, which is outside the purview of thermodynamics' macroscopic measurements—but Lewis worked to come up with an empirical equation that would explain the data. He finally succeeded with his concept of *ionic strength*, which is expressed by a simple formula involving the concentrations and charges of all the ions involved. He provided no theoretical justification for this formula, but he validated it experimentally and published it in 1921.[6] It provided the first practical solution to the problem of strong electrolytes. Two years later, Peter Debye and Erich Hückel provided the theoretical justification. Their theory predicted the same behavior—Lewis's empirical formula for ionic strength followed directly from their equations.[7]

Lewis used his book *Thermodynamics* to make what he claimed to be a final statement of the third law of thermodynamics, which was more restrictive than Walther Nernst's claims that the law applied to all liquids and solids. According to Lewis, the third law applied only to regular crystals of pure compounds:

> If the entropy of each element in some crystalline state is taken as zero at the absolute zero of temperature: *every substance has a finite positive entropy, but at the absolute zero of temperature the entropy may become zero, and does so become in the case of perfect crystalline substances.*[8]

By 1923, Lewis was growing weary of thermodynamics and had decided to move on; he stopped all work in that area after the publication of *Thermodynamics*. He might have continued his work on the chemical bond and applied the new quantum mechanics, then being developed in Germany, to chemical problems. Instead, Lewis published *Valence and the Structure of*

Atoms and Molecules, a summary of his ideas on the chemical bond, and effectively left that field, as well.

Lewis's and Irving Langmuir's model for the chemical bond was called "static"—the electrons did not orbit the nucleus as they did in Bohr's model. Both Lewis and Langmuir had attempted to reconcile Bohr's theory with theirs, and Lewis continued to do this in his book *Valence*. He came up with a thought experiment that may be seen as related to the uncertainty principle—that it is not possible to simultaneously determine both the position and the velocity of any particle. This uncertainty becomes more pronounced the smaller the particle, and the electron is the smallest subatomic particle. Werner Heisenberg did not publish the uncertainty principle until 1927. Here is Lewis's 1923 thought experiment:

> Let us in [fig. 7-1] represent a hydrogen atom according to Bohr with an electron in the first orbit, that is to say in the most stable state, and let us represent by AA′ a small wire which may be brought next to the hydrogen atom. Now if the electron in the orbit exerts any sort of a force at a distance, when the electron is in position X there will be a slight flow of positive electricity toward A, and when the electron is at X′ there will be a slight flow toward A′. Indeed at any finite distance of the wire from the atom there should be set up in the wire a finite alternating current which would continue indefinitely. Such a current should generate heat, but since the atom is supposed to be in the state of lowest possible energy, there appears to be no source from which the heat could originate. In other words, we must conclude that...such an alternating current is not produced....If these considerations are correct, we must conclude that an electron, in a Bohr orbit, exerts upon other electrons no force which depends upon its position in the orbit. In other words it seems that we should add another assumption to those of Bohr, namely, that while the orbit of one electron may as a whole affect the orbit of another electron, we should look for no effects which depend upon the momentary position of any electron in its orbit.[9]

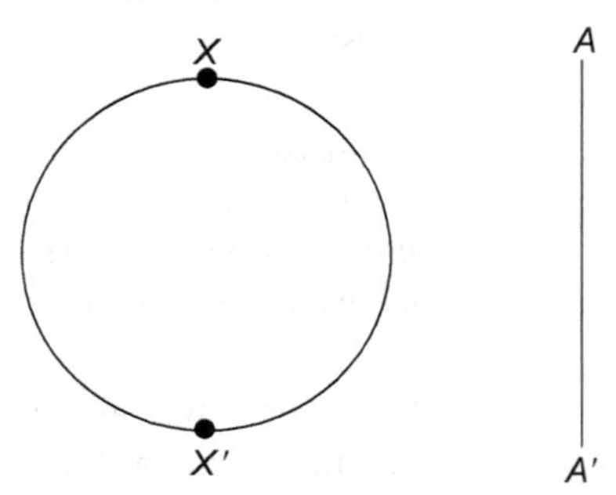

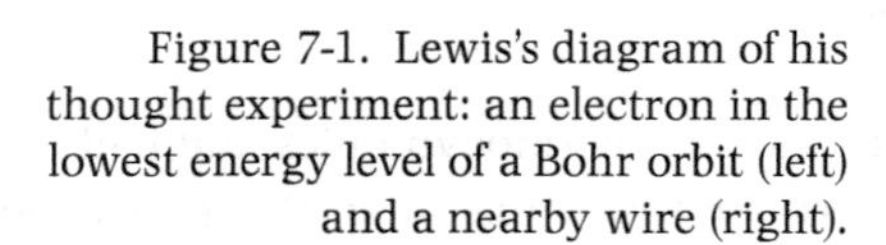
Figure 7-1. Lewis's diagram of his thought experiment: an electron in the lowest energy level of a Bohr orbit (left) and a nearby wire (right).

Lewis's thought experiment should not be viewed as a historical precursor of the uncertainty principle, which came out of an exchange among the

German physicists who were developing the new quantum theory. But it shows that Lewis shared the skepticism of Heisenberg and Wolfgang Pauli about the usefulness of talking about electron positions in a Bohr orbit.[10] If Lewis had chosen to continue his work on the chemical bond, he would still have had a great deal to offer.

Lewis knew that quantum mechanics would soon provide theories that would supplant his own work on the chemical bond, but said that "it is ... not unlikely that some of the things that are said in this book may soon have to be unsaid, but I trust that these may be matters of detail rather than essence."[11] And that is what they were. Linus Pauling published his own theories of the chemical bond based on the new quantum mechanics beginning in the late 1920s, and made it clear that he saw his work as an extension, rather than a replacement, for Lewis's ideas.[12] When Pauling published *The Nature of the Chemical Bond* in 1939, he dedicated it to Lewis.

Into the Unknown—Theories of Radiation

When Lewis left his work on thermodynamics and the chemical bond, he decided to attack the most pressing problem in theoretical physics—the dual wave–particle nature of light. This problem dated back to the late 1600s, when Newton and Christiaan Huygens had proposed competing theories of light. Newton had described light in terms of particles, and Huygens in terms of waves. In many experiments (e.g., diffraction, interference, refraction), light seems to behave as an electromagnetic wave, and in the latter part of the 19th century the experiments and theories of Thomas Young, Augustin-Jean Fresnel, and others seemed to have firmly established that light could best be described in terms of waves.

But by the 1920s, things were less clear. Much new experimental work seemed to indicate that light behaved as a particle when it interacted with electrons, which had a known mass and were therefore particles themselves. Einstein's 1905 explanation of the photoelectric effect, in which electrons are emitted from a substance upon irradiation with ultraviolet light, had been awarded the 1921 Nobel Prize in Physics. Bohr explained the light emitted by a hydrogen atom as resulting from a shift in energy of the atom's electron. X-rays had been discovered in 1895; like light, they propagate in straight lines and are diffracted. Arthur Compton observed in 1923 that X-rays were scattered by electrons, that the electrons changed direction and gained momentum in the scattering, and that the wavelength of the X-rays was lengthened and their path changed. Compton pointed to the problem, then typically ignored, of the dual nature of light.

Einstein had been working on the problem since at least 1905, and Hendrik Lorentz said in 1923 that "it must be possible after all to reconcile the different ideas. Here is an important problem for the physics of the immediate future."[13] Lewis felt that he had a chance at solving the great riddle. He was not satisfied with being one of the most prominent physical chemists—he wanted to be a Newton or an Einstein, discovering the equations that drove the universe. He attacked the problem with a theory that physicist and science historian Roger Stuewer has called "the most original and distinctive of the pre-complementarity period."[14]

Lewis started with what he knew—the theory of equilibrium, which is the essence of chemical thermodynamics. Take a reaction where a substance A is transformed to substance B, but after some time the progress of the reaction slows and there is no detectable change in the concentrations of either A or B—it seems as if everything has stopped. What is actually happening is that both the forward reaction, $A \rightarrow B$, and a reverse reaction, $B \rightarrow A$, continue to occur, but at rates such that A is consumed in the forward reaction at exactly the same rate as it is formed in the reverse reaction, and that B is formed at exactly the same rate in the forward reaction as it is consumed in the reverse reaction. The chemical reactions have not stopped at all—they are in a state of balance, or equilibrium. A shorthand notation for this equilibrium is as an equality sign, $A = B$. But what if there were three substances present that could be transformed into each other by reactions between A and B, B and C, and C and A? When the system reaches equilibrium, it might be that each of the three reactions reached equilibrium separately with a forward and reverse reaction occurring at the same rate; or it might be that a cyclic equilibrium exists, such that A is being transformed to B, B to C, and C to A, all at rates such that the concentrations of A, B, and C are unchanging. Lewis argued that cyclic equilibria did not exist and that each of the three reactions had to be in equilibrium separately. If a cyclic equilibrium did exist, it would be possible in principle to disturb it by introducing a suitable catalyst. A catalyst would change the forward and reverse rates of one of the reactions but leave its equilibrium point unchanged; because the hypothetical cyclic equilibrium depended only on the forward rates, the catalyst would in that case disturb the cyclic equilibrium. Because this had never been observed experimentally, Lewis concluded that cyclic equilibria were not possible. He put it this way:

> Suppose that during a period in which there are no births or deaths the population of the several cities in the United States remains constant, the number leaving each city being balanced by the number entering it. This stationary condition would not correspond to our case of thermal equilibrium. We would require further, to complete the analogy, that as many people go from New York to Philadelphia as from Philadelphia to New York. If there were three railroad lines between these two cities we should require that the number of passengers going by each line be equal in both directions.[15]

He called this conclusion the "law of entire equilibrium," asserting that *any* physical process must proceed in the same manner in both the forward and reverse directions. When he proposed this law he believed he was the first to do so, although a number of others had proposed or assumed something similar.[16] He corrected his claim of priority in a subsequent paper.[17] But he took this principle much further than others had, asserting that it applied to radiation—that if an atom of a substance in a higher energy state A^* emitted a quantum of radiation hν to move to a lower energy state A, there must also exist an exact reverse process where an atom in the lower energy state would absorb a quantum of radiation from the electromagnetic field to move to the higher energy state, and that if the substance were in radiative equilibrium, the emission and absorption processes must be occurring at exactly the same rates. The two processes may be written as

$A^* \rightarrow A + h\nu$ spontaneous emission

$A + h\nu \rightarrow A^*$ stimulated absorption

The emission process is called spontaneous, because the higher energy state needs no interaction with the field, while the second process is stimulated by the electromagnetic field.

In the late 19th century, Wilhelm Wien had proposed a formula, based on classical electromagnetic theory, that described the intensity of emission of radiation from what was called a black body—a perfect absorber of radiation at any frequency. Wien's law described emission well for radiation of long wavelengths, but failed at short wavelengths—the failure was known as the "ultraviolet catastrophe." In 1900, Max Planck proposed a formula that fit the experimental data much better but seemed to imply that energy was quantized—that it occurred in discrete packets rather than continuously (although Planck resisted assigning this physical interpretation for years). He derived his 1900 law based on an electromagnetic theory of absorption and emission of radiation, and in 1917 Einstein rederived Planck's formula based on statistical mechanics. In order to end up with Planck's law, Einstein assumed the existence of "stimulated emission" in addition to the stimulated absorption and spontaneous emission processes. In stimulated emission, an atom in a higher energy state emitted radiation through an interaction with the energy field, while in spontaneous emission the intervention of the field was not part of the process:

$A^* + h\nu \rightarrow A + 2h\nu$ stimulated emission

If he did not include stimulated emission, Einstein ended up not with Planck's law but Wien's. For Lewis, stimulated emission caused a problem—it seemed to violate his law of entire equilibrium. He saw spontaneous absorption and stimulated emission as inverse processes. Einstein had not proposed an inverse process for stimulated emission, so Einstein's theory seemed to Lewis to involve a cyclic equilibrium.

Lewis's approach to deriving equations for radiation was an almost exact mirror of his successes in physical chemistry: he applied his concept of activity, which he had used to reformulate chemical thermodynamics (chapter 2). The ideal gas law, like the radiation law, can be derived from statistical mechanics, but the ideal gas law ignores some of the interactions of the particles and is only approximate. Lewis had been able to make the gas law exact by replacing concentration (matter density) with activity, which then had to be determined experimentally. Lewis applied the same approach to radiation. Considering only the stimulated absorption and spontaneous emission processes, Einstein had derived Wien's law for the radiation density of a black body, so Lewis decided that Wien's was the distribution law for *ideal* radiation, and that *actual* radiation would follow Wien's law if the energy densities were expressed as activities. Just as for chemical substances, the activities would be identical to the densities only at infinite dilution—in the case of radiation, for a single atom and a single quantum of energy. At all real radiation densities, the activities would differ from the ideal densities, which explained to Lewis why Wien's law did not hold experimentally.

As Lewis was developing these ideas, he corresponded with a number of his friends, asking their opinions. R.H. Fowler, a physicist at Cambridge, wrote Lewis two letters[18] in 1925 saying that he was not sure that Lewis was on the right track and suggesting that he read two recent papers, one by John van Vleck and one by Satyendranath Bose. The latter set out what later came to be known as Bose-Einstein statistics, which describes the behavior of radiation and other particles now called bosons. Lewis unfortunately read neither, or he might have realized that his theories had problems. The initial reaction from most physicists to Lewis's papers was cautious interest mixed with skepticism, and Lewis decided to seek the opinion of the master—Einstein. Einstein's reply to Lewis, which began this chapter, was deeply disappointing. Einstein told Lewis that he had misunderstood his theory. There were not three processes, but two: stimulated emission and spontaneous emission were two parts of a single process, and its inverse process was stimulated absorption. Einstein endorsed Lewis's law of entire equilibrium, but not the conclusions he had drawn from it.

Lewis continued to theorize and defined a particle that he called a "photon," which served as the carrier of radiation. Lewis's name has stuck, but his photon was not the photon of modern physics. Lewis's photons were neither created nor destroyed—they were like atoms of radiation, and a constant number populated the universe. When a substance emitted radiation, Lewis saw it as filling an already-existing photon with the right amount of energy and sending it out. Again, a wider reading of the literature could have kept him out of trouble. Bose and Einstein had expressly assumed that the quanta of radiation were *not* conserved but were dynamically created and destroyed.[19] At Compton's suggestion, the Fifth Solvay Conference on Physics in 1927 adopted Lewis's term "photon" for the particle that carried radiation, but as used it did not signify Lewis's atom of radiation but rather Einstein's quantum.

Lewis's insistence that every physical process have an exact inverse process led to some strange conclusions. If everything is reversible, things must work in the same fashion even if time is assumed to run backward. For matter, this is in accord with Newton's laws. Think of a billiard table with perfectly elastic collisions between the balls and the bumpers and without any friction between the rolling balls and the table. Once the balls are started moving, they will continue rolling, rebounding, and colliding forever. Watching a film of the balls on this table, it would be impossible to tell whether the film were being played forward or in reverse. Lewis argued that this was a general principle, which he called "the principle of the symmetry of time," and that it applied to radiation as well as to matter. When we see films of actual physical events, we can usually tell the difference between forward and backward time—a vase is broken, for example, and breaks into a hundred pieces. When we view the film of the vase in reverse and watch it reassemble itself, it seems comical. Lewis said that vases break rather than repair themselves not because of a one-way direction of time, but because of probability (a point made mathematically half a century earlier by Ludwig Boltzmann)—there are many more broken states of the vase than the one unbroken state. But on a microscopic level, Lewis asserted, all physical processes including radiation are completely reversible in time, and time runs equally well in both directions, even for radiation. He took this to its logical conclusion: for every emitter atom, there must also be an absorber atom, and this process must be completely reversible in time, such that the absorber becomes the emitter and vice versa. This led him to a paradox: "The light from a distant star is absorbed, let us say, by a molecule of chlorophyll which has recently been produced in a living plant. We say that the light from the star was on the way to us a million years ago. What rapport can there be between the emitting source and this newly made molecule of chlorophyll?"[20] Lewis based his argument on extensions of Einstein's theory of relativity, which he was proud to have been the first to present to the American Physical Society in 1909.[21] Just before Lewis arrived at MIT in 1906, Einstein had published his special theory of relativity, and Lewis had become fascinated with the subject. Lewis and his MIT colleague Richard Tolman had spent many nights trying to find contradictions in the theory. Eventually they gave up, accepted the theory, and began publishing on the subject.[22] Lewis arranged to meet Einstein in Zurich for a beer to discuss physics sometime in 1910 or 1911.[23] Many of the conclusions of relativity were paradoxical (and many more paradoxes would soon be arriving from quantum mechanics). Lewis was willing to accept his new paradox, which seemed to deny the commonsense idea of causality—it seemed that light had to "know" where it was to be absorbed even before it was emitted. He came up with the idea of "virtual contact," asserting, "I may say that my eye touches a star, not in the same sense as when I say that my hand touches a pen, but in an equally physical sense."[24] Regarding the paradox, he quoted James Boswell quoting Samuel Johnson: "Sir, to leave things out of a book, merely because people tell you that they will not be believed, is meanness."[25] Lewis's ideas ended up in the popular press as humorous articles under such headlines as "Two-Timing."[26] Most physicists saw this as "philosophic speculation," saying that they did not accept it as a "law of physics, of exceptional scope and power."[27] In 1930, Langmuir wrote Lewis expressing his enthusiasm for Lewis's ideas on the symmetry of time, and Lewis responded with a friendly letter.[28] For a moment, it looked as though Lewis might give up his grudge.

Figure 7-2. Lewis in 1928, courtesy of the Bancroft Library, University of California, Berkeley.

The Anatomy of Science

Lewis was invited to give the Silliman Lectures at Yale in 1925, which were published under the title *The Anatomy of Science*.[29] The Silliman Lectures, given annually by a prominent scientist, consist of a series of popular talks. Lewis talked about his own philosophy of science, beginning his lectures as follows:

> The strength of science lies in its naïveté. Science is like life itself; if we could foresee all the obstacles that lie in our path we would not attack even the first, but would settle down to self-centered contemplation. The average scientist, unequipped with the powerful lenses of philosophy, is a nearsighted creature, and cheerfully attacks each difficulty in the hope that it may prove to be the last. . . . I take it that the scientific method, of which so much has been heard, is hardly more than the native method of solving problems, a little clarified from prejudice and a little cultivated by training. . . . I have no patience with attempts to identify science with measurement, which is but one of its tools, or with any definition of a scientist which would exclude a Darwin, a Pasteur, or a Kekulé. . . . The scientist is a practical man and his are practical aims. He does not seek the *ultimate* but the *proximate*. The theory that there is an ultimate truth, although very generally held by mankind, does not seem useful to science except in the sense of a horizon toward which we may proceed.[30]

These talks gave Lewis a platform to speak on any subject, and he did not hold back. A good deal of his lecture series was occupied with his current interest—the nature of radiation and of the symmetry of time. But he spoke not only about things that he knew—chemistry, physics, and mathematics—but about biology, where he espoused a Lamarckian belief that offspring could inherit acquired characteristics from parents.[31] His biological philosophy was close to vitalism, a belief that the processes of life are quite different from physical and chemical processes, and that animate beings may have some way to cheat the second law of thermodynamics.[32] He made an interesting biological suggestion: he observed that many organic molecules could occur in two isomers, which may be thought of as mirror images of each other—they are like left- and right-handed gloves (chapter 1). If these are synthesized in the laboratory from simple starting materials, the result is a mixture of equal amounts of the left- and right-handed isomers. In living creatures only one form is usually found, and it is puzzling to see how this separation could have been originally made by nature. Lewis suggested that it was by autocatalysis—that an initial microscopic fluctuation in the balance between the two isomers was magnified by a process in which the momentary presence of one isomer in excess catalyzed the formation of more of itself.[33] This may be the first suggestion of this sort,[34] and it is now accepted as a possible explanation.[35] He suggested that a similar process may be responsible for inheritance of acquired characteristics.

Lewis continued to work on his physical theories through the rest of the 1920s, as well as offering his ideas on economics[36] and on "ultimate rational units,"[37] which was an almost mystical belief that all physical constants could be reduced to dimensionless units leading to scientific formulas consisting only of integers and the constant π. In the right units, he believed that numbers like the Boltzmann constant, Planck's constant, the Rydberg constant, and Newton's gravitational constant would all become integers. He found some accidental relationships among the physical constants, but as William Jolly has said, it was all coincidence—"like coming up with your phone number after taking the square root of the logarithm of your street address."[38]

Although Lewis's theories of thermodynamics and the chemical bond had been successful, he was in danger of being viewed as a crank by scientists outside physical chemistry. He spent ten of what might have been his most productive years doing what Arrhenius had earlier done with immunochemistry—dabbling in fields in which he had little experience.

Langmuir Continues with What He Knows

While Lewis was off working on theoretical physics, Langmuir continued on the same path that had made him so successful. He worked on the problems that concerned General Electric—the behavior of filaments, cathode screens, electrons, gases, electrical discharges, chemical reactions at high temperatures and low pressures, and high-vacuum phenomena, all the while looking for underlying theories. While none of this was as revolutionary as his earlier work, they were all solid extensions in the fields that he had pioneered. He was as much an engineer as a scientist. He loved solving problems, and GE seemed to be able to keep him busy. His colleague Saul Dushman said of him, "Langmuir is a regular thinking machine. Put in facts, and you get out a theory."[39] In attacking a problem, Langmuir wanted to understand what was happening at a fundamental level, not merely to fix an isolated case. He thought of science as the ultimate fun—his 1946 Hitchcock Lecture at Berkeley would be titled "Science for the Fun of It"—and he was happy to engage in scientific discussion at any time. After his successes in surface chemistry and the theory of the chemical bond, industrial chemists, who had little interest in the abstract theories of academics like Lewis, considered Langmuir to be a celebrity. Langmuir was a hypnotizing public speaker, enjoyed travel, liked meeting new people, loved the camaraderie of scientific discussions, and was willing to take on leadership positions, serving as president of the American Chemical Society in 1929.

In 1917, Langmuir had hired as an assistant Katherine Blodgett, who had earned a master's degree in chemistry from the University of Chicago at age nineteen. Her dissertation had been on the adsorption of gases on charcoal, a subject of clear interest to Langmuir and to GE. She left GE briefly in 1924 to study with Ernst Rutherford at Cambridge (with Langmuir's recommendation), becoming in 1926 the first woman to earn a doctorate there, and her dissertation was on another subject of interest to GE—the behavior of electrons in ionized mercury vapor. She and Langmuir maintained a long working relationship. Their collaboration is best known in Langmuir-Blodgett films, the deposition of multiple layers, each one molecule thick, on a substrate. (Using this technology, Blodgett was responsible for the development of nonreflective glass in 1938, which would make GE a fortune. In a remarkable example of corporate sexism even for that time, she was entirely omitted from a corporate history of GE that was published in 1953.)[40] One of Langmuir's senior collaborators, Kenneth Kingdon, said, "Langmuir certainly knew how to make the best use of people. He let them do the things they could do best, and he never tried to make anybody over."[41]

In the 1920s, Lewis turned down almost all honorary degrees because he did not want to spend time traveling and because he disliked public speaking. Langmuir, on the other hand, enjoyed his role as a prominent scientist. He went on long speaking tours, which he combined with vacations. He began receiving awards and honorary degrees both in America and in Europe. He and his wife Marion went to Europe in 1921 where, with almost no conditioning and then forty years old, Langmuir decided to climb the Matterhorn. This was a peak so difficult that it had been the last of the Alps to be conquered, in 1865. His diary records his climb from a base camp at 11,000 feet (3,350 meters):

> I was awakened at 1:50 AM and we left at 2:40 AM. Beautiful night with thin crescent moon. I found it hard to climb so fast in the dark, soon began to be winded easily. At about 4:30 arrived at Solvay refuge and we stayed there about 3/4 hr. Suffered greatly from lack of breath and once twice from faintness and sudden dizziness. Thought at one time that I could not reach the top. Climbing very uniformly steep, using hands 95%–99% of the time. But no really difficult places except above the shoulder, where we used the numerous ropes provided. Intense and sudden gusts of wind on top, making it necessary to lie down and hold onto each other. Temp probably about 20°F. Very hazy. Arrive on top at 8:20 am.... I was very tired (in arms) and winded all way down peak and very stiff in thighs.... Next Day: extremely stiff, especially towards night. To bed right after dinner but could not sleep because of pain in right leg above knee. Knee swollen and very sore.[42]

(Compare with Lewis: Max Born recalled that, while relaxing in a mountaineering outfit at a scientific meeting in the Alps, he met Lewis, who was wearing "a black suit, gray striped pants, shiny black shoes, and a bowler."[43])

Langmuir went from the Alps to a conference in Edinburgh in 1921, where he gave a lecture, received another honorary degree, and participated in a discussion on quantum theory hosted by John Nicholson. (Nicholson would marry Dorothy Wrinch, whom we will meet again in chapter 9, the following year.) Langmuir's diary reports that he and his wife stopped in London, Paris (the Folies Bergère: "Fine dancing, apparently naked women"), Brussels, Amsterdam, Berlin, and a side trip to Göttingen, where Langmuir had studied with Nernst. He was able to show Marion his old room because the "present occupant [was] out fighting his duel."[44]

Langmuir's diaries began to be filled with famous names.[45] Herbert Hoover and Orville Wright were acquaintances, J.J. Thomson and Guglielmo Marconi (the radio pioneer) were houseguests, and Lord Rutherford and Max Born hosted him in their homes.

Langmuir had corresponded with Bohr about his atomic theories beginning in 1921, but he and Bohr first met in 1923 in Amherst, Massachusetts, where Bohr was speaking. Langmuir had driven the 120 miles from Schenectady, New York, to hear his talk. The two spoke privately for two hours, and Langmuir convinced Bohr to return with him and visit GE. They became close friends, and Langmuir wrote, "Never before have I met anyone who impresses me more and inspires me more—scientifically—than Bohr. He is the most marvelous man I know."[46]

The Solvay Conference

The Fifth Solvay Conference on Physics—where Lewis's name "photon" was taken up for the quantum of energy—was held in Brussels in 1927 and had a remarkable attendance, as can be seen in the photo in fig. 7-3. Each Solvay Conference focused on a different problem, and the focus for 1927 was the new quantum mechanics and radiation, the very problem that had been occupying Lewis. The conference was by invitation only, and Lewis was not invited, probably because his theory was viewed as a dead horse by all participants by 1927. This must have been an intense disappointment to Lewis. But, out of the blue, a week before the conference was to begin in Brussels, Langmuir received an invitation. He was surprised—he was the only chemist to be invited—and wrote to his mother, "This year the subject is the 'Quantum Theory,' a field in which I have contributed nothing so I can't see why I have been invited."[47] Langmuir was in Europe on vacation and was hobbling around on crutches after breaking his foot on the voyage from America. He was recuperating with his wife in Italy when a telegram arrived from Lorentz, who was in charge of the conference's organization. Leon Brillouin, the French physicist and an invitee to the conference, had been in

Figure 7-3. The Fifth Solvay Conference, Brussels, 1927, with Langmuir holding his cane at the left of the front row. Back row: A. Piccard, E. Henriot, P. Ehrenfest, E. Herzen, T. De Donder, E. Schrödinger, E. Verschaffelt, W. Pauli, W. Heisenberg, R.H. Fowler, L. Brillouin. Second row: P. Debye, M. Knudsen, W.L. Bragg, H.A. Kramers, P.A.M. Dirac, A.H. Compton, L. de Broglie, M. Born, N. Bohr. Front row: I. Langmuir, M. Planck, Mme. Curie, H.A. Lorentz, A. Einstein, P. Langevin, C.E. Guye, C.T.R. Wilson, O.W. Richardson. Photograph by Benjamin Couprie, Institut International de Physique Solvay, courtesy of the AIP Emilio Segrè Visual Archives.

contact with Langmuir's boss, Willis Whitney, trying to negotiate a consulting contract for himself with GE, and Whitney had asked Brillouin to connect with Langmuir during his European trip. Brillouin was unable to meet Langmuir at an earlier conference in Como,[48] so he apparently fudged things a bit to get Langmuir an invitation to the Solvay Conference. He led Lorentz to believe that Langmuir was going to be in Brussels anyway that week[49] and asked him to extend Langmuir an invitation. Langmuir jumped at the chance and later sent Whitney a report on the conference:

> We had planned to leave Cortina on Oct. 23rd and go to Eindhoven, but on Oct. 19th I got a telegram from Prof. H.A. Lorentz (Haarlem) inviting me to attend the Solvay Congress which was to meet in Brussels for 1 week beginning Oct 23rd to discuss the quantum theory. L. Brillouin had written to Lorentz suggesting that I be invited. I had heard about the Solvay Congress in Como in Sept. and had inquired about it even then from Lorentz, but was told that only those invited specially could attend and the membership of about 25 was limited to the most active workers in the quantum theory—so

I was surprised to receive the invitation.... Strenuous meetings—8 hours per day in lecture room and all evening (often till 1 A.M.) in hotel where we all stayed, discussing quantum theory. The wave mechanics is the universal mode of expression today. For the 1st time the quantum theory can be formulated in apparently complete form so that the discrepancy between the classical wave theory and the quantum phenomena (and even the photon theory) seems to be clearing up. Hope rather than bewilderment characterized the meeting. Bohr and Heisenberg were full of a new principle "Das Ungenauigkeits-princip" [*sic*] [uncertainty principle] as broad as the Relativity Theory or the Corresponding-Princip [*sic*] [correspondence principle] by which they interpret nearly all quantum phenomena. It denies the reality of the position and velocity of an electron:—these things depend on the point of view just as the velocity of the earth in space depends on the point of view. If s is a coordinate giving the position of an electron and p is its momentum and Δs and Δp are the limits of accuracy with which s and p can be determined by any given experimental conditions then $\Delta s \times \Delta p = h$, the quantum constant. Thus if we know the velocity of an electron very accurately it has no meaning whatever to specify the position more accurately than $\Delta s = h/\Delta p$, thus the electron has a large extension in space. But if the position is known accurately then the velocity becomes indeterminate. The principle takes many different forms and is very powerful in yielding useful results, but sounds very queer.[50]

More than half the attendees of the conference had received or would eventually receive Nobel Prizes. The new quantum theory was revolutionizing physics, and this was the meeting where Einstein, who resisted the statistical interpretation of quantum mechanics, supposedly said, "God does not play dice," and Bohr replied, "Einstein, stop telling God what to do." The Solvay Conferences emphasized discussion rather than presentation, and in 1927 English was not yet the international language of science. Langmuir, who spoke fluent French and German from his days as a schoolboy in Paris and his studies in Göttingen with Nernst, fit right in. He took home movies of the attendees on their breaks—he did work for a technology company, after all. His movies have been preserved and may be seen on the Internet,[51] where Langmuir can be seen pantomiming to Paul Ehrenfest how he managed to get around on crutches. After the conference, the Langmuirs continued their European tour to Holland, spending time with the Lorentzes and the Ehrenfests; to Berlin; to Göttingen; to Copenhagen as houseguests of the Bohrs; and to Cambridge. In 1930, Langmuir returned to Italy, walking and climbing in the Apennines, and gave a lecture in Berlin in flawless German to an audience of 500, including Einstein, Planck, Nernst, Debye, Michael Polanyi, and Lise Meitner. The heads of the two major German electrical companies, AEG and Siemens, delivered talks where they justified their own research programs by pointing to the success GE had had with Langmuir.[52]

Stereo Sound and Flying

While Lewis kept to himself in Berkeley, Langmuir made friends outside of science. In 1930, Langmuir became friendly with Leopold Stokowski, the conductor of the Philadelphia Orchestra, and visited Philadelphia as his guest. The two of them were interested in improving the quality of sound recording and playback, including stereo sound, which was a natural extension of Langmuir's World War I work on submarine detection (see chapter 4). With GE's permission, the two of them began a collaboration that included RCA Victor.[53]

Beginning in 1928, Langmuir became seriously interested in flying, and became friendly with Charles Lindbergh when Lindbergh visited the GE laboratories in 1929. In 1930, Langmuir took flying lessons (including flying with Lindbergh), purchased a small Waco plane, and received his license as a private pilot. From Langmuir to his mother:

> Last Friday evening Marion and I had dinner with Col. and Mrs. Lindbergh (with only 3 other people) in the Lindbergh's suite at the Van Carlin Hotel. He makes a better impression than her—and she is very delightful and pretty (which one wouldn't think from her pictures). After dinner we all went to the airport and after Mr. and Mrs. L took a couple of loops to try out some special kinds of lights that G.E. has installed, Lindbergh asked me to take a trip with him over the lighted city—a beautiful night—in Mrs. Lindbergh's plane, one almost exactly like mine.[54]

Langmuir's enthusiasm for flying infected others at GE, and Langmuir began agitating for GE to purchase a company airplane. Management was not happy. At the end of 1928, E.W. Rice, GE's past president and then honorary chairman, had written to Gerard Swope, GE's president, to voice alarm. Rice copied Langmuir, Langmuir's boss Whitney, William Coolidge (who would become Langmuir's boss when Whitney would retire in 1932), and Ernst Alexanderson (the inventor of the alternator that allowed radio to be broadcast as voice rather than Morse code). Rice told the four, "While I admire your enthusiasm, I do hope that you will refrain from taking undue risks by flying in machines which are not considered first-class and handled by competent pilots. When the Company gets its machine, I hope that you will confine your flying to the machine and pilot which will be supplied, and which I assume will be the best that money and the wisdom of the best experts can provide."[55] Langmuir, of course, paid no attention. GE was not about to fire him.

Nobel Consideration

By the 1920s, Lewis and Langmuir were well known to American chemists. Lewis's public image was that of a theoretician and Langmuir's that of an experimentalist and inventor, a worthy successor to Edison. Neither of these was entirely accurate: Lewis worked in the laboratory until the day he died, and Langmuir's theories of surface chemistry, adsorption, and thermionic emission were as mathematically sophisticated as was Lewis's work. Where they differed was in their personalities—Lewis was unsure and mistrusting around those whom he did not know, while Langmuir was gregarious and friendly; and in their attitudes toward industrial work—Lewis banned any consulting or industrial grants at Berkeley, while Langmuir saw GE's research as his path to the most important unsolved problems.

Industrial chemistry had undergone an enormous expansion in America, especially when the United States was cut off from European suppliers during World War I. Membership in the American Chemical Society had increased from 1,715 in 1900 to 15,582 in 1920.[56] Most of the new chemists were in industry, and Langmuir was the most prominent industrial chemist.

In 1916, when he was only thirty-seven years old, Langmuir received a single nomination for the Nobel Prize in Physics, but nominations in quantity, for both physics and chemistry, began to arrive in 1927. His diary contains an entry from 27 November 1927: "Cambridge: Arrive 9:50. . . . See Lord Rutherford and have tea at his home. He tells me that he has proposed my name for the Nobel prize and expects me to get it next year."[57] Lewis's nominations began in 1922, when he was nominated by his old dissertation adviser Theodore Richards, who had received the chemistry prize for 1914. Richards had apparently forgiven Lewis for their falling-out in 1901–2 (see chapter 2). Lewis's nominations were for his work on thermodynamics and the chemical bond, but by the time Lewis began receiving nominations in numbers, he had already left both of these fields and was developing his ideas on the nature of radiation.

In 1924 the Nobel Committee for Chemistry asked Arrhenius to write a special report assessing Lewis's work.[58] Arrhenius, as the originator of the dissociation theory and at that time Sweden's only Nobel chemistry laureate, was the *éminence grise* of Swedish science. He had dominated the Nobel awards for years, arranging awards to his friends Richards, Jacobus van't Hoff, and Wilhelm Ostwald and withholding them as long as he could from his foes Nernst and Paul Ehrlich.

Arrhenius dismissed Lewis's theory of the chemical bond in one sentence: "[I]t is rather insignificant; and moreover the major part was done by Langmuir, and it is in opposition to the theory of Bohr, which is probably correct."[59] Arrhenius was wrong on all points: even Langmuir would not have claimed to have done the major part of the work, and because Bohr's theory did not explain the chemical bond in any general sense, it could not be considered "correct" for chemical purposes. Arrhenius's analysis of Lewis's thermodynamics was more positive, but he criticized Lewis for *not* doing something that Lewis had in fact done. In his report, Arrhenius described Lewis's concepts of fugacity and activity and his early work on strong electrolytes, but entirely omitted Lewis's 1921 formula for ionic strength. He then complained that Lewis had not come up with a simple equation for the behavior of electrolytes. He drew an analogy with the van der Waals equation, an empirical equation that describes the behavior of nonideal gases:

> It is clear that the goal for Lewis's work should be to find more simple laws for the activity of substances, with which one could treat the whole area, not only the most dilute solutions or the lowest pressure gases. Some simple laws for concentrated solutions analogous to van der Waals' equation for strongly compressed gases would be of value.... [T]hrough Lewis's work, a large amount of material has been collected through the work of students and others, but still without succeeding in finding any law which can explain all the data.... Through his careful and systematic work in this area, Lewis has made a great contribution to our knowledge of physical chemistry....However it is difficult to consider [Lewis's] work as a new discovery or invention, when the authors themselves say that they have proceeded on the "broad highway of thermodynamics." No new principle has been given by Lewis in this area, but he has used methods that long before had been worked out by Kirchoff, Guldberg, Gibbs, Helmholtz, Clausius, and Planck, and others of the leaders in mathematical physics over the past 70 years. As outstanding and important as this work is, it does not come up to the demands that Alfred Nobel has set for the prize he has established.[60]

Arrhenius's assessment was clear; Lewis had done good work, but it was not worthy of a Nobel Prize, because he had not done what was needed—develop a simple empirical formula that would describe the behavior of electrolytes. The implication is that such a "discovery" would be worthy of a Nobel Prize. What is puzzling is that Lewis *had done exactly that*—his theory of ionic strength, published three years earlier in 1921 and also described in his 1923 book, involved only simple calculations and had given a practical solution to the determination of the activity of strong electrolytes.

Author Robert Friedman sees Arrhenius's opposition to Lewis as spiteful, based on Arrhenius taking offense at the attempts by Lewis and others to modify Arrhenius's dissociation theory.[61] Possibly. Arrhenius was a curmudgeon who thought that chemical thermodynamics had ended with the work done by himself and van't Hoff thirty years earlier, and Lewis had said in print in 1907 that the equations developed by Arrhenius, Nernst, van't Hoff, and Ostwald were "approximate" and would "no longer suffice"[62] (see chapter 2). But Arrhenius was unlikely to resent Lewis in particular. Lewis had been the student of Arrhenius's friend Richards, whom Arrhenius had successfully supported for the 1914 chemistry prize; Lewis was an enemy of Arrhenius's enemy Nernst; and Lewis had gone out of his way to repeatedly praise Arrhenius's contributions.[63] Another explanation for Arrhenius's criticism is possible. His report had omitted Lewis's formula for ionic strength and then criticized him for the lack of that achievement. This omission is so egregious that it is unlikely that it was intentional. It seems to be a sufficient explanation that Arrhenius was sixty-five years old at this time (he was to die three years later) and was just not keeping up with the literature.

In 1926, the Nobel chemistry committee, perhaps aware of the inadequacies of Arrhenius's 1924 report, asked Theodor Svedberg to prepare a special report on Lewis's work in thermodynamics.[64] Svedberg was a rising star in Swedish physical chemistry. Only forty-two years old, he had done extensive work on proteins and macromolecular systems. Svedberg's report was much more positive than Arrhenius's. Svedberg began with a description of the theory behind Lewis's formulation of thermodynamics. When he came to the treatment of strong electrolytes, Svedberg was explicit in his assessment of the importance of Lewis's work:

> Some as yet unsolved problems arise when one calculates the treatment of electrolytes. Thermodynamically-defined data is not enough.... The area of which we are speaking is totally inaccessible to numerical calculations if we stay on the ground of the classical theory. But with the aid of Lewis's methods [of ionic strength], and by using the table that he has assembled, calculations are now possible with great exactitude. From the above investigations, it should be clear that the service that Lewis has made to chemical science through the creation of exact methods to determine the changes of free energy in chemical reactions has to be acknowledged as of the greatest value. Through his work, research has made a huge step forward in the exactitude with which one can calculate in advance how a certain reaction will take place. In this regard, his contribution can best be compared with that of Richards's determination of atomic weights or Siegbahn's determination of X-ray wavelengths. He has further, through his welcome analysis of the meaning of the third law experimentally as well as theoretically, helped give an answer to this question more than any other after Nernst. One can question whether he should not in fact be placed next to Nernst without wishing to detract from Nernst, since he [Nernst] before anyone else understood the importance of the third law and before anyone else organized experimental research into this important area, but one still has to admit that Nernst did not explain it sufficiently. Clarity was first won by Lewis and his coworkers. So it seems to the undersigned [Svedberg] that Lewis's work on chemical affinities is of such importance that it would deserve to be honored with a Nobel prize in chemistry. Several important points will become clearer in the near future, because some of the ongoing or expected work will have been completed. It seems to me that it may be advisable to postpone the award of the prize for a few years.[65]

Svedberg's suggestion that the committee open up the third law for reevaluation was unlikely to be welcome. The award of the chemistry prize for 1920 to Nernst had been one of the most contentious battles in the history of the

Nobel Prizes, and the committee almost certainly wanted to put that behind it. Svedberg's suggestion that the award be delayed to await further developments ignored the fact that Lewis had left thermodynamics with the publication of his book in 1923, so any further clarification of Lewis's theories would need to be done by others. Although the chemistry committee acknowledged Svedberg's recommendation that Lewis's thermodynamic work was worthy of the chemistry prize, the committee's 1926 report to the Royal Swedish Academy of Sciences was not quite as certain as Svedberg in its assessment of Lewis:

> Lewis has earlier been proposed for the prize in chemistry in 1922, 1924, and 1925. This year's proposal to award him the prize considers especially the theoretical and experimental work that he has done in the creation of tables of the free energy of substances.... From [Svedberg's] report, it seems that the service that Lewis has provided the chemical community in the ability to determine the changes in a range of chemical reactions' free energy must be regarded as extremely important, because through his work research has taken a great step forward, as a reaction's progress may now be calculated beforehand. He can also be said, through his welcome analysis of the third law of thermodynamics, to have given not only experimental but also theoretical solution to this problem more than anyone but Nernst. Even if one admits on the grounds earlier mentioned that Lewis's work on the determination of chemical affinities is of the sort that would deserve a Nobel prize in chemistry, it seems that a postponement of the question would not be unmotivated, because the importance of Lewis's research can be expected to be clearer in the near future, since part of the work is close to termination.[66]

It is possible that Svedberg's suggestion to delay the award to Lewis may have had a personal component: the recipient of the chemistry prize for that year, 1926, was Theodor Svedberg. But it was quite common for the committee to want to evaluate the work of a nominee over a period of several years.

Wilhelm Palmaer

Although the committee delayed an award for Lewis, Svedberg's 1926 report put Lewis on the list of those whose work had been judged as of the first rank and worthy of a Nobel Prize, and Svedberg was now himself a Nobel laureate and Sweden's foremost young physical chemist. In the years to come, anyone who opposed awarding the prize to Lewis for his work in thermodynamics would need to rebut Svedberg's 1926 report. As we shall see, later reports on Lewis's thermodynamics did attempt to rebut Svedberg's report, and were all written by another committee member, Wilhelm Palmaer, whom we have seen in chapter 6 as a friend of Nernst and as a key supporter of Nernst's Nobel

Prize. I have uncovered circumstantial evidence that Palmaer used his position on the committee to manipulate the nomination and reporting process in order to block the award of a prize to Lewis. To substantiate a circumstantial case, an accuser at law must show that the accused had both motive and opportunity and that other reasonable explanations are inconsistent with the accused's actions. I will now attempt to make that case.

First, Palmaer's motives: Lewis had studied in the Göttingen laboratory of Nernst, the dean of German physical chemists, in 1901 after receiving his doctorate at Harvard, and Lewis and Nernst seemed to have developed a lifelong aversion to each other.[67] No correspondence between the two of them has been preserved, and for the next twenty years, whenever possible, Lewis published Nernst's errors. In his 1923 book, *Thermodynamics*, Lewis gave himself free rein. He started in on Nernst on page 5 in the introduction: "Nernst

Figure 7-4. Wilhelm Palmaer. Courtesy of the Royal Institute of Technology, Stockholm.

and his associates have made remarkable contributions both to theory and practice. Some arithmetic and thermodynamic inaccuracy occasionally marring their work is far outweighed by brilliancy of imagination and originality of experimentation." After this backhanded compliment, Lewis proceeded to cite errors by Nernst or his students, and called Nernst's "chemical constants," which he had devised in an effort to make his heat theorem predictive, "a regrettable episode in the history of chemistry."[68] Nernst did not respond in kind, but simply ignored Lewis completely in his own publications.

Palmaer, a Swedish electrochemist, had been a student of Arrhenius's. When the Nobel Prizes were first organized, he was appointed secretary to the Nobel committees for chemistry and physics, and he remained in that position from 1900 to 1925. After losing to Svedberg for election to a vacant position on the chemistry committee in 1924,[69] he was elected to that committee as a voting member in 1926, a position he held until his death in 1942. Palmaer's own list of scientific publications was limited, consisting of twenty-three publications on electrochemistry. Palmaer and Nernst were close friends. Palmaer had repeatedly nominated Nernst for the Nobel Prize[70] and, together with Svedberg, had led the fight against Arrhenius in supporting the award of the prize to Nernst, including submitting a thirteen-page memorandum on his behalf to the academy after the chemistry committee decided not to propose Nernst for the prize in 1920.[71] Palmaer was personally cited and thanked by Nernst in his Nobel address.[72] Their friendship would persist until Nernst's death. As mentioned in chapter 6, Nernst's two daughters, who had married Jews, would be forced to emigrate to Britain and to Brazil during the Nazi era. With the beginning of hostilities in 1939, Nernst would write to Palmaer as a resident of a neutral country, addressing him as "Lieber Freund Palmaer" ("Dear friend Palmaer," a very intimate form of address for a German professor of that time), asking for help in forwarding letters to his daughters, to which Palmaer agreed[73] (see chapter 6). In addition to his friendship with Nernst, Palmaer may have had a personal motive to oppose Lewis. Palmaer had made his career as an electrochemist, which is closely related to thermodynamics through Nernst's equation for the determination of free energy from voltage. Lewis's reformulation of thermodynamics had implicitly criticized the work of Palmaer and Nernst, and older scientists often react negatively to the iconoclastic ideas of younger scientists.

By 1932, the nominations for Lewis were starting to pile up. In 1929 alone, Lewis received six nominations, and Svedberg's 1926 report was still on file, calling Lewis's work on thermodynamics prize-worthy but suggesting that the committee wait a few years. It is clear from the committee's report for 1932 (quoted further below) that the two leading candidates for the year were Lewis and Langmuir. By mid-January 1932, Lewis had received four nominations and Langmuir two.[74] If Palmaer were to block Lewis's award, 1932 seemed to be the time to act.

Second, Palmaer's opportunity: As a member of the five-person Nobel Committee for Chemistry, he would have only a single vote. If he were to block the award of a prize to Lewis, his best hope was to rebut Svedberg's 1926 report with the hope of influencing others to vote against Lewis. The usual practice within the committee was for a committee member who nominated a candidate to write a report on the candidate's work if the candidate were under active consideration. The committee could assign the report to someone else, but that would be an exceptional circumstance.[75]

Third, Palmaer's actions: *Two days* before the deadline for nominations, Palmaer nominated Lewis for the 1932 Nobel Prize in Chemistry for his work on chemical affinity and standard potentials—that is, for his work on thermodynamics.[76] He then accepted the task of investigating Lewis's work and submitted a negative report.[77] If he were to rebut Svedberg's positive 1926 report, he would need to submit more than a supplementary update on Lewis's activities in thermodynamics from 1927 to 1932 (which activities were virtually nonexistent, as Lewis had left the field). Palmaer was not a thermodynamicist, but he began his report by explaining why he found it necessary to cover the same ground that Svedberg had covered earlier—that is, by describing the theory behind Lewis's thermodynamics:

> The research by Lewis regarding the magnitude of chemical affinities has already been the subject of two reports by the committee, in 1924 by Arrhenius and in 1926 by Mr. Svedberg. In these reports, the theoretical part of Lewis's work in question has been examined. Therefore it might seem sufficient to now limit the report to deal only with the experimental work. However considering the fact that the two previous reports draw different conclusions regarding Lewis's merit for a Nobel prize in chemistry, I consider myself obliged to treat the theoretical results.[78]

Palmaer then began to explain Lewis's concepts of fugacity and activity as they applied to gases, but he did not summarize Lewis's own explanations of these concepts, as Arrhenius and Svedberg had done earlier; he instead explained Lewis's concepts in terms of Nernst's older formulation of thermodynamics. He eventually arrived at the conclusion that Lewis's thermodynamics of gases was simply a recapitulation, with minor extensions, of Nernst's and van't Hoff's conclusions drawn thirty years earlier.

Having finished with gases, Palmaer moved on to Lewis's treatment of electrolytes. Again, he did not follow Lewis's formulation, but rather explained Lewis's concept of activity of a solvated electrolyte in his own way. He did admit that his statement of the problem did not follow Lewis's presentation: "Lewis does not reason exactly in this way, and he does not even present Guldberg's formula, but I however believe that the [explanation of Lewis's work that Palmaer had given in terms of Nernst's formulations] above can comfortably explain the path."[79] Palmaer then suggested that Lewis had disparaged the work of Arrhenius, Sweden's first Nobel laureate, without justification:

> In this connection it seems to me with regard to Lewis's work, it is fitting to remember in what sense remarks can be put forward regarding the validity of Arrhenius's formula for calculation of the electrolytic degree of dissociation.... Arrhenius's formula, next to van't Hoff's law regarding osmotic pressure and Nernst's formula for electromotive force, ought to always keep its importance and should be viewed in the same class as the ideal gas laws. [Lewis's] assumption [that ion mobility varies with concentration, in opposition to Arrhenius's theory] rests upon rather loose grounds, and such are not given at all by Lewis. I cannot see that Lewis has any use for the hypothesis other than it provides the possibility of blaming existing discrepancies on the assumed circumstance.[80]

Palmaer moved to ionic strength, but faulted Lewis for giving no theoretical support for his concept. Palmaer made no mention of the theoretical justification that Debye and Hückel had given for ionic strength in 1923, nine years earlier:

> [Lewis] does not give any theoretical motivation for this remarkable statement [his formula for ionic strength]. The experimental proofs are seemingly weak, and the assumptions concerning ionic strength cannot be considered proven. Lewis admits as much himself. The extremely difficult calculations have, as Mr. Svedberg has shown, provided a step forward. On the other hand, one must in my opinion agree with Mr. Arrhenius that "the ideal goal for Lewis's work should be to find simple laws for the activities of substances and that some simple laws for concentrated solutions would be of great value, as the van der Waals equation is for strongly compressed gases." In Mr. Arrhenius's statement [one sees] that Lewis and his coworkers have put together a far-reaching compendium of material without succeeding in finding a law tying it together which could give an overview of it all. In reality one cannot in Lewis's work find any overall theory or simple hypothesis of larger scope. His proposal of the concept "activity" is an expression that had been used even earlier and cannot be counted for anything, because it only means a formulation of the earlier-known fact that concentrations alone cannot be used with success to calculate free energy. The concept of ionic strength is newer, but of doubtful value.... Any overall theory or simple hypothesis in this direction is not available in Lewis's work, nor has it been proposed by anyone else.[81]

All of this led Palmaer to the conclusion that Lewis's work in thermodynamics was not worthy of the award of a prize: "On the basis of what I have now said, I have come to the conclusion that Lewis's contributions in the above-stated thermodynamic work, in spite of their undisputed merits, are not of the sort that can motivate the award of the Nobel prize."[82]

After completing his assessment of Lewis's theoretical thermodynamics, Palmaer went on to deal with Lewis's contributions to electrochemistry, specifically to standard electrode potentials. This was a subject in which Palmaer, an electrochemist, was competent. He examined Lewis's work in minute detail, citing any conflicting results by other researchers over the past twenty years. He complained that even when Lewis's work was the most accurate available, Lewis had reported his results to too many significant figures. Finally, he complained about Lewis's definition of the zero point reference for electrode potentials, which was different than that employed by Nernst and by Palmaer himself.

By the end of his report, Palmaer had disparaged Lewis's work on the theories of both gases and electrolytic solutions, his work on electrochemistry, his work on the third law, and a great number of additional minor points. He nonetheless gave his report a neutral conclusion, leaving it to the committee as a whole to decide:

> I have hereby tried to provide a report considering Lewis's work on the magnitude of chemical affinities, as well as the work of his coworkers, whose merits to a high degree ought to be ascribed to Lewis. Any decisive opinion whether this work merits the Nobel prize or not, I have not wanted to express, but in order to more clearly present my opinion, I want however to say that Lewis together with Langmuir seem to me to be those that stand foremost regarding this year's Nobel prize.

Svedberg's Report on Lewis's Electronic Theories

Lewis had been nominated several times for his theory of the chemical bond. In addition to Palmaer's report on Lewis's thermodynamics, in 1932 the committee asked Svedberg for a special report on Lewis's work in this field.[83] After tracing the evolution of Lewis's ideas, Svedberg dealt with Lewis's intuition that the electron pair was central to chemical bonding:

> [Lewis] claims that the formation of electron pairs can be conceived as a connection between two electrons' orbits in such a way that their magnetic fields are neutralized and that the magnetic moment disappears. That the occurrence of electron pairs within molecules is a factor of the greatest significance is statistically evident by inspection of the known types of chemical bonds. The few molecules with an odd number of valence electrons are all more or less reactive, showing a tendency to associate with each other or with other "odd molecules" to form molecules with an even number of valence electrons. Without doubt, Lewis's discovery of the role of electron pairs in molecular structure, and his identification of an

electron pair with a chemical bond, must be considered an outstanding achievement.

Lewis's suggestion explaining the cohesion of the electron pair by an interaction between the magnetic fields of their orbits was not at first considered a happy one. In Bohr's atomic model, there was no place at all for Lewis's electron pair, and it is therefore completely natural that [Lewis's] thoughts on atomic and molecular structure were not held in any high esteem during the heyday of [Bohr's] theory. The extension of our atomic concepts, brought about by wave mechanics, has however led to a renaissance for Lewis's ideas. In the most recent atomic physics we find again that Lewis's paired electrons occur in a variety of forms. Pauli's principle stipulates that within one molecule, two electrons which are identical with respect to the main quantum number, and with respect to the two quantum numbers of the orbital momenta, have opposite spin. They form an electron pair as envisaged by Lewis.[84]

But Svedberg's overall conclusion was that Lewis's theory of the chemical bond was not of great significance:

There is every reason to admire the intuition which led Lewis to the idea of the concept of the electron pair as being responsible for the chemical bonding; but on the other hand one should not be blind to the fact that Lewis's hypothesis has not had much significance for later research. Spectroscopy and wave mechanics together go far beyond Lewis's idea.... Also bond directionality can be calculated by wave mechanics.... Quantitative statements of this sort cannot be made by means of Lewis's theory of molecular structure.

As a concluding judgment, it seems justified to say that Lewis's theory of valence neither has been nor can become of such importance for chemistry that an award of a Nobel prize should be motivated. At the same time, it should be said that the theory gives an impressive proof of the originator's clear-sighted intuition and that this achievement therefore contributes to the strengthening of Lewis's candidacy for a Nobel prize motivated by his work in the field of experimental and theoretical thermodynamics.[85]

Svedberg, writing in 1932, was evaluating Lewis's 1916 work using hindsight based on the advances in quantum physics and chemistry that had been made beginning after Lewis left the field in 1923. His statement that "Lewis's hypothesis has not had much significance for later research" was from the point of view of a physical chemist. Many organic chemists would certainly not have agreed with Svedberg, as Lewis had explained the chemical bonds that they had been drawing for more than half a century: the bonds were electron pairs, and that understanding allowed organic chemists to explain how bonds could be formed and broken in reactions.

Langmuir, who had received his Ph.D. under Palmaer's friend Nernst, was the other leading candidate for 1932. Although Svedberg said in his 1932 report on Lewis's valence theory that he still believed that Lewis deserved a prize for his work in thermodynamics, Svedberg was backing Langmuir for the 1932 chemistry prize. He apparently had no idea of what Palmaer was up to with his nomination for Lewis, believing that Palmaer was supporting Lewis for the 1932 prize, as is shown by this exchange of letters between Svedberg and Arne Westgren, the secretary to the committee:

From Svedberg to Westgren, 10 May 1932:

I submit my relatively modest contribution to this year's Nobel prize awards. I have made a brief supplementary investigation on Langmuir, since I did not think it makes sense to make all the arguments again, when it is after all probably in vain, since it seems to me as if S. will support P. Well, I do not have anything against the idea of Lewis receiving the prize, but it is somewhat annoying when one has been working on behalf of Langmuir. ["S." is likely Soderbaum, another committee member, and "P." is likely Palmaer.][86]

From Westgren's reply, 14 May 1932:

I do understand your emotions regarding Palmaer's action on Lewis. It comes to the worst, since by all means Palmaer does not have a clue about Lewis. In any case, so far I have not received a report from him, and he will probably be long about it.[87]

Westgren's comment that "Palmaer does not have a clue about Lewis" is interesting. Palmaer's report on Lewis's theoretical concepts of activity and fugacity showed an extremely close reading of Lewis's work, and it is hard to believe that Palmaer had the theoretical background to criticize Lewis's thermodynamics in the detail supplied in his report. Did he have help from Nernst in writing the thermodynamics section of his report? His report on Lewis's thermodynamic theory carefully followed Lewis's work, noting inconsistencies and errors, while ignoring the achievements; it almost seems that Palmaer (perhaps with Nernst's help) had done to Lewis what Lewis had earlier done to Nernst—spent a year finding all the errors. While Palmaer used Nernst's notation and nomenclature, and while there are a number of references to Nernst's work in the report, I have found no evidence that Nernst was involved in the preparation and editing of Palmaer's report, or even that he knew of the report at all. The letters from Nernst to Palmaer that have been preserved do not mention Lewis, and Palmaer's letters to Nernst have not been preserved. However, if Nernst were involved in preparing the report on Lewis, in violation of the Nobel standards of confidentiality, it is likely that

Nernst and Palmaer would have been careful to conceal any evidence. The submission of reports with a signature other than the author's had happened before—biographer Elisabeth Crawford believes that Arrhenius had been the secret author of reports opposing Ehrlich and supporting Ostwald (see chapter 1).

Because the Nobel committees preserve no tally of votes or minutes of their meetings, there is no record of whether Palmaer voted for Lewis or supported him during committee discussions. The committee's 1932 report to the academy did discuss the possibility of awarding the prize to Lewis.[88] After first dismissing Lewis's work on valence theory, which had been covered by Svedberg's report, the committee moved to Lewis's work on thermodynamics:

> Greater importance should be placed upon [Lewis's] work in thermodynamics, and his closely related work on the magnitude of chemical affinities, including careful determinations of the standard potentials of a great number of chemical elements, on which Mr. Palmaer has given a report.... After having reviewed the two reports on Lewis's work, the committee has however not been able to find that Lewis's work is weightier than Langmuir's, and it has often been said that when comparing the work of these two, the balance swings to Langmuir's advantage.[89]

On 10 November 1932, Langmuir received two telephone calls from Swedish newspapers, saying that he had been awarded the Nobel Prize in Chemistry. At 5 P.M. he received a telegram, "Nobel prize for chemistry awarded to you. Please wire whether you can be present at Stockholm on Dec. 10. Signed Secretary, Academy of Sciences."[90]

Langmuir took his wife and children with him to Stockholm, sailing on the *Bremen*. Before leaving, he was invited to give a radio talk on NBC and to address the New York Chemistry Club. He described the Nobel ceremony to his mother in a letter from Copenhagen, where he and his family stayed at Bohr's home.

> When we arrived at Stockholm we were met by a Dr. von Euler, a Nobel prize winner and a member of the Nobel foundation who escorted us to the Grand Hotel where we had three rooms facing the King's Palace across a beautiful body of water. The next day, Dec. 9, ... we attended a formal dinner given by the American Ambassador Morehead in his very beautiful home.... The award of the Nobel prizes took place the following day, Dec. 10, at 5 PM in the large Concert House holding about 2000 people with every seat filled. It was the most impressive ceremony I have ever attended. At exactly 5 PM the King and the Royal Family, etc. (in all about 25 people) entered the hall and were seated in the front row of the auditorium.... The King's arrival was announced by trumpet calls with trumpeters on each side of huge columns on the platform:—trumpets 6 ft. long like in Aida,

> pointed upwards by trumpeters in silver and black uniforms.... After a beautiful musical selection the President of the Foundation gave a 15 minute speech about Alfred Nobel and the Nobel prizes. Then another selection and a talk by Prof. Soderbaum, Secretary of the Swedish Academy of Sciences (in Swedish) about my work. Then in English he addressed me for 2–3 minutes while I stood and as soon as he was through I went down the steps from the platform with more trumpet calls and flag lowering to stand in front of the King while the whole audience arose. The King presented me with a gold medal and a large hand decorated diploma and assignation for the prize.[91]

Lewis was at home in Berkeley while Langmuir was receiving his prize. Lewis, of course, did not know about Palmaer's report or the recommendations of the committee, but Langmuir's Nobel Prize must have been galling to him.

Palmaer Repeats Himself

Was Palmaer's 1932 nomination of Lewis dishonest? Viewing his actions in the most charitable light, one might say that Palmaer nominated Lewis in 1932 in good faith, reviewed Lewis's work carefully, came to the conclusion that he was mistaken in his nomination, and then submitted his negative conclusions to the committee. But if this were the case, why did Palmaer do what he did next, which was to nominate Lewis again for 1933? He appears to have wanted to continue to control the reporting process and to block the award of the prize. Palmaer did not write another report on Lewis for 1933, but Palmaer was committee chairman for 1933 and presumably was responsible for the committee's summary report[92] on all nominees.[93] The committee's reports to the academy are not signed by an individual member. But the verbose, convoluted prose style of this report seems to be the same as in Palmaer's signed 1932 special report on Lewis. (My Swedish-born translator had a great deal of difficulty with Palmaer's prose, saying it was "more like German than Swedish.") The committee recommended that the academy reserve the chemistry prize for that year, and the discussion of that decision seems to point directly to Lewis:

> The attention of the Academy cannot avoid being drawn to the recurring candidates who appear year after year, almost touching the goal. Yet time after time they have had to step back for others who have come ahead of them. It would undoubtedly be most comfortable for the committee to suggest these reserve candidates for the Nobel prize to the Academy. So why not do it in chronological order, because it would be hard to find another norm for the candidates. Such a method would probably not offend anyone or provoke any serious opposition. It is dangerous in a way, yes one could

with near certainty predict that it would lead to a cheapening of the Nobel prize. Such a devaluation of the prize should be forcefully opposed by the prize-givers.

The committee intends that the prize should be given to groundbreaking discoveries that are fundamental. These discoveries should be in the theoretical or practical areas that give new views that are fruitful for research. On the other hand, it is necessary for the committee to act with great caution regarding such work, which even if brilliant with regard to the result, depends more or less directly on the work of previous research in the same or closely related areas....

Such a case presents itself this year, and thus the majority of the committee has decided to recommend the reservation of this year's Nobel prize in chemistry and to wait for next year.[94]

Lewis seems to be the one referred to as "depend[ing] more or less directly on the work of previous research in the same or closely related areas." He had been reduced to a second-rate imitator of Nernst, and an award to him would lead to a "cheapening of the prize."

Despite the committee's assessment that Lewis should be passed over for the chemistry prize for 1933, Palmaer nominated him yet again for 1934, and he did write a supplementary report on Lewis for 1934.[95] Although Lewis had not been active in thermodynamics since 1923, Palmaer began with the observation: "I have not been able to find any new work in this field where Lewis is given as an author. On the other hand, there is some work from the university at Berkeley done by Lewis's coworkers or followers."[96] Palmaer made a grudging admission that his 1932 statement, that Lewis's concept of ionic strength had no theoretical foundation, was not correct, possibly because Svedberg had called him on this: "I also want to say that Lewis's assumptions regarding ionic strength..., which is a term that he has coined, in some regard seems to have won support through theoretical work by Debye and Hückel."[97]

Debye and Hückel had published their theory eleven years earlier, and Palmaer had cited the lack of theoretical support for ionic strength as one of foundations of his 1932 recommendation that Lewis's work on strong electrolytes was not worthy of the prize, but Palmaer did not change his assessment in his 1934 report. In the same way he had done in 1932, he concluded his supplementary 1934 report with a neutral finish:

Considering what I have said in the previous report about the work on standard potentials, I want to say in conclusion that any eventual award of a prize for Lewis's work on the magnitude of chemical affinities should be motivated partially by his thermodynamic input, and partially by his and his coworkers' research on chemical affinity in reactions using inert electrodes for the determination of the free energy of formation of chemical compounds.[98]

Palmaer was also the committee chairman in 1934. The committee's 1934 report to the academy, presumably written by Palmaer, includes the following: "In consideration of the question as to whether Lewis should be awarded the prize for his work on chemical affinities, there is a supplementary report by Mr. Palmaer. (Report 6). Very little has been added since it was originated in 1932. But in certain areas, several remarks have been added so that the hesitation from the committee for the award of the prize has increased rather than diminished."[99] Lewis's chances for a Nobel Prize for his work in thermodynamics (although not in other areas) were finished.

I have found no correspondence that indicates that any of the other committee members suspected dishonesty behind Palmaer's repeated nominations. One would think that after his third nomination of Lewis and second negative report, it would have been evident that things were not as they seemed. It may have been that in the committee's deliberations on the award of the prize, Palmaer initially supported Lewis and then deferred to other members who were basing their assessments of Lewis on Palmaer's reports. I believe, however, that the consistently negative tone of the body of his reports, their neutral conclusions (which would allow Palmaer to pretend to a continued interest in supporting Lewis), and the pattern of repeated nominations lead to the conclusion that Palmaer used his position as nominator and reporter to block the award of a prize to Lewis.

Did Lewis Deserve a Nobel Prize for Thermodynamics?

But even if Palmaer acted dishonestly, and even if he did so with the complicity of Nernst, it does not necessarily follow that Lewis deserved an award for his thermodynamics. In many ways, Lewis was the Henry Ford of thermodynamics:[100] just as Ford did not invent the automobile but made it readily available to the public, Lewis did not invent chemical thermodynamics but made it a standard tool for chemists. He reformulated chemical thermodynamics rigorously in terms of measurable quantities (pressure and concentration) while preserving the form of the equations then in common use; he explained thermodynamics in terms that a working chemist could follow; and he compiled the most extensive collection of chemical thermodynamic data to date. Lewis's work might have been viewed as a great "improvement" and given an award in the same spirit in which Richards was awarded the 1914 chemistry prize for his accurate determination of atomic weights or as Manne Siegbahn was awarded the 1924 physics prize for his measurement of X-ray spectra; this is what Svedberg suggested in his 1926 report. And yet, as Arrhenius pointed out in his 1924 report, all of this might not have been enough to merit the award of a prize—there was no single "discovery" there. But the great unsolved problem of physical chemistry in the period 1887–1923 was the anomalous behavior of strong electrolytes, which was not explained by Arrhenius's dissociation theory. Lewis was the first to provide an empirical solution with his 1921 concept of ionic strength, which was given theoretical support by Debye and Hückel in 1923. Ionic strength was ignored in Arrhenius's 1924 report, but was noted favorably in Svedberg's 1926 report recommending Lewis for the prize, only to be subsequently disparaged in Palmaer's 1932 report and then grudgingly acknowledged in Palmaer's 1934 report. Although the problem of strong electrolytes was the most pressing problem in physical chemistry in the early 20th century, no Nobel award was ever made in this area. An award of a chemistry prize to be shared among Lewis, Debye, and Hückel would have been appropriate. The committee's problem was that a great many people had worked in this area. Ludwig Ramberg, a committee member who supported an award in the field, wanted to make sure that the Danish scientists Niels Bjerrum and Johannes Brønsted, who had developed theories on the behavior of strong electrolytes, were not excluded,[101] and the Nobel statutes limit the award to a maximum of three people. Debye did receive a chemistry prize in 1936, but not for his work on strong electrolytes, and Hückel, like Lewis, never received the prize. In 1939, the Royal Swedish Academy of Sciences did award Lewis a consolation prize—the first Arrhenius Medal, for his "epoch-making contribution to dissociation theory, based on the application of exact thermodynamic methods."[102] It is likely that Lewis had mixed feelings about this award.

8

Nuclear Chemistry

Lewis, Urey, and Seaborg

NOBEL PRIZE MAY GO TO DR. LEWIS OF U.C.

Dr. Gilbert Newton Lewis of the University of California, and Dr. Harold Clayton Urey, former student at the university, were mentioned today in Stockholm dispatches as likely recipients of this year's Nobel Prize in Chemistry.

Dr. Urey, now professor of chemistry at Columbia University, attracted wide attention by his discovery of "heavy water," and Dr. Lewis has developed notable methods of concentrating this water. Dr. Urey was a student under Dr. Lewis while attending the university at Berkeley.

Dr. Lewis, one of the world's outstanding chemists, has been mentioned for the Nobel Prize before. His scientific works have been widely acclaimed.

During the war he was chief of the defense division of the gas service.

This newspaper article appeared on the front page of the Oakland, California *Post-Enquirer* on 25 October 1934 (Oakland is a larger city adjacent to Berkeley).[1] Whoever wrote it had likely based it on a leak from the Nobel committee, whose internal reports and deliberations were supposed to be secret—Theodor Svedberg had, only a few months earlier, given the Nobel chemistry committee a report recommending that Urey and Lewis share the 1934 prize.

After spending a decade in the wilderness of theoretical physics, Lewis returned to physical chemistry around 1932, beginning to work in a new field within physical chemistry—nuclear chemistry, the chemistry of what went on *inside* the atom's nucleus. Most scientists are finished after leaving a field for ten years, but some of Lewis's most productive work in physical chemistry was still ahead after 1932, although he seemed to become even more isolated from the larger scientific community.

His cigar smoking was on the increase, as well. By this time, Lewis was chain-smoking a great number of Alhambra Casino cigars, imported from the Philippines, every day. His son Edward puts the number at 12–20 per day,[2] but this may have been what Lewis told his family to assuage their concerns (he also claimed that he did not inhale). The Berkeley graduate students were able to measure his cigar consumption a few years later. Art Wahl used Lewis's empty cigar boxes to store his plutonium samples, and empty boxes that had once held fifty cigars became available from Lewis's secretary almost once a day.[3] (Fifty cigars a day works out to one every fifteen to twenty minutes, given eight hours a day for sleep. Difficult, but it can be done, and Lewis was the man to do it.) The Monday after Pearl Harbor, Lewis and his coworkers were discussing the Japanese attack when David Lipkin, Lewis's research assistant, said, "Professor Lewis, what are you going to do about your Philippine cigars?" After a long pause, Lewis said, "Lipkin, this is more serious than I realized." He had his secretary buy up the local cigar wholesaler's entire supply of his Philippine cigars, but it was still not enough. Lewis was forced to ration them and mix them with other brands until Douglas MacArthur's return.[4] When Lewis was preparing to travel to Philadelphia for a talk on acids and bases at the Franklin Institute in 1938, he and his research assistant, Glenn Seaborg (more on him later in this chapter), planned a demonstration experiment that required him to bring chemicals and glassware with him on the train. Seaborg took the equipment for the experiment to Lewis's room at the Berkeley faculty club to pack for the train trip and found Lewis absent, but he saw a large open suitcase lying on the floor, into which he carefully packed everything. When Lewis arrived, he took one look at what Seaborg had done and said, "No, no! That suitcase is for the cigars!" He made Seaborg remove everything and filled the suitcase with cigar boxes.[5]

Nuclear Chemistry and Heavy Water

John Dalton's 1803 atomic theory had joined the ancient idea of the atom with Antoine Lavoisier's concept of the chemical element—each element consisted of a different kind of atom, and different atoms had different atomic weights. More than a century after Dalton, it became evident that a particular element might consist of atoms with similar chemical properties but different atomic weights—these were called an element's *isotopes*. The new field of nuclear chemistry dealt with the identification and separation of isotopes, and a few years later, scientists at Berkeley would be the first to transform an isotope of one element into an isotope of an entirely new element. A modern introductory chemistry text might explain isotopes this way: "Each atom's nucleus contains a certain number of protons, which determine its nuclear charge and its electronic configuration, and thus fix its place in the periodic table. Isotopes are atoms with the same number of protons but different numbers of neutrons within the nucleus, giving them different weights." But this explanation is historical revisionism if viewed from the 1920s—the name "isotope" was proposed by Frederick Soddy in 1913, but the neutron was not discovered until 1932. Soddy defined different isotopes simply as atoms that had different atomic weights but that were chemically identical. For some elements, two or more isotopes are abundant. Thus, naturally occurring chlorine, which has a measured atomic weight of approximately 35.5, is a mixture of two isotopes, ^{35}Cl with atomic weight 35 and ^{37}Cl with atomic weight 37, in a ratio of approximately 3 to 1, respectively. For other elements, such as hydrogen, whose dominant isotope ^{1}H constitutes almost all of the naturally occurring material (99.98%), it was difficult to determine even that other isotopes existed. The chemical differences among isotopes of the same element are usually small—the number of protons is much more important chemically than the number of neutrons. Isotopes were very difficult to separate chemically, and it was not until the advent of radiochemistry that their existence was even determined. For chlorine, for example, the relative increase in atomic weight from ^{35}Cl to ^{37}Cl is only about 5%, and the chemical effects of such a small relative change are usually insignificant. But for hydrogen, the change between the atomic weight of the common hydrogen isotope with an atomic weight of 1 and of a (then hypothetical) isotope with an atomic weight of 2 would be 100%, and the chemical effects would be expected to be significant. For hydrogen, discovering a new isotope would be almost like discovering a new element. (Soddy, in fact, did not think that the heavy isotope of hydrogen fit his definition of an isotope—a substance with different atomic weight but identical chemical properties.)

The "heavy water" mentioned in the newspaper article that begins this chapter is water whose two hydrogen atoms have an atomic weight of 2 rather than the normal mass of 1—heavy water would have a molecular weight of 20 rather than 18. There was speculation in the literature about the existence of a heavier isotope of hydrogen as early as 1928. Lewis had reportedly first become interested in a possible heavy hydrogen isotope in the late 1920s. During a seminar at Berkeley, Raymond Birge, the chairman of the physics department, mentioned that he was getting inconsistent values for the atomic weight of hydrogen depending on its source. He said that hydrogen prepared from water in an electrolytic cell that had been in use within the physics department for years seemed heavier than hydrogen from other sources. Lewis wondered whether a heavier isotope of hydrogen might exist that had become concentrated there over time through the repeated process of distillation. He asked Simon Freed, who was an instructor at Berkeley from 1927 to 1929, to compare the density of the water from Birge's still with normal distilled water. When Freed asked Lewis to what accuracy he should measure the density, Lewis replied, "Oh, four decimal places should do it." That was not quite enough. If Freed had gone to five decimal places, he might have found evidence of the existence of heavy hydrogen.[6]

Ferdinand Brickwedde, one of Urey's collaborators in discovering heavy hydrogen, recalled the interest in isotopes of hydrogen:

> I remember a conversation in 1929 with Urey and Joel Hildebrand, a famous professor of chemistry at Berkeley. It took place during a taxi ride between their hotel and the conference center for a scientific meeting we were attending in Washington. When Urey asked Hildebrand what was new in research at Berkeley, Hildebrand replied that William Giauque and Herrick Johnston had just discovered that oxygen has isotopes with atom atomic weights 17 and 18 [in addition to the most common isotope with atomic weight 16].... Then Hildebrand added, "They could not have found isotopes in a more important element." Urey responded, "No, not unless it was hydrogen."[7]

Harold Urey

Harold Urey was born on an Indiana farm in 1893 and moved to Montana as a child when his parents became homesteaders there. After graduating from high school, he taught school for three years and then entered the University of Montana, where he received a degree in zoology with a chemistry minor. He lived and studied in a tent during the school year (in Montana!) and worked on a railroad gang during the summer. After graduating in 1917, he worked in industry for a few years and entered the Berkeley graduate program in chemistry in 1921. Lewis was Urey's nominal research director, but Urey chose his own line of research, the rotational contribution to heat capacity and entropy of molecules consisting of two atoms—a subject that most people would have seen as part of physics rather than chemistry. When he published his doctoral work,[8] apparently neither he nor Lewis thought that Lewis's name should be attached. Urey believed that the future of physical chemistry was quantum chemistry, and after receiving his Ph.D. from Berkeley in 1923, he went off to study for a year with Niels Bohr in Copenhagen. Lewis and Urey corresponded after Urey left Berkeley. In 1923, Urey wrote from Copenhagen asking for Lewis's help in finding a permanent academic position, dropping strong hints that he would like to return to the Berkeley chemistry department.[9] Lewis likely saw Urey as too mathematically oriented for his liking as a chemist. (Although Lewis was mathematically skilled himself, he always disliked presentations that he thought were excessively mathematical. Lewis once publicly dressed down Kenneth Pitzer, then a young faculty member, at the end of a seminar full of equations that he had given, saying, "I never want to see a seminar like that again."[10]) Lewis tried to arrange a faculty position for Urey at Berkeley in the physics department. Urey seemed unenthusiastic, and the appointment apparently fell through in any case.[11] Their correspondence moved into a more collegial relationship after Urey managed to find positions at Johns Hopkins in 1924 and then Columbia in 1929, apparently without much help from Lewis. In 1930, Urey and Arthur Ruark, a physicist, wrote *Atoms, Molecules, and Quanta,* which became the standard American text on chemical quantum theory for some years, firmly establishing Urey's professional reputation.

A Heavy Isotope of Hydrogen?

Within the Berkeley physics department, Birge and Donald Menzel continued to puzzle over the discrepancies in the atomic weight of hydrogen and found a slight discrepancy between its measurement by mass spectrometry—which examined a single hydrogen atom at a time—and by bulk weight of hydrogen, the bulk atomic weight being slightly heavier. They took this to indicate that a heavy isotope of hydrogen might exist, and that if it had atomic weight 2, about 1 in 4,500 hydrogen atoms would be of the heavy form.[12]

After Urey read Birge and Menzel's paper in July 1931, he immediately began to develop a plan to find the heavy isotope of hydrogen. Working with his assistant George Murphy, he looked at the atomic spectrum of ordinary hydrogen using a grating twenty-one feet long to diffract the light so that individual frequencies could be identified with precision. Urey found very faint lines where he had predicted they would appear for an isotope of hydrogen of mass 2. But he was not ready to publish—these lines might arise from impurities in the hydrogen or from ghost lines due to instrumental artifacts. He decided that he needed to enrich the hypothetical heavy isotope's concentration; if the lines then became more intense, this would confirm the isotope's existence. Using calculations based on statistical mechanics and on Peter Debye's theory of solids, he predicted that distillation of liquid hydrogen would enrich the heavy isotope. He expected that normal hydrogen would boil off more quickly, leaving an excess of the heavy isotope in the liquid phase. But in order to obtain a small amount of concentrated heavy liquid hydrogen, Urey calculated that he would need to start with a large amount of ordinary liquid hydrogen—five or six liters—boiling more than 99.9% off until only two milliliters were left. In 1931, there were only two laboratories in America capable of generating five or six liters of liquid hydrogen—William Giauque's laboratory in the Berkeley chemistry department and the National Bureau of Standards laboratory in Washington, D.C. Urey enlisted Brickwedde at the National Bureau of Standards, who provided the hydrogen. When they distilled it, it showed the spectral lines intensified six- or sevenfold, establishing the existence of an isotope with mass 2.[13] It was Thanksgiving Day, and Urey arrived home late to find his wife and guests waiting. The only excuse he could offer his wife was "Well, Frieda, we have it made."[14] He announced the discovery at the end of 1931 at a scientific meeting in New Orleans and published the results shortly thereafter.[15]

Just before Urey was awarded the Nobel Prize for 1934, Birge and Menzel discovered that they had made an error in the calculations that had led them to suggest the existence of the new isotope. Given their corrected results, the difference in the values of the atomic weight as determined by mass spectrometry and bulk measurement would have been within the experimental error, and Urey would not have had the evidence he had used to plan his approach. Urey noted this in an addendum to his Nobel presentation speech, saying that without the incorrect value, he would not have made a search for the heavy isotope, and its discovery might have been delayed for some time.[16] This sort of thing, an important discovery based on incorrect data, is more common than one might think. Lewis once noted that Dalton had based his 1803 atomic theory upon the definite proportions among elements, and that the experimental data that had convinced him of the general nature of the law was the oxides of nitrogen. Lewis said, "The crudity of the experiments upon which he based this law, and the fact that his analysis of one of the oxides of nitrogen was entirely erroneous, indicate a strong predisposition toward the conclusion which he reached."[17]

Urey suggested the name "deuterium" for the isotope of hydrogen with mass 2. It was not usual to name isotopes—the different isotopes of oxygen did not have separate names other than ^{16}O, ^{17}O, and ^{18}O—but the heavy isotope of hydrogen, which was expected to be quite different chemically from ordinary hydrogen because of its double mass, would be an exception.

About this same time, Lewis had returned to experimental chemistry. Lewis and his research assistant Ronald T. Macdonald began working on isotopes in December 1931, attempting to separate the isotopes of lithium and oxygen. Although deuterium was a more interesting problem, Lewis might have decided not to pursue it because he thought that the natural occurrence of deuterium, which he believed to be one part in 30,000,[18] was too low for easy concentration and detection.

Although Urey had shown that deuterium existed in nature, no one had produced it in a pure enough form to allow its chemical properties to be measured. Birge and Menzel had originally estimated that one hydrogen atom in 4,500 occurred naturally as deuterium, and Urey's work roughly supported this estimate (the modern value is one in 7,000). Even if the deuterium concentration of hydrogen were increased 100-fold—to about 2% deuterium—it would not be expected to show measurable chemical difference from normal hydrogen. To characterize deuterium chemically, it would be necessary to synthesize compounds with almost all the significant hydrogen atoms replaced by deuterium. After both the paper by Birge and Menzel and the paper by Urey and coworkers had estimated that the normal concentration of deuterium was much higher than his earlier estimates, Lewis decided to enter the race to be the first to make pure heavy water. Once he had the heavy water, he could use it to synthesize a variety of deuterium-substituted compounds.

A rumor circulated at Berkeley about Lewis's motivation—and it is only a thirdhand rumor at that: Franklin Long, who received his Ph.D. from Berkeley in 1935, told Lewis's graduate student Jacob Bigeleisen that Wendell Latimer, Lewis's colleague and friend, was responsible for Lewis's decision to pursue heavy water. Long said that Latimer told him that he had advised Lewis that this was the short and sure route to the Nobel Prize.[19]

Distilling liquid hydrogen was an awkward method for purifying deuterium, and Edward Washburn at the National Bureau of Standards suggested that electrolysis of distilled water would work as well or better. When water is electrolyzed (when the terminals of a battery are inserted into water), hydrogen gas forms at one of the terminals. A smaller percentage of heavy hydrogen moves into the gas phase, thereby concentrating heavy water in the liquid phase. This is what Lewis and Freed had assumed in their earlier investigation of the water remaining in Birge's still. Working with Urey, Washburn published his electrolytic method for concentrating heavy water, although the first results produced only a small amount of water containing only a few percent deuterium.[20] Urey credited Washburn with the method's discovery, because he had suggested it.[21]

Lewis, along with other scientists, immediately adopted the Washburn-Urey purification technique and began producing heavy water. Lewis had an advantage over other scientists, however—Giauque, a faculty member at Berkeley, had for years been producing hydrogen from two large electrolytic cells, essentially applying Washburn and Urey's electrolytic purification method without intending to do so, and Lewis and his research assistant Macdonald confirmed that the water remaining in these cells was significantly denser than normal distilled water. Lewis and Robert Cornish investigated other methods of isolating heavy water,[22] of which the most successful was fractional distillation—essentially the same process by which wine is distilled to brandy—although Washburn and his coworkers scooped them by publishing similar results on fractional distillation a week earlier.[23]

Distillation columns are most effective in separating substances when the ratio of their height to their diameter is as large as possible and they are filled with small inert particles—small glass beads are commonly used—upon which the material being distilled can condense. But while very tall, narrow columns give the highest purity, they cannot handle much material. It is common to use distillation columns in stages—use fatter columns at first to give an initial separation, then move the concentrated product from those columns to narrower columns that will further increase purity. Lewis enlisted Merle Randall, his colleague at Berkeley and his former student, to build fractional distillation columns designed to work under reduced pressure, including one a foot in diameter and seventy-two feet high, and another column of the same height with a diameter of two inches.[24] Lewis said of these columns, "Owing to circumstances which are too universal to dwell upon, [the largest] column had to be filled, not with the best, but with the cheapest available packing material. This proved to be waste aluminum turnings, which have caused some difficulty through the formation of aluminum hydroxide."[25] Using fractional distillation followed by electrolytic concentration, Lewis and Macdonald were soon producing more than a gram per week of heavy water at greater than 99% purity, allowing them to measure its physical and chemical properties and to employ it in chemical reactions.[26] Despite his disdain for chemical engineering, Lewis had built the world's first heavy water factory.

In the space of sixteen months, Lewis published twenty-six communications on heavy water and deuterium. He measured its boiling point, melting point, vapor pressure, dielectric constant, refractive index, and so forth. Lewis also investigated heavy water's biological properties—would it support life? He found that tobacco plants (an interesting choice for Lewis, given his consumption of cigars) would not grow if they were given only heavy water, that yeast cultures did not prosper, and that flatworms died. And then the ultimate test:

> Finally, I wished to test the effect of heavy water upon a warm-blooded animal. For this purpose I obtained three young white mice of respectable ancestry, weighing approximately ten grams apiece, and kept them in the laboratory for several days while their normal habits were being observed. Then, after they had all been deprived of water over night, two of the mice were given ordinary water while the third was given heavy water.... Nevertheless, the experiment was a very costly one and I regret that since it was undertaken solely to ascertain whether the heavy water would be lethal, no preparation was made for a careful clinical study of the effects produced.
>
> The answer to the main question was decisive.... The mouse survived and on the following day and thereafter seemed perfectly normal. Nevertheless, during the experiment he showed marked signs of intoxication. While the control mice spent their time eating and sleeping, he did neither, but became very active, and spent much of the time, for some mysterious reason, in licking the glass walls of his cage. The more he drank of the heavy water, the thirstier he became, and would probably have drunk much more if our supply of heavy water had not given out. The symptoms of distress that he showed seemed more pronounced after each dose but not cumulative with succeeding doses, which leads me to suspect that the heavy water was being rapidly eliminated by the mouse. This could have been ascertained if suitable preparation had been made.[27]

The mouse had consumed the world's supply of heavy water and done nothing more than run around his cage, licking the walls and urinating into his straw. Within the Berkeley chemistry department, this became known as "the mouse that made a monkey out of G.N. Lewis." From all of this biological work, Lewis concluded that heavy water was not in itself toxic, but that if supplied to the exclusion of normal water, it caused changes in rates of biological functions such that life would not be supported.

Lewis provided heavy water to Ernest Lawrence to produce deuterium nuclei to be accelerated in his Berkeley cyclotron. Lewis and Lawrence were on good terms; Lewis had chaired a faculty committee that had voted to overrule the budget committee and promote the twenty-nine-year-old Lawrence to full professor when he was considering moving to Northwestern in 1930.[28] But Lawrence had to wait for his supply of deuterium; he had planned to use the heavy water that Lewis had fed to the mouse and was apoplectic when he found what Lewis had done.[29]

Naming the Baby

Lewis was generous with his production of heavy water, providing it to investigators around the world, including Otto Stern. Stern used it to measure the magnetic moment of the proton, for which he would receive the Nobel Prize in Physics for 1943. Lewis also pushed to name the deuterium nucleus the "deuton," which was stretching things a bit. Urey, not Lewis, was the discoverer of deuterium and had already decided on the isotope's name. Based on Urey's name "deuterium," the logical name for the nucleus would be "deuteron," which is what it is now called. Lewis had apparently forgotten that when Irving Langmuir had named Lewis's electronic theory the "octet theory," he had complained, "Sometimes parents show singular infelicity in naming their children, but on the whole they seem to enjoy having the privilege."[30]

Urey and Lewis corresponded about names for the new hydrogen isotope's nucleus; Urey was polite but was clear that while he was willing to listen to Lewis, he felt that it was his prerogative to choose the name.[31] Lewis preempted things, writing to Urey that

> I am sorry that the physicists with whom I am working...rather forced an early decision regarding the name of the nucleus of the hydrogen isotope..., so when [we] sent off an abstract to the Physical Review we used the word "deuton," but we should still have had time to change if Lawrence had not given the paper in Pasadena, which, by virtue of their extraordinary publicity department, was telegraphed to all parts of the world, and in which this name was used.[32]

Lord Rutherford also got involved with his idea for the name for the hydrogen isotope and its nucleus—"diplogen" and "diplon," respectively[33]—which Urey resisted, copying Lewis on Rutherford's letter to him and his own communication to the editors of *Nature* arguing against Rutherford's names.[34] The naming dispute with Rutherford made it into an article in *Time* magazine.[35] In the end, Urey's name prevailed.

The 1934 Nobel Prize

In 1934, the Nobel Committee for Chemistry said that deuterium was "a recent discovery that is seemingly of such quality that there is no doubt that it deserves a Nobel prize."[36] The only question was the prize's recipients. There were three principal candidates: Urey, who had been the first to obtain spectroscopic evidence of the deuterium isotope; Lewis, who had been the first to purify heavy water in chemically significant amounts and to measure the properties of deuterium-containing compounds; and Washburn, who had led the collaboration with Urey to develop a method for isolating heavy water by electrolysis. Washburn had died between the time of his nomination and the committee's report to the Royal Swedish Academy of Sciences. While Washburn was still technically eligible (candidates who were alive at the time of nomination could be awarded the prize that same year even if subsequently deceased), the committee was not enthusiastic about posthumous awards, and in any case Washburn's role seemed minor when compared to Urey's and Lewis's. It was likely to come down to Urey alone, or to Urey and Lewis jointly.

In Lewis's favor was that this was the chemistry prize, and Lewis's work had been of more direct chemical interest than Urey's. Once again, Svedberg was called on for the report.[37] His report of 18 May 1934 describes the contributions of Urey, Lewis, and Washburn in detail and then concludes with the following:

> If one tries to picture what part each of the researchers Urey, Washburn, and Lewis have played in the discovery of the heavy hydrogen isotope [deuterium] and in its isolation, and also in the research into its chemical and physical chemical qualities, one is met with great difficulty. The main merit for the discovery is without doubt ascribed to Urey. It is not only that the research done by him and by his coworkers ended with a positive result, but also that this research was done with a rational plan, based on statistical thermodynamical calculations regarding the qualities of the unknown element, and this must be taken into account. It is considerably more difficult to decide how much weight one must ascribe to Washburn's and Lewis's contributions. The method that Lewis worked out regarding electrolytic separations has made possible the production of heavy hydrogen in quantities large enough for chemical research. More or less the entire collection of

experimental material regarding the qualities of heavy hydrogen is therefore thanks to this method. Lewis has also, more than any other researcher, been working on the chemistry and physical chemistry of heavy hydrogen. He has, before anyone else, made heavy water in a pure state, and he has studied its qualities and reactions and has made several compounds containing heavy hydrogen. I do not know how large a part he had in the earlier experiments in nuclear fission that were done at Lawrence's laboratory in Berkeley using a high voltage accelerator. His name is on these publications; one publication even has his name as first author. After having purified heavy water, he gave samples to several researchers in America as well as in Europe. By doing this, he has supported other pioneer work in this area. . . . Therefore the possibility remains of dividing the prize between Lewis and Urey. Against this lies the fact that Lewis's work rests upon Washburn's ideas. On the positive side lies the fact that Lewis has been and still is the leading researcher in the field and more than anyone else has contributed to results won to date. If one does a thought experiment to erase Lewis's work from the history of heavy hydrogen, what is then left at present is so fragmentary and the meaning of the discovery so unclear that it could be questioned whether the award of a Nobel Prize in chemistry at the present time would be motivated. Considering this, the undersigned (Svedberg) is more inclined to dividing the prize.

Although Svedberg's report recommended sharing the prize between Urey and Lewis, it was clear that he was struggling with second thoughts. Svedberg's cover letter to Arne Westgren, the committee's secretary, that accompanied Svedberg's report makes this even clearer: "I enclose the report concerning heavy hydrogen. I have, after a lot of hesitation, recommended a division of the award between Urey and Lewis. Should the committee be of the opinion that Urey alone should be the only candidate, I will agree. What is your opinion?"[38]

The newspaper article that began this chapter was based on old news. By the time it appeared, Svedberg had already changed his mind. Other workers were beginning to purify heavy water, and Lewis's monopoly on the chemical characterization of deuterium was vanishing. In a supplementary report dated 18 August 1934, Svedberg reviewed the latest work in the field and concluded:

The above-mentioned works regarding heavy hydrogen illustrate the great importance that one must ascribe to the discovery of this isotope. They also help to clarify the importance of the earlier works. The prime importance of Urey's work comes forward even more clearly now. One also finds Washburn's initiative to use electrolysis to have been of decisive importance. On the other hand, the work by Lewis now seems perhaps rather less when compared to others', especially the performance of the Princeton school. One even gets the impression that the research by Lewis in some

> measure has the character of a speed record. It seems not improbable that workers in Urey's own laboratory as well as those at Princeton could have achieved the same results if they had only used Washburn's suggestions unconnected with Lewis's work regarding of heavy hydrogen.[39]

It is unclear whether Svedberg revised his conclusions on his own or in consultation with other committee members, but the 1934 chemistry prize was awarded to Urey alone. Because of the birth of his child, Urey was unable to attend the award ceremony, and his Nobel presentation speech was delivered by that year's chairman of the Nobel chemistry committee, Wilhelm Palmaer.

Urey was known for his generosity. The Nobel Prize was awarded to Urey alone, a decision that his collaborator, Brickwedde, endorsed, writing, "Although George M. Murphy and I co-authored the papers reporting the discovery; it was Urey who proposed, planned, and directed the investigation. Appropriately, the Nobel prize for finding a heavy isotope of hydrogen went to Urey."[40] Nonetheless, Urey gave half of his Nobel Prize money to be divided between Brickwedde and Murphy. When Urey was awarded a Carnegie Institution research grant shortly after receiving the Nobel Prize, he shared it with Isidor Rabi, also a professor at Columbia, who used it to build his first molecular beam machine to investigate the magnetic properties of atomic nuclei.[41] Rabi was awarded the Nobel Prize in Physics for 1944. Mildred Cohn, who was a graduate student working for Urey at Columbia in the 1930s, remembers Urey as "the only professor in the chemistry department at Columbia who was concerned with the welfare of the graduate students in those depression years. Were they paid enough as teaching assistants? Were the long hours they worked interfering with their research?"[42] When Cohn was without summer financial support, Urey approached her and said, "Miss Cohn, what are you doing for money?" When she told him she was borrowing it, he told her, "Ever since I got the Nobel prize, I've wanted to use some of that prize money to help my students. So, why don't you let me lend you some money, and some day when you have a job, you can pay me back." She had difficulties finding a job—she had two strikes against her as a Jewish woman, as many of the job postings specified "Christian male"—and Urey worked hard to find her a position.[43] Urey felt that Columbia was particularly anti-Semitic. David Altman, a young Jewish chemist from Cornell, applied to both Berkeley and Columbia for graduate school. Berkeley accepted him and was pressing him to either accept or decline, but he preferred Columbia because he wanted to stay on the East Coast. At Lewis's recommendation, he caught a midnight bus to New York and appeared in Urey's office. Urey closed the door, took a deep breath, and advised him that given what he knew of Berkeley and the attitudes toward Jews of some Columbia faculty members, Altman might do better to choose Berkeley.[44]

Figure 8-1. Harold Urey measuring isotope effects with a mass spectrometer in the Columbia chemistry department, sometime between 1934 and 1939. Courtesy of the College of Chemistry, University of California, Berkeley.

Did Lewis deserve a Nobel award for his work on heavy water? Jacob Bigeleisen, Lewis's student and research associate from 1941 to 1943, who spent his career in isotope chemistry, believes not:

> If the 1934 Nobel Prize was to be awarded for the discovery of deuterium, the record is clear—it goes and went to Harold Urey. . . . Lewis jumped on the bandwagon. He had a large supply of enriched water from the electrolytic cells from which Giauque generated hydrogen gas for liquefaction. Should the scope of the prize have been enlarged to include work beyond the discovery [of deuterium], Lewis would not have been the only one to consider. Of course there was the work by Urey beyond the discovery. There was a legion of people who published on the properties of deuterium by 1934. What was unique about Lewis was he published 26 papers in one year! But he did this because he had a supply based on the work of Washburn and Urey and he enlisted the "collaboration" of everyone in the [Berkeley chemistry] department who could make measurements relative to the difference between protium [the normal hydrogen isotope] and deuterium.

> In my view, Lewis did not warrant a share of the 1934 Nobel Prize even if [the subject of the award were] more broadly defined than the discovery of a heavy isotope of hydrogen.[45]

The award to Urey made it clear that how much the prize mattered to Lewis: he stopped all work on deuterium. He published nothing at all for eighteen months, the longest publication break since his time as an instructor at Harvard. Seven years later, when Bigeleisen found the samples and notebooks from Lewis's earlier research with Macdonald on lithium and oxygen isotopes, Lewis told Bigeleisen to destroy them. (Bigeleisen did as instructed, but not before reading the notebooks.)[46] The fractionation columns, seventy feet high, that Randall had built sat in front of the Berkeley chemistry building, still producing heavy water until they were dismantled years later.

Urey was the founding editor of the *Journal of Chemical Physics*, which began publication in 1933. Lewis had been supportive of the new journal, which was seen by Lewis and others as a remedy for the deficiencies of Wilder Bancroft's *Journal of Physical Chemistry*,[47] which Lewis had shunned since Bancroft had snubbed him in a review of his dissertation thirty years earlier (see chapter 2). Lewis published a paper titled "The Chemical Bond"[48] in the first (January 1933) issue of Urey's new journal and published his first comprehensive report on concentration of heavy water in the June issue that same year.[49] After Urey's Nobel Prize, Lewis did not submit another paper to the journal, although Lewis's colleage Melvin Calvin and Lewis's research assistant Michael Kasha submitted a paper there after Lewis's death in 1946 with Lewis's name as first author.[50] A letter dated 1 June 1933, which discusses the name for the new hydrogen isotope, is the last correspondence from Lewis to Urey that has been preserved in the Lewis papers; there is no subsequent letter or telegram from Lewis congratulating Urey on his Nobel Prize. There are letters preserved from Urey to Lewis dating after Urey's prize, in which Urey informs Lewis of various symposia or exhibits on heavy water or deuterium. Urey's language suggests that Lewis's replies, if any, were brusque refusals to participate.[51] Urey, along with Langmuir, Walther Nernst, and Theodore Richards, had joined those from whom Lewis had estranged himself, but the feeling was not reciprocated by Urey. A few months after Lewis's death, Urey asked Seaborg for a picture of Lewis.[52]

Lewis Drops Out

Urey's Nobel Prize was announced on 15 November 1934. Six weeks later, on 29 December, Lewis abruptly resigned from the National Academy of Sciences. The entire text of his letter read, "It is with regret that, after careful consideration, I have decided to withdraw from membership in the National

Academy. May I ask that this letter be taken as a formal resignation?"[53] Membership in the National Academy is considered a great honor. Lewis had been elected a member in 1913, and his resignation was a shock to the entire American scientific community. The academy organized Lewis's friends to ask his reasons and to persuade him to change his mind, but he refused to give any explanation. There has been speculation that he saw the academy as an East Coast old boys' club because the candidates that he had nominated had been rejected: Edwin Bidwell Wilson's letter to Lewis after his resignation seeks to assure him that William Giauque, a Berkeley faculty member, would be elected soon,[54] and Lewis's son believes that Lewis was upset because George Ernest Gibson, another prominent Berkeley faculty member, had not been elected.[55] While Lewis may have had such reasons, the timing of his resignation suggests that his disappointment over the 1934 Nobel Prize was the proximate cause. When Lewis felt unappreciated or insulted, his reaction was to leave the scene, as he had done at Harvard years earlier.

After an eighteen-month publication break following Urey's Nobel Prize, Lewis picked himself up once more and got back to work, starting with some unproductive and disappointing work on neutrons in 1936 and 1937. Lewis theorized that slow neutrons were captured by an interaction between the spin of the nucleus and the spin of the neutron and went into quantized orbits outside of the nucleus; from there they then collapsed into the interior of the nucleus in a secondary process.[56] He believed that he had succeeded in diffracting neutrons.[57] This was to be Lewis's last attempt at physics, and it was not well received by the physicists. Physicist Hans Bethe reviewed one of Lewis's manuscripts and wrote, "I think it is an extremely instructive example of the dangers of purely qualitative arguments."[58] One of the researchers in Lawrence's laboratory wrote of Lewis's work, "The effect is so feeble, and the instruments so barbaric (he doesn't want to hear about counters) that no one believes him here."[59] Seaborg, then a graduate student at Berkeley working on nuclear chemistry under George Gibson, worked up the nerve to tell Lewis where he thought his work on neutrons was going wrong. From Seaborg's journal on 20 April 1937:

> This afternoon Professor Lewis called me into his laboratory... where he is working with Phil Schutz studying what they believe is a new phenomenon—the diffraction of slow neutrons and their focusing through the use of paraffin lenses.... He asked me my opinion of their work; I gulped, looked at him and said, "I think it is wrong." He then asked how I explained their observations if they are not due to "neutron diffraction." I said that I believe their maxima in the slow neutron activations... are spurious and caused by scattering of neutrons and not [by] focusing with their paraffin lenses. Lewis thanked me; I left the room with some misgivings.[60]

And a journal entry a month later: "Professor Lewis and Phil Schutz, perhaps as the result of my conversation with Lewis last month, repeated their slow neutron focusing experiments with their setup outdoors in order to minimize the spurious scattering of neutrons from extraneous material. Today they sent [a] retraction to the Editors of the *Physical Review*."[61] Lewis was impressed enough with Seaborg to offer him a position as his research assistant upon his graduation in 1937, a job that paid $1,800 a year and would allow Seaborg to stay in Berkeley. Seaborg accepted enthusiastically.[62]

Glenn Seaborg

Glenn Seaborg was born in 1912 to Swedish parents who had immigrated to Ishpeming, Michigan. His father was a mechanic, and the family lived in an all-Swedish neighborhood. He was so shy as a child that his mother arranged with his teachers for him to go to the bathroom without asking permission because he was too embarrassed to raise his hand. But shyness did not seem to be a problem once he reached maturity. These are the opening paragraphs from his autobiographical Web site:

> When the phone rang in the Radiation Laboratory one January afternoon I recognized the Boston accent from the newscasts. President-elect John F. Kennedy wanted to know: Would I join his administration as head of the Atomic Energy Commission?
>
> I said that I needed to think it through.
>
> Take your time, he replied, I'll call you tomorrow morning.
>
> Was I ready to leave the laboratory for government—and national politics? Could I adapt my scientific skills to managing a bureaucracy? Was I ready for the burdens of overseeing weapons testing and development, nurturing a fledgling nuclear power industry, and encouraging international cooperation?[63]

When Seaborg was ten years old, his family moved to the suburbs of Los Angeles because his mother wanted greater opportunities for the children than Michigan offered. Once in California, his Swedish-speaking father never managed to find permanent employment, and the family was pressed financially. Seaborg became interested in science in his junior year of high school and enrolled as an undergraduate at UCLA in 1929. He worked his way through school as a stevedore and enrolled in the graduate chemistry program at Berkeley, where he received his Ph.D. under George Gibson in 1937. As Lewis's research assistant, Seaborg would work on what would become one of Lewis's most lasting achievements—a reformulation of the theory of acids and bases.

Acids and Bases

In his 1923 monograph *Valence and the Structure of Atoms and Molecules*, Lewis had suggested almost in passing that the electron pair could be used to define a theory of acids and bases. He returned to this idea in 1938 after his work on heavy water. Acidity is one of the oldest concepts in chemistry, dating back to alchemy. Before the 1880s, the definitions of acids and bases were empirical—acids were water solutions that tasted sour, turned litmus red, etched or dissolved some metals, and neutralized bases to form salts; similarly, bases were water solutions that felt slippery, tasted bitter, turned litmus blue, etched glass, and neutralized acids to form salts. One of the puzzling things about acids and bases had been that the heat of neutralization of strong acids with strong bases seemed always to be the same. For example, the reaction of hydrochloric acid with sodium hydroxide evolved the same measured heat per equivalent as did the reaction of sulfuric acid with potassium hydroxide. Svante Arrhenius's 1887 dissociation theory was able to explain this: strong acids and bases were dissociated into ions. Acids dissociated to a positively charged hydrogen ion (H^+, or a proton) and a negatively charged counterion, and bases to a negatively charged hydroxyl ion (OH^-) and a positively charged counterion. Arrhenius proposed that the reaction that occurs in the neutralization of a strong acid and strong base was always the same—a reaction between the hydrogen ion and the hydroxyl ion:

$$H^+ + OH^- \rightarrow H_2O$$

The counterions (chloride ion and sodium ion, for example, in the reaction of hydrochloric acid with sodium hydroxide) remain solvated after the neutralization and do not contribute to the heat of reaction. So Arrhenius's dissociation theory provided a theoretical foundation for acids and bases: an acid is a substance that dissolves in water to provide hydrogen ions, and a base is a substance that dissolves in water to provide hydroxyl ions.

Arrhenius's theory was sufficient to explain acid–base reactions in water solution. But what of reactions such as that in the equation below, where a proton in an acid reacts with pure ammonia, and no water or hydroxyl ions are present?

$$H^+ + NH_3 \rightarrow NH_4^+$$

In 1923, Johannes Brønsted and Thomas Lowry defined a new concept of acids and bases that included bases such as ammonia. According to Brønsted and Lowry, acids were substances that donated protons, bases were substances that accepted them, and the above reaction was an acid–base reaction.

When Lewis published *Valence* in 1923 (at about the same time as Brønsted and Lowry published their theory), he suggested that his concept of the electron pair could provide a framework for a theory of acids and bases in completely nonprotonic solvents. In terms of Lewis's electron-dot structures, the Arrhenius and Brønsted-Lowry neutralization reactions show some similarities:

$$\mathrm{H{:}\underset{..}{\ddot{O}}{:}^{-}} + \mathrm{H^{+}} \longrightarrow \mathrm{H{:}\underset{..}{\ddot{O}}{:}H}$$

$$\begin{array}{c}\mathrm{H}\\ \mathrm{H{:}\ddot{N}{:}}\\ \ddot{\mathrm{H}}\end{array} + \mathrm{H^{+}} \longrightarrow \begin{array}{c}\mathrm{H_{+}}\\ \mathrm{H{:}\ddot{N}{:}H}\\ \ddot{\mathrm{H}}\end{array}$$

In each of the above reactions, the base (the molecule on the left) can be thought of as donating a lone (unbonded) electron pair to a proton, and the electron pair becomes shared between the base and the proton in the salt (the molecule on the right). Lewis pointed out that this concept could be extended to other reactions that did not involve protons at all, such as the reaction between ammonia and boron trifluoride (the additional electrons surrounding the fluorine atoms have been omitted in the diagram below for clarity):

$$\begin{array}{c}\mathrm{H}\\ \mathrm{H{:}\ddot{N}{:}}\\ \ddot{\mathrm{H}}\end{array} + \begin{array}{c}\mathrm{F}\\ \mathrm{\ddot{B}{:}F}\\ \ddot{\mathrm{F}}\end{array} \longrightarrow \begin{array}{c}\mathrm{H\ F}\\ \mathrm{H{:}\ddot{N}{:}\ddot{B}{:}F}\\ \mathrm{\ddot{H}\ \ddot{F}}\end{array}$$

Lewis saw this as an acid–base reaction, as well. To Lewis, "*a basic substance is one which has a lone pair of electrons* [an electron pair that is not part of a bond] *which may be used to complete the stable group of another atom,* and...*an acid is one which can employ a lone pair from another molecule* in completing the stable group of one of its own atoms."[64] In the example above, ammonia was what would later be known as a Lewis base, and boron trifluoride a Lewis acid.

In May 1938, Lewis agreed to accept an honorary degree and to give a talk at the Franklin Institute in Philadelphia, when the large new building there was to be dedicated. He decided to make the subject of his talk his 1923 ideas for redefining acids and bases, and he began looking for experimental evidence to support his ideas. Most people who have taken a chemistry course remember using two burettes to titrate a water-based solution of an acid against a solution of a base, and that a single drop changed the color of the indicator phenolphthalein from clear to red. Lewis began examining these same sorts of reactions in nonprotonic solvents (solvents such as acetone where the solvent itself provides no protons). He wrote:

> I had hoped by these experiments to show the changes in indicator color to be as sharp and the titrations as precise as in corresponding neutralizations in water. I was, however, unprepared for the really astonishing resemblances between the effects produced by H-acids and the other more generalized acids. I venture to say that any one who has carried out for himself a few of these titrations will never again think of acids in the present restricted way.[65]

Lewis showed that his generalized acids and bases behaved almost exactly as did conventional acids and bases. Even in the absence of water or another protonic solvent, the process of neutralization was rapid, with no activation energy; a stronger acid or base would displace a weaker acid or base from its compounds; acids and bases could be titrated against each other, often employing the same colorimetric indicators as in protonic solvents; and both acids and bases played an important role in promoting chemical reactions by acting as catalysts. It was this last point that was to prove so useful to organic chemists, who usually work with nonprotonic solvents. Lewis's theory gave organic chemists a framework for explaining catalytic effects and could suggest alternative catalysts: if one Lewis acid or base worked to catalyze a reaction, try another and see if it worked better. Lewis's acid–base theory did not receive widespread recognition at first, but in 1947 W.F. Luder and Saverio Zuffanti published *The Electronic Theory of Acids and Bases*, explaining the applicability of Lewis's theory to much of chemistry.[66] Lewis read the manuscript of this book in 1946 and provided the authors his suggestions and criticisms shortly before his death.[67] They dedicated the book to Lewis.

Lewis's acid–base experiments were simple and inexpensive and took only a few months. Seaborg recalls:

> I was immediately struck by the combination of simplicity and power in the Lewis research style, and this impression grew during the entire period of my work with him. He disdained complex apparatus and measurements. He reveled in uncomplicated but highly meaningful experiments. And he had the capability to deduce a maximum of information, including equilibrium and heat of activation data, from our elementary experiments. I never ceased to marvel at his reasoning power and ability to plan the next logical step toward our goal. I learned from him habits of thought that were to aid continuously my subsequent scientific career.[68]

Bigeleisen made similar comments about Lewis: "Lewis was a very original thinker, with a fundamental grasp of nature. He was very methodical and clever in planning his research and designed experiments very economically.

A Lewis experiment was designed so that a positive outcome would confirm a theory and so that a negative outcome would suggest a pathway for future work. He was not interested in designing an elaborate experiment for its own sake."[69] Today, Lewis acids and bases are one of the dominant motifs in organic chemistry (see, for example, *Lewis Acids in Organic Synthesis*,[70] a 1,200-page two-volume publication).

Seaborg Joins the Berkeley Faculty, and Work on Plutonium Begins

In June 1939 Lewis invited Seaborg to join the Berkeley faculty as an instructor, and in August 1940 Lewis told him that he would be promoted to assistant professor. By this time Seaborg was working on the creation of new elements. In 1939, the periodic table ended with uranium, with atomic number 92. In 1940, Edwin McMillan and Phillip Abelson from the Berkeley physics department produced neptunium (Np, atomic number 93) by irradiating the nonfissionable isotope of uranium U^{238} with neutrons to produce a highly radioactive isotope of uranium, U^{239}, which had a half-life of only twenty-three minutes. This quickly emitted a beta particle (an electron) to gain a nuclear charge and produce Np^{239}, also highly radioactive with a half-life of twenty-three days. McMillan suspected that this then decayed by beta emission to a radioactively stable isotope of an element with atomic number 94 and atomic weight 239, now called plutonium (Pu^{239}), but he did not have time to confirm this before leaving for MIT for war work. Seaborg wrote McMillan that he, Joe Kennedy (an instructor at Berkeley), and Art Wahl (Seaborg's student) would be happy to collaborate and extend his work, to which McMillan agreed. Seaborg and his collaborators used deuterons rather than neutrons to bombard U^{238}; they succeeded in producing a different highly radioactive isotope of neptunium (later shown to be Np^{238}), which they suspected decayed to a different isotope of plutonium (later confirmed as Pu^{238}). Wahl managed to chemically separate neptunium and confirm that irradiation by neutrons and deuterons had produced different isotopes. He eventually managed to separate plutonium based on a method suggested by Wendell Latimer (who did not have the proper security clearance to even know of the work).[71] Seaborg's diary entry for 10 January 1941 says, "I met with Professor Lewis and told him about our probable discovery of an alpha-particle emitting isotope of element 94 [Pu^{238}]. He was both interested and intrigued."[72] It was not until late February 1941 that the discovery was confirmed. Seaborg wrote McMillan, "These results came just in time to me (!) in the 94^{239} project which I (!) am doing for the Uranium Committee."[73] The Seaborg-Wahl-Kennedy crew, augmented by Emilio Segrè, an Italian-Jewish physicist whom Lawrence had welcomed to his laboratory, bombarded uranium with neutrons in Lawrence's cyclotron and produced enough plutonium for experiments to demonstrate that it was fissionable—that it could support a nuclear chain reaction even more readily than uranium.

This was the eve of World War II, and the government was becoming aware that this sort of work might have military applications. Seaborg and his collaborators wrote up the plutonium experiments and submitted them to *Physical Review*, but they were not published until after the war for national security reasons.[74] Some elementary security precautions were imposed within the Berkeley chemistry department, but security does not seem to have been taken too seriously. Seaborg's diary for 1 March 1941 reports, "I saw Professor Lewis and told him about our success in oxidizing our element 94 isotope. Although Professor Lewis and Professor Latimer are not, strictly speaking, cleared for such information, I am keeping both of them informed about our work—I am confident that they will treat it as secret."[75] Lewis also served a more practical purpose. Seaborg later noted that "I presented [one of the original plutonium samples] to the Smithsonian Institution at a public ceremony on March 28, 1966, the 25th Anniversary of the first date on which the slow neutron fission of $^{239}94$ [plutonium] was demonstrated. It had been stored in one of G. N. Lewis's cigar boxes most of the intervening years."[76] Seaborg's diary of 3 June 1941 reads, "I saw Professor Lewis today and told him about our demonstration of the fissionability of $^{239}94$ [plutonium]."[77] Lewis and Seaborg were named as principal investigators on a government contract to search for naturally occurring isotopes of neptunium and plutonium, a contract that would expire in December 1941.[78]

Naturally occurring uranium consists of 99.3% ^{238}U and only 0.7% ^{235}U, the isotope that is fissionable. If a uranium bomb were to be constructed, the fissionable ^{235}U isotope would need to be concentrated. This would not be easy, as the two isotopes are almost identical chemically and would need to be separated based on physical properties—the different rates of diffusion of their compounds, for example. But if the fissionable isotope of uranium *could* be concentrated in the midst of the surrounding inert isotope, it could be used to start a controlled chain reaction and produce neutrons to transform the remaining inert uranium into plutonium. This is the way that a breeder nuclear reactor works today—plutonium, which can be used as fuel in a reactor, is a by-product made during the normal operation of a uranium reactor. Once it was decided to build atomic bombs, it made sense to build both a uranium and a plutonium bomb simultaneously.[79]

Seaborg and the Manhattan Project

President Franklin D. Roosevelt gave the order to proceed with development of the bomb about 9 March 1942, and the Manhattan Project was born. The physicist Arthur Compton was put in charge of one of its most important branches, which had the cover name "Metallurgical Laboratory" and came to be known as the Met Lab. Located at the University of Chicago, the Met Lab had the charter of producing and purifying plutonium for a bomb. On 19 April 1942, Seaborg took over leadership of the purification of plutonium at Compton's request.[80] It was his thirtieth birthday.

Separating the isotopes of uranium could not be done chemically, but plutonium purification was a relatively simple chemical problem. It was complicated, however, by the radioactive mess in which plutonium was embedded after its production in a nuclear pile. Seaborg later wrote:

> The problem of separating the new element plutonium from uranium and fission products might not at first seem difficult, for it was primarily a chemical problem. However in many ways it differed from ordinary chemical problems....From the beginning, our limited time seemed the most nearly insurmountable difficulty. It was impossible to complete the design and testing of the process before it had to be put into operation. Even a simple chemical process usually requires a much longer time to place in large-scale operation than did the plutonium separation process, although the latter cannot be regarded as either simple or short.
>
> The problem that had to be solved during fall 1942 was to develop a separation process that would [separate] plutonium in high yield and purity from many tons of uranium in which the plutonium would be present....At the same time, the radioactive fission products produced along with the plutonium...had to be separated from the...final product from the process. This requirement made it safe to handle the plutonium.*...Thus a unique feature of the process was the necessity of completely separating a wide variety of elements from the final product and doing so by remote control behind large amounts of shielding to protect operating personnel from the radiation....[This] made it necessary to thoroughly test the process in advance to minimize the possibility of errors in the design of the equipment....In all of these considerations, it was obviously desirable to keep the process to be operated as short and simple as possible. Similarly, the process from the standard of plant design should consist of steps rendering the same sort of equipment rather than steps that were so fundamentally different as to require many different types of equipment. At the same time, it seemed advisable to design the process and the equipment in such a way as to facilitate changes in case of failure.[81]

Seaborg was put in charge of Group C-1 of the Met Lab, which was eventually given the codename "Separation Studies and Basic Chemistry of the Heavy Elements." Throughout the project's lifetime, Seaborg kept an active group working on plutonium purification at Berkeley's chemistry department in addition to

---

* Plutonium is an alpha-particle emitter and is not dangerous unless it is ingested or inhaled or enters the body through a wound.

his larger group in Chicago. Michael Kasha worked on plutonium purification while at Berkeley and remembers that the safety standards were minimal:

> There were only about 14 graduate students [in Berkeley's chemistry department] during the war. I got a 1-A [draft] classification several times, and if I had worked just with Lewis, it wouldn't have counted for a deferment, but I was working on the separation of plutonium. I got several pure plugs of plutonium. We were working with hot concentrated perchloric acid—explosive—and with nitrates, another explosive. We were working with concentrated solutions of plutonium, which were beautifully colored. We had no rubber gloves and used wooden lab benches. We had a thing called a "poppy" constantly taking readings—filter paper and a Geiger counter. It would go "pop, pop, pop" and then suddenly go "vrrrm".... We'd all run out of the room. Something had leaked. But none of us suffered any apparent ill effects. There was one death—a graduate student who was doing dry chemistry breathed plutonium dust. He died in a month. I had one accident where I cut myself and had to go to the hospital. We were not allowed to say anything. They removed a sliver of flesh, but it's probable that some radioactive material diffused into my bloodstream. Robert Connick ran the project—a wonderful person to work for—and we were not allowed to leave the lab if we were still hot. So at the end of the day he would check our hands with the Geiger counter. We didn't wear gloves. He devised a method. We had chromic acid cleaning solution around. We dipped our hands in that. When we pulled them out, we felt nothing. But when we put them under the tap, the heat of dilution was murder. Once a month we had a safety meeting, but it was low on the agenda. We had urine tests, but we didn't carry radiation badges. We mouth-pipetted concentrated plutonium solutions, but we never got a plug. We worked with concentrated perchloric acid, and if you put anything in that, it's explosive, but we knew the right amounts of nitrates to add. We knew what we were doing. The really dangerous work was done at Los Alamos.[82]

In 1941, Lewis was forced to retire from his positions as dean and department chairman because he had reached the mandatory age. In 1943, Seaborg recommended to Compton that Lewis be hired as a consultant to the Manhattan Project,[83] but nothing came of this—perhaps Compton or others did not want to deal with Lewis's sometimes difficult personality. Professors Latimer, now dean, and Emmon Eastman were put in charge of Berkeley's government contract on plutonium.[84] Although they did not formally report to Seaborg, these two were now effectively working for a thirty-year-old who had been a Berkeley graduate student only a few years earlier.

When Seaborg started on the plutonium purification project, the only source of plutonium was from bombardment of uranium in a cyclotron—the uranium nuclear pile that would produce plutonium in quantity did not yet exist. Only two milligrams of plutonium were made in a cyclotron, but this turned out to be enough to plan the scale-up to industrial quantities. When the Hanford nuclear plant was built in Washington State, the process that Seaborg's group developed was successfully scaled up by a factor of a billion, which Seaborg notes was

> surely the greatest scale-up factor ever attempted. . . . The chemical plants were massive structures ingeniously engineered to fit the grave problems inherent in handling the extremely high levels of radioactivity. It is self-evident that no one ever saw the plutonium entering the plant. It is also true that no one saw it until just before it finally emerged as a relatively pure compound. In the meantime, it had passed through a maze of reaction vessels via thousands of feet of piping with only instruments and an occasional sampling to chart its progress.[85]

Environmental Problems

Seaborg's praise for the Hanford plant omits a few things. In retrospect, there were serious environmental problems at Hanford, problems that were well known even to the public by the time Seaborg wrote the above quote. He might have chosen to discuss them, even though decisions as to plant and process design were not all under his control. While Hanford was quite successful in purifying plutonium, it has been an environmental disaster. The bismuth phosphate method that was used for plutonium purification had a yield of only about 90%, and much of the remaining plutonium and the waste fission products ended up in the Columbia River, the groundwater, and the air. While this may have been understandable under the pressure of development during World War II, it has caused long-term problems. There were massive releases of radiation at Hanford for years, and Seaborg was involved in management of plutonium production well into the 1970s. From the Washington State Department of Health:

> Because of the secrecy surrounding nuclear weapons production, the public did not know much about Hanford's operational details until 1986. By February of that year, citizen pressure had forced the U.S. Department of Energy to release 19,000 pages of Hanford historical documents that had been previously unavailable to the public. These pages revealed there had been huge releases of radioactive materials into the environment that contaminated the Columbia River and more than 75,000 square miles of land. . . .
>
> The documents revealed that Hanford was key to U.S. participation in the nuclear arms race. In 1943 the federal government had selected Hanford as the site for the world's first large-scale nuclear production plant. Hanford produced the plutonium for the bomb dropped on Nagasaki, Japan, during

> World War II. About half of all U.S. nuclear weapons were made with plutonium from Hanford. Hanford officials cited national security considerations as a justification for the secrecy.
>
> Contained in the documents were descriptions of how Hanford operations had released radioactive materials. The plutonium was produced in nuclear reactors along the Columbia River. The reactors needed large amounts of water from the river for cooling. Materials in the river water were made radioactive when they passed through the reactors. After passing through the reactors, the water and the radioactive materials it carried were put back in the river. The radiation contaminated the water and aquatic animals downstream as far as Pacific oyster beds along the Washington and Oregon coasts. The highest releases to the Columbia were from 1955 to 1965.
>
> After the plutonium was removed from the reactors, it had to be separated and purified for use in nuclear weapons. Separating the plutonium resulted in radiation being released into the air. Winds carried Hanford's airborne radiation throughout eastern Washington, northeastern Oregon, northern Idaho and into Montana and Canada. Food grown on contaminated fields, and milk cows grazing there, transferred the radiation to people who ate the food and drank the milk. The years of highest releases to the air were 1944 through 1951, with 1945 being the largest.[86]

The Smyth report, the official history of the Manhattan Project published in 1945 immediately after the end of the war,* had this to say about to say about environmental concerns at Hanford:

> [T]he really troublesome materials are the fission products, i.e., the major fragments into which uranium is split by fission. The fission products are very radioactive and include some thirty elements. Among them are radioactive xenon and radioactive iodine. These are released in considerable quantity... and must be disposed of with special care. High stacks must be built which will carry off these gases along with the acid fumes..., and it must be established that the mixing of the radioactive gases with the atmosphere will not endanger the surrounding territory. (As in all other matters of health, the tolerance standards that were set and met were so rigid as to leave not the slightest probability of danger to the health of the community or operating personnel.)
>
> Most of the other fission products can be retained in solution but must eventually be disposed of. Of course, possible pollution of the adjacent river must be considered. (In fact, the standards of safety set and met with regard to river pollution were so strict that neither people nor fish down the river can possibly be affected.)[87]

---

* The Smyth report is a remarkable document, providing a very thorough account of the scientific and engineering decisions made in the development of the bomb. It almost certainly has been invaluable to other nations' bomb projects.

The Bomb

As the war and the Manhattan Project drew to a close, the subject of patent rights for plutonium became important, especially for the work done on the discovery of plutonium before the project got under way. The government was pushing the inventors and the University of California to give up their patent claims. In view of the projected demand for peacetime nuclear energy, there might be a lot of money involved, and Seaborg was not rolling over; his patent rights were a continuing subject of concern in his journal entries of the time. The negotiations and litigation dragged on well into the 1950s and involved a number of scientists.[88]

In June 1945, a month after the end of the war against Germany, both the uranium and the plutonium bombs were ready. The bomb had been developed at the urging of Einstein and others because of fear of a German atomic bomb, and many exiles from Europe had taken leadership roles in the Manhattan Project. Now Germany was out of the war, and the only possible target was Japan. James Franck, one of the Jewish exiles from Germany, chaired the Manhattan Project's ad hoc Committee on Political and Social Problems, which recommended demonstrating the bomb rather than using it against Japan. Seaborg was a member, and the committee issued its report on 11 June. Seaborg quoted the report's summary in his journal for that day:*

> The development of nuclear power not only constitutes an important addition to the technological and military power of the United States but also creates grave political and economic problems for the future of this country.
>
> Nuclear bombs cannot possibly remain a "secret weapon" at the exclusive disposal of this country for more than a few years. The scientific facts on which their construction is based are well known to scientists of other countries. Unless an effective international control of nuclear explosives is instituted, a race for nuclear armaments is certain to ensue following the first revelation of our possession of nuclear weapons to the world. Within ten years other countries may have bombs, each of which, weighing less that a ton, could destroy an urban area of more than ten square miles. In the war to which such an armaments race is likely to lead, the United States, with its agglomeration of population and industry in comparatively few metropolitan districts, will be at a disadvantage compared to nations whose population and industry are scattered over large areas.

---

* The quotations from Seaborg's wartime journals (notes 89, 90, 91, and 93 in this chapter) are postwar reconstructions by Seaborg and his staff from his contemporaneous journals and his correspondence. The contemporaneous journals themselves are retained by the United States government on the grounds that they contain classified material.

We believe that these considerations make the use of nuclear bombs for an early unannounced attack on Japan unadvisable. If the United States were to be the first to release this new means of indiscriminate destruction upon mankind, she would sacrifice public support throughout the world, precipitate the race for armaments, and prejudice the possibility of reaching an international agreement on the future control of such weapons.

Much more favorable conditions for the eventual achievements of such an agreement could be created if nuclear bombs were first revealed to the world by a demonstration in an appropriately selected uninhabited area.[89]

Seaborg's journal entry of 13 June shows that he agreed with the committee's suggestion, but President Harry S. Truman did not. Fifty years after the war, Seaborg seemed to have modified his conclusions. He said, "The Japanese, after all, started it all with Pearl Harbor.... So you can't really regard this as the U.S. attacking Japan. The U.S. was in a sense defending itself, trying to bring the war to an end as soon as possible." Seaborg talked about the Franck report's recommendation that the bomb be demonstrated to the Japanese and others, saying, "That report received quite a bit of attention, but somehow the other side prevailed." He then pointed out that the United States had only enough uranium and plutonium for three bombs: one to test, and two to use: "That's one of the reasons they didn't take our recommendation for a demonstration first, because if that demonstration had failed, they wouldn't have had enough plutonium for another bomb for a while yet, and that could have led to a continuation of the war."[90]

The uranium and plutonium bombs had different triggering mechanisms. The uranium bomb used a simple trigger: one part of the uranium was fired into the other, bringing the bomb to critical mass. The plutonium bomb used conventional explosives to compress it to critical mass through an implosion, and this needed to be tested. The "Trinity" test on 16 July 1945 was a successful test of the plutonium bomb. Seaborg had spent the last three years working on the chemistry of plutonium, and one might have thought he would have been emotionally involved in the outcome of the test. He was based in Chicago and did not attend the New Mexico test, but Joe Kennedy, with whom he shared credit for the discovery of plutonium at Berkeley, was stationed in Los Alamos and sent him an account. Seaborg's journal entry for that day shows him to be at least as concerned with his negotiations with Berkeley and Chicago for an academic position after the war as he was with the bomb test:

At 5:30 AM mountain time, the first explosion of an atomic bomb occurred at Alamogordo, New Mexico, about 100 miles southwest of Albuquerque. Later I read the account of that event that Joe Kennedy wrote in his diary. [Seaborg here inserted Kennedy's account of the test.]

I received and read Latimer's reply to my letter of July 9. He says, "You have put your finger on our most important problems. In general, the

University departments of chemistry and physics have not kept pace with the tremendous developments which have occurred in these subjects. Now on top of this we have the developments in the nuclear field. Chancellor Hutchins at Chicago seems to have grasped the situation. However, when he is willing to increase his offer to you from $8,000 to $10,000 and wants to add a dozen new assistant and associate professors in chemistry, I can't help but be encouraged. I am confident that the logic of the situation is such that the proper presentation will achieve the same results here... He adds that he thinks much of the present project properly belongs to the industrial field. Latimer also mentions that he will be in Tennessee on July 23 and hopes to see me then.[91]

The first American bomb, dropped on Hiroshima on 6 August 1945, was a uranium bomb, but the second bomb, dropped on Nagasaki on 9 August, used Seaborg's plutonium. Seaborg, of course, had no control over the decisions as to whether, where, and when the bombs would be used. There has been a great deal of discussion as to whether these bombings were justified and, in particular, whether there should have been a longer delay between the Hiroshima and Nagasaki bombs to give the Japanese more time to consider surrender. But whatever the merits of these arguments, approximately 70,000 people died either at once or from the immediate aftereffects of the Nagasaki bomb, with many more injured or to eventually suffer radiation poisoning.[92] Most of these were civilians. One can imagine that Seaborg might have had any of several reactions to the news of the bomb—triumph, doubt, regret, or horror. Seaborg's journals are extremely detailed (to the point of recording what he had for lunch) and, as published, fill entire bookshelves. This, however, is his entire entry for 9 August 1945:

I am still in Berkeley staying with the Calvins.

A plutonium bomb was dropped on Nagasaki shortly before 11:00 AM Japanese time, August 9 (yesterday our time). The announcement was the top headline in this morning's paper here.

I had a meeting with Provost Deutsch (President Sproul is out of town) [both of Berkeley] to tell him about my offer of a professorship at the University of Chicago (at $10,000 per year) with the additional inducement of other academic staff as well as graduate students in the Department of Chemistry to work with me. Deutsch was encouraging that the University of California's offer could be improved to reduce the discrepancy between it and that from the University of Chicago. Because I held the rank of Assistant Professor, in which I served only one semester before going on leave to work on the Manhattan Project, it is not surprising that the Administration at Berkeley finds it difficult to go much beyond the direct promotion to Full Professor that they already provided me last month.

I talked at the regular weekly meeting of Latimer's group.

Figure 8-2. Seaborg listing the elements in whose discovery he had participated—plutonium, americium, curium, berkelium, and californium—in 1951, the year he won the Nobel Prize in Chemistry. Courtesy of the College of Chemistry, University of California, Berkeley.

> The Soviets have declared war on Japan and have made their first attack on the eastern border of Manchuria.[93]

In his negotiations with the University of California, Seaborg got everything he asked for—full professorship, salary, staff, facilities, and budget. He later served as associate director of Berkeley's Lawrence Radiation Laboratory, as chancellor at Berkeley, as chairman of the Atomic Energy Commission, and as president of the American Chemical Society. He won the Nobel Prize in 1951, and was the only then-living scientist to have an element named for him—seaborgium, element 106. He worked with every president from Truman to Richard M. Nixon, and he gave his autobiography the title *A Chemist in the White House.*

Urey and the Manhattan Project

Harold Urey was also a leader in the Manhattan Project. Unlike Seaborg, Urey had been politically active long before the war. As early as 1932, Urey had been staunchly anti-Nazi and antifascist and had been an advocate of world government. As the European scientists fled Mussolini and Hitler, Urey and his wife welcomed them to America. Like Einstein, Urey was fearful that Hitler would be the first to develop the atomic bomb and supported an American bomb program. As World War II approached, Urey took responsibility for separation of uranium isotopes with essentially the same goal as Seaborg's parallel program for plutonium—to concentrate the radioactive isotope in a form where it could be used in a bomb. Under Urey's direction, a number of approaches were examined for separating the isotopes. In May 1942, Urey advised using countercurrent centrifuges (the preferred method for most nations producing enriched uranium today), which he had invented in 1938, but he was overruled. The uranium separation program seemed to have lacked effective management. Rather than choose one separation method and make it work, the project's management pursued them all. At the end of 1943, more than 700 people worked for Urey on the gaseous diffusion process, enriched uranium which provided enriched uranium that was fed to Lawrence's electromagnetic process for final enrichment for the Hiroshima bomb.[94] Ten thousand workers began to build a thermal diffusion plant at Oak Ridge, Tennessee, but that process, headed by Abelson, was quite inefficient, although it did provide some low-enriched feedstock for the electromagnetic process as well.

Urey later said of this time, "I was most unhappy during the war. I had bosses in Washington who didn't like me, and I had people working for me who didn't like me. Imagine a more miserable situation—where you can't resign, but nobody wants you around! When the war was over, I got out. I was very close to a nervous breakdown during the war."[95] His biographers wrote, "Early in 1944 . . . it was clear to Urey that the [thermal] diffusion plant would have little relevance to the war effort. . . . Urey remained nominal director of the Columbia [Manhattan Project] laboratories until 1945, but his heart was not in it. From that time forward his energies were directed to the control of atomic energy, not its applications."[96]

Urey told Mildred Cohn, his former student, that he was so upset by the use of the bomb against Japan that he would never again work on military research.[97] After World War II, he served as vice chairman of the Emergency Committee of Atomic Scientists, of which Einstein was chairman, opposing the use of atomic weapons and advocating peaceful use of atomic energy. In 1948, the chairman of the House Un-American Activities Committee, J. Parnell Thomas, attacked Edward Condon, who had played a leading role in the Manhattan Project and was then head of the National Bureau of Standards, saying he was "one of the weakest links in our atomic security."

Urey and Einstein supported Condon, saying such accusations were "a disservice to the interests of the United States."[98] When Julius and Ethel Rosenberg were sentenced to death for atomic espionage, Urey sent a telegram to President Dwight D. Eisenhower, pleading for him to block their execution; Urey later said, "I doubt seriously that justice had been done."[99] Urey was treated as a Communist fellow traveler (someone who was not a Communist but nonetheless shared views held by Communists), and in 1953 was called to testify before the Army-McCarthy hearings, the nationally televised congressional inquiry into charges by the U.S. Army against McCarthy and his chief counsel, and vice versa.

9

The Secret of Life

Pauling, Wrinch, and Langmuir

In early 1938, Linus Pauling became involved in a scientific controversy with Dorothy Wrinch, an English mathematician turned biologist whose "cyclol theory" of protein structure was opposed to his own ideas on the subject. Warren Weaver, who was managing the natural sciences grants for the Rockefeller Foundation, had awarded grants to both Pauling and Wrinch. Weaver asked Pauling for a private report on Wrinch's work (but did not ask her for a reciprocal report on Pauling's work). Pauling told Weaver that Wrinch's papers were "dishonest," that she was "facile in the use of terminology of chemists and biologists, but her arguments are sometimes unreliable and her information superficial."[1] He did say that reviewing her work had led him to think that it was time to publish his own ideas on protein structure, and Weaver encouraged him to do so. The resulting paper, coauthored by Carl Niemann, was titled "The Structure of Proteins,"[2] but one of Pauling's correspondents more accurately called it "The De-bunking of Wrinch."[3] After Pauling's report and paper, Wrinch's career was in ruins. Weaver refused to renew her grant, and she was unable to find other research funding.

Linus Pauling

Linus Pauling is one of the few chemists widely recognized by the general public, although he is perhaps better known generally for his political activities and for his claims that vitamin C is a cure for the common cold than he is for his chemistry. He won two Nobel Prizes, for chemistry in 1954 and the Nobel Peace Prize in 1962.

Pauling was born in 1901 in Portland, Oregon. His mother, Belle, had psychological difficulties throughout her life, and she and he had a troubled relationship. His father, Herman, was a druggist who was often absent from the family as he attempted to start new businesses in smaller towns in Oregon. Herman died when Linus was nine years old, leaving Belle with little money. In an attempt to make ends meet, she sold Herman's pharmacy and invested the money in a boardinghouse in Portland. A few years later she developed pernicious anemia, and she depended on the children to help financially. At age thirteen, Linus was sent to work after school to help support the family, and his mother urged his sister Pauline, then seventeen, to date a well-to-do man in his thirties. This stopped when Pauline complained to the police. Linus was an excellent student and was interested in science, particularly chemistry, but when he attempted to enroll at Oregon Agricultural College, his mother objected that she needed him to continue working to support her. He enrolled anyway, working in the summers to pay his college expenses. The summer before his junior year, his mother confiscated the funds he had saved, leaving him unable to pay his tuition.[4]

The college saw his promise and offered him a position as an instructor in quantitative analysis with an office next to the library, where he read the papers on the chemical bond by Irving Langmuir and Gilbert Lewis. He was fascinated with chemical theory, and in his senior year he wrote to Berkeley, where Lewis was dean, to apply to enter the graduate school. Unfortunately for Lewis and Berkeley, Pauling's application letter was misplaced, and he enrolled at CalTech, where Arthur Noyes, Lewis's old boss at MIT, was now running the chemistry program. (After Pauling's letter was lost, Lewis decided that perhaps the application process needed to be better organized and turned graduate recruiting over to Wendell Latimer.)[5] Before leaving for CalTech, Pauling married his college sweetheart, Ava Helen Miller. She became the central person in his life, and he was to rely on her for the next sixty years until her death in 1981.

Student Years at CalTech

Noyes suggested that Pauling work in Roscoe Dickinson's laboratory in the new field of X-ray crystallography, which promised a way to determine the arrangement of atoms in a crystal. X-rays are light waves with very short wavelengths—approximately the length of the spacing between atoms in a crystal. If X-rays were directed at a crystal from different directions, the waves' diffraction patterns—the patterns in which the X-rays were scattered by the atoms in the crystal—could give information on the distances and angles among the atoms. But the information was not simple to extract, especially before the invention of the computer, and in complicated crystals the problem could become practically insurmountable. By the nature of the experiment, it was impossible to precisely locate the atoms—it was as if the pictures were out of focus and blurry. Furthermore, the X-ray images lost what was called the "phase" information—one could determine the intensity of the waves that had been diffracted, but not the positions of the crests and valleys of the waves. It was as if scientists were watching silent films of shadow projections of actors performing an unknown stage play. They could project the light from as many different directions as they liked, but it was still not easy to determine the play's plot. And just as the same shadow projection may correspond to more than one reality—we have all seen hand motions whose shadows look like a rabbit or an elephant—more than one structural arrangement could lead to the same X-ray image. Henry and Lawrence Bragg, a father-and-son team who had developed X-ray diffraction as a technique at Cambridge and Manchester, had been able to determine the structure of simple crystals such as sodium chloride and diamond, but the difficulty of analysis increased dramatically as the crystal structure became more complicated.

Pauling made an immediate splash at CalTech, but he showed that he had a few rough edges. After his initial attempts at determining crystal structures failed, Dickinson suggested that he try a particular crystal, molybdenite (MoS_2), and it worked beautifully. Pauling then wrote up the results for publication in the *Journal of the American Chemical Society*, listed only himself as an author, and gave Dickinson a copy. Noyes called Pauling into his office and explained how scientific credit was assigned: Dickinson had suggested molybdenite, trained Pauling in crystallography, and supervised and reviewed the work, so he therefore deserved primary credit. The paper appeared with Dickinson and Pauling as coauthors, listed in that order. Svante Arrhenius, Lewis, and Langmuir had all had difficulties with their research advisers, as well, but Pauling's dispute with Dickinson was different. The solution of the structure of molybdenite was a straightforward application of a technique in which Dickinson was expert—similar to the first part of Lewis's thesis on zinc and cadmium electrodes, where Theodore Richards directed the work, and where Richards was listed appropriately as first author in the publication. Pauling later said, "I think it was a good experience, in that it pointed out to me how easy it is to underestimate the contributions that someone else has made."[6] But throughout his career, many of Pauling's colleagues and coworkers, in particular CalTech's chief Robert Millikan, would see him as self-aggrandizing and autocratic.[7]

Pauling continued to solve crystal structures and began to see patterns in the ways that bond distances and angles changed depending on the influences of other nearby atoms. He applied Lewis's idea of the covalent bond, noting a gradual transition between covalent and ionic bonds depending on the bonded atoms' relative positions in the periodic table; he found that covalent bonds were generally stronger and shorter. He found that bonds between the same atoms often exhibited the same lengths and angles even in different crystals. These generalizations would become the basis for his later successes in inorganic chemistry, organic chemistry, and biology.

Still a graduate student, Pauling began working with Richard Tolman, CalTech's leading theoretical chemist. Tolman had worked with Noyes and Lewis at MIT, earning his Ph.D. there in 1910, and along with Lewis had become fascinated with Einstein's special relativity theory; he published on the subject jointly with Lewis and on his own. When Lewis had left MIT for Berkeley in 1912, he had taken Tolman with him, but Noyes had drawn him to CalTech in 1922 with a high salary. Tolman was an expert in statistical thermodynamics, which gives a way to calculate entropy—a measure of the disorder within a substance—from first principles. Walther Nernst's third law of thermodynamics can be formulated in terms of entropy—that the entropy of any substance to which it applies will disappear as the substance is cooled toward absolute zero. Nernst had claimed the third law applied to all liquids and solids, while Max Planck and Lewis, among others, had provided theoretical and experimental evidence that it applied only to pure crystals and not to glasses or solutions (see chapters 2–4 and 7). Pauling and Tolman jointly published a paper in 1925 on the third law, using statistical mechanics to calculate the difference in entropy at absolute zero of an imperfect crystal or glass as compared to that of a perfect crystal.[8] When Pauling insisted that his name should appear first in the list of authors, Tolman agreed but apparently took offense and declined to publish with Pauling again, although they were colleagues at CalTech for years.[9]

With Tolman's guidance, Pauling had begun to develop the physical and mathematical skills to understand the physicists' theories of the atom. Tolman was tracking the new field of atomic and molecular quantum mechanics, which had begun with Niels Bohr's work on the hydrogen atom in Manchester in 1912; Bohr was now professor in Copenhagen. Arnold Sommerfeld in Munich had extended Bohr's model to explain many-electron atoms. Bohr's and Sommerfeld's models involved dynamic electrons moving rapidly in orbits in space, while Lewis had developed static-electron models to explain the chemical bond. Erwin Schrödinger, Werner Heisenberg, Wolfgang Pauli, Max Born, and others in Germany were developing more general mathematical formulations for quantum mechanics, but their new formulations still had little appeal to chemists in the early 1920s—there was as yet no role for the electron pair, which was the essence of Lewis's idea of the chemical bond. With Noyes as his champion, Pauling obtained a National Research Council fellowship to do postdoctoral study in quantum mechanics.

By this time Noyes had a high opinion of Pauling and wanted him for the CalTech faculty when he finished his doctorate. The biggest potential obstacle to Noyes's plan was Lewis. Pauling intended to spend his fellowship time at Berkeley, where Noyes was sure Lewis would offer him a faculty position. Pauling later said of this time:

> When I was approaching the completion of my work for the Ph.D. degree there were many National Research Council fellows in Pasadena, and

I applied for a National Research Council fellowship, saying in my application that I would go to Berkeley and work with Gilbert Newton Lewis. There was a rule that a National Research Council fellow had to leave the institution from which he had received his doctorate....

When I prepared to go to Berkeley, after receiving the NRC fellowship, Noyes said, "You have done a large amount of x-ray work that you haven't yet written up for publication. That work could more conveniently be done here in Pasadena, so why don't you just stay here and complete writing the papers?"

After I had held the fellowship for about four months, Noyes told me, "There are some new fellowships...called the John Simon Guggenheim Memorial fellowships. You ought to go to Europe, which is the center of scientific work now, so why don't you apply for a Guggenheim fellowship for the coming year?" I applied for the fellowship, and again my wife and I prepared to go to Berkeley. Noyes then told me that the Guggenheim fellowships are not decided until the end of April, but he said, "You are sure to get a Guggenheim fellowship, so I'll give you enough money to pay the fare to Europe and to support you from the end of March until the beginning of the Guggenheim fellowship. It really isn't worth your while to move to Berkeley and then to make another move to Europe." I agreed....

It wasn't until about twenty years later that I realized that Noyes had been afraid that I would become a member of the staff at Berkeley if I took my fellowship at Berkeley.... In fact, Gilbert Newton Lewis had made a visit to Pasadena a few months earlier, about the same time that I got my doctorate. Many years later I learned that he had come to offer me an appointment, but that Noyes had talked him out of it.[10]

When Lewis came to Pasadena to try to hire Pauling, Noyes may have reminded Lewis of the debt he owed from 1905, when Lewis had returned from the Philippines and Noyes had hired him at MIT.

Pauling Makes His Mark in Quantum Chemistry

Pauling spent his Guggenheim fellowship in Munich with Sommerfeld, and in brief trips to Copenhagen and Zurich, where he met Bohr and Schrödinger. Pauling realized that although he was a competent applied mathematician, he was not in the same league with the German physicists who were building the new quantum mechanics. But he had something that they did not—a deep understanding of chemistry. He began to tie together the things that he knew—the new formulation of quantum mechanics, knowledge of experimental bond lengths and angles as determined experimentally from X-ray diffraction, and a sense for the important chemical problems. When he returned to the CalTech faculty, he began publishing papers that took experimental chemical properties and explained them in terms of the new quantum theory.

Schrödinger had expressed the possible energy states of an electron moving around a nucleus in terms of "wave functions," and Pauling reformulated Schrödinger's equations in terms of "hybridized orbitals" that explained the known directionality of chemical bonds—and matched the experimental bond angles in water, hydrogen cyanide, methane, ethane, ethylene, and acetylene, some of the simplest molecules.

Pauling viewed his ideas as an extension of Lewis's electron-pair bond. In the abstract to a 1931 paper that set out the essence of his ideas, he said that his work "provided a formal justification of the rules set up in 1916 by G. N. Lewis for his electron-pair bond," and "permit[ted] the formulation of an extensive and powerful set of rules for the electron-pair bond supplementing those of Lewis."[11] Pauling found that he could partition the binding energy of an entire molecule (as represented by the heat of formation from its elements) among the individual chemical bonds, and that adding the individual bond energies would often allow him to predict the binding energy of a new molecule. But the experimental binding energy of a molecule with several alternating double and single bonds was often lower than that predicted by adding the individual bond energies. Pauling took this to mean that the molecule's energy was stabilized by freeing the binding electrons to move among the different bonds. Lewis had discussed similar ideas in his 1923 book *Valence and the Structure of Atoms and Molecules*, but Pauling put it all on a more quantitative basis. He called the difference between the actual binding energy of a molecule and the sum of his standard bond energies the molecule's "resonance energy."

Lewis had said that there was a continuum between ionic bonds like those in sodium chloride and covalent bonds like those in chlorine molecule. There was an old concept in chemistry, *electronegativity*, which Pauling revived. Electronegativity had referred to Jöns Jakob Berzelius's idea that molecules were held together by electrical attraction between positive and negative atoms and was associated with the ability of an atom to take on a negative electrical charge. Combining thermodynamics and quantum mechanics, Pauling developed a quantitative scale of electronegativity based on the ability of a particular atom to pull an electron pair in a bond toward itself. His concept of electronegativity allowed chemists to assign percentages to the ionic and covalent character of a particular chemical bond. The most electronegative elements are those toward the top right of the periodic table, typically oxygen, nitrogen, fluorine, chlorine, and sulfur (see fig. 1-4). When bonded to an atom with lower electronegativity, an atom with higher electronegativity carries a partial negative electrical charge, and the atom with lower electronegativity carries a partial positive charge—an extension of Lewis's ideas on the polarizability of the chemical bond.

Pauling was particularly interested in the *hydrogen bond*, which would be fundamental to his later ideas on biological systems. As the name implies, the hydrogen bond involves a connection to a hydrogen atom. If a hydrogen bond is to form, the hydrogen must be covalently bonded to an atom more electronegative than hydrogen—typically oxygen, nitrogen, or sulfur in biological systems—which draws the electron pair away from the hydrogen, leaving the hydrogen with a partial positive charge. The hydrogen atom is then attracted to a different electronegative atom (which may be in a different molecule or in the same molecule) that bears a partial negative charge. It is this electrostatic attraction that forms the hydrogen bond. Pauling and others explained the ordering of water molecules in water and ice by hydrogen bonding (see fig. 9-1). The hydrogen bond had first been proposed in 1920 by two colleagues of Gilbert Lewis at Berkeley, using Lewis's theory of the chemical bond,[12] but Pauling expanded upon the idea.

As Pauling successfully extended Lewis's theories, and as Lewis's dreams of solving the problem of the dual wave–particle nature of light faded (see chapter 7), Lewis made an attempt to get back into the game that he had left in 1923—the chemical bond. In May 1932, Lewis received a letter from his friend William Noyes at the University of Illinois suggesting that Lewis collaborate with Pauling on a book on electronic theory.[13] Lewis had formed a

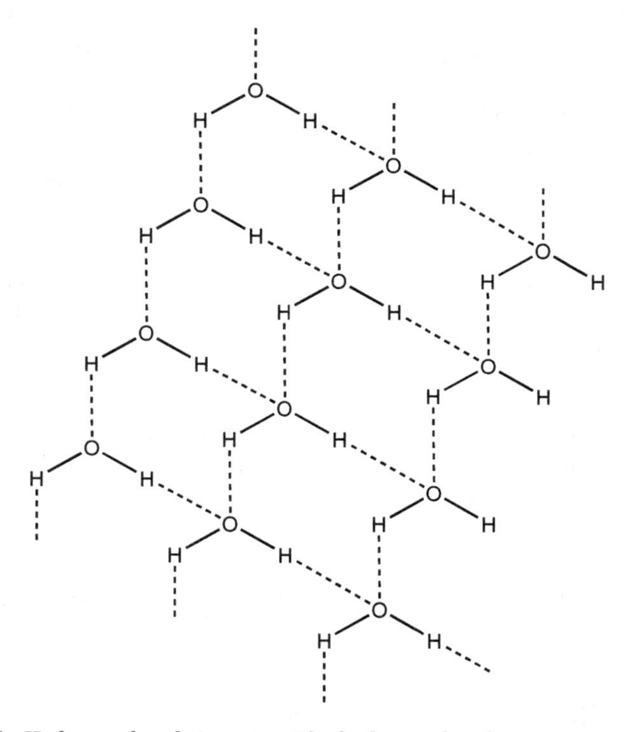

Figure 9-1. Hydrogen bonds in water. The hydrogen bonds are represented as dashed lines and connect the more electronegative oxygen atoms, with partial negative charge, to the less electronegative hydrogen atoms, with partial positive charge. (This drawing is a schematic only; hydrogen bonding in water is three-dimensional.)

close relationship with Pauling, who had spent several summers teaching at Berkeley. Lewis responded enthusiastically to Noyes, saying, "[T]here is no one I know with whom I would prefer to collaborate,"[14] and then sent Pauling copies of both the letter from Noyes and his own reply to Noyes.[15] Pauling returned a note saying no more on the matter than, "I was interested to read your letters to and from W.A. Noyes."[16] This exchange of correspondence may be significant in explaining Lewis's paper "The Chemical Bond," submitted in November 1932 for inclusion in the first issue of Harold Urey's *Journal of Chemical Physics*.[17] This paper is a rambling attempt by Lewis to place his ideas on the chemical bond in the perspective of the new quantum theory. It may have had an intended audience of one—Pauling—and been meant to persuade him of the value of collaboration. Pauling did not accept Lewis's 1932 suggestion of coauthorship, but when Pauling published his book *The Nature of the Chemical Bond* in 1939, he dedicated the book to Lewis. Pauling's book did for quantum chemistry what Lewis's *Thermodynamics and the Free Energy of Chemical Substances*, coauthored with Merle Randall in 1923, had done for chemical thermodynamics—it made an abstruse subject accessible to working chemists. Like Lewis's book, Pauling's was idiosyncratic, presenting the work from the author's own perspective, but it also collected the results of others. It was firmly rooted in experimental values of bond lengths, angles, and energies. Its publication brought Pauling recognition as Lewis's natural successor. As chemist Pierre Laszlo has noted, Pauling's success was due to his use of the language of chemists rather than that of physicists or mathematicians:

> Pauling became such an influential chemist because *he explains things*. . . . There is total continuity between his work and earlier Lewis structures for molecules, with the notion of valence, etc. Pauling tackled *well-established phenomena*, such as tetrahedral carbon, the bent water molecule, the linear hydrogen cyanide molecule, free rotation around single bonds, and so on. . . . And all these qualities are capped by a superb teaching talent: Pauling is easily understandable, he skips entirely sophisticated mathematics, the rules he gives are simple and they are widely applicable, to atoms in general.[18]

Pauling's new ideas explained the properties of ordinary chemical compounds, but Pauling saw that they might also apply to something bigger—the chemistry of life.

The Rockefeller Foundation and Molecular Biology

By the 1930s, the Rockefeller Foundation had become a major funding source in American science. Before World War II, federal funding of science was less important than it is today, and private foundations provided much of the money. In 1932, Weaver, a protégé of Millikan at CalTech, was named director of the natural sciences division of the Rockefeller Foundation, putting him in charge of a large pot of money. In 1936, Weaver invented the term "molecular biology." He believed that the application of chemical, physical, and mathematical techniques could transform biology and even psychology, enabling laboratory study of the very processes and constituents of life. He called his new program "the science of man" and focused all the Rockefeller funding there. This caused a number of physicists, chemists, and biologists to find biological relevance in their work; Pauling had become dependent on Rockefeller funding and followed the money. Although he knew little organic chemistry and almost no biology, he plunged into the new field and impressed Weaver by studying the magnetic properties of hemoglobin, the protein molecule that binds oxygen in the blood. Weaver and Pauling twisted the arm of Millikan, CalTech's director, to provide Pauling with increased funding, and Weaver provided him with a three-year grant, as well. Pauling and Weaver were now working in tandem.[19] With hemoglobin, Pauling had begun to use his ideas on the chemical bond to explain proteins, which are fundamental building blocks of life.

Proteins and the Secret of Life

Proteins are found almost everywhere in life—in muscle, feathers, fingernails, hair, blood, brain cells, and reproductive cells, among others. In the 1930s proteins were suspected to be the carrier of genetic information, which we now know to be encoded in DNA. Proteins are a primary constituent of the chromosome, and DNA was at that time thought to provide only structural support for proteins. If the structures of proteins could be determined, it was thought that it might lead to an understanding of "the secret of life."

Determining the nature, structure, and function of proteins was the most pressing problem on Weaver's agenda for the new molecular biology. Scientists had been speculating about the nature of proteins for years. Some supported a vitalist attitude that proteins were somehow different from other molecules and were endowed with a principle of life. Emil Fischer, the German organic chemist, had done fundamental research on proteins between 1899 and 1908 and believed that their nature was firmly rooted in chemistry. He determined that proteins could be decomposed into a few α-amino acids [organic acids that contained an amine ($-NH_2$) group and a carboxylic acid ($-COOH$) group separated by one carbon]; twenty of these are known today to occur naturally, and many of them were identified by Fischer. He showed that the amino acids were differentiated only by their "side chains," the organic groups represented as R_1 and R_2 in fig. 9-2, and that amino acids could be joined into polypeptide chains by a peptide bond, also shown.

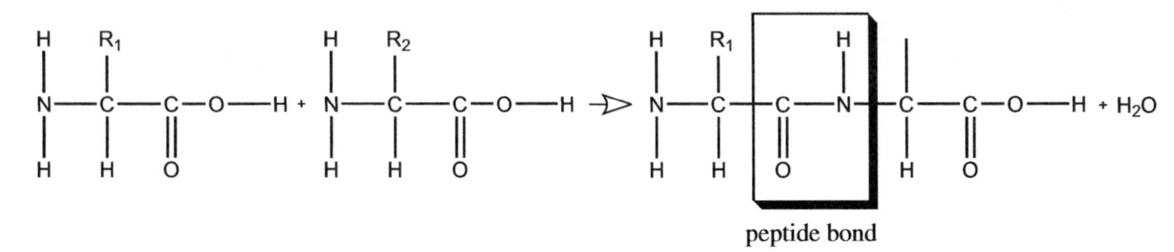

Figure 9-2. Two amino acids with different side chains R_1 and R_2 can unite by forming a peptide bond. The resultant molecule can react with another amino acid with side chain R_3, then another with side chain R_4, and so forth, forming a molecule consisting of a chain of amino acids, all linked by peptide bonds. The resultant molecule is called a polypeptide.

Fischer synthesized a great number of polypeptides in his laboratory, eventually linking eighteen amino acids in such a chain. He proposed that proteins were nothing more than long polypeptides. Proteins were known to be highly specific in their physiological actions—two particular proteins will combine strongly with each other, but not with most other proteins. Fischer was the first to propose the "lock and key" model for proteins, suggesting that the biological specificity of protein–protein interactions is due to specific chemical interactions between their side chains. By the 1930s, there was general agreement that the peptide bond was important in the formation of proteins but not that Fischer had been right in his belief that proteins were long polypeptides. His was only one of the possible models.

Organic chemists, physical chemists, and biologists had taken different paths in the study of proteins. When measured by the usual chemical techniques, such as osmotic pressure, the molecular weights of proteins can be quite high, sometimes exceeding one million—molecules this large are called "macromolecules." A subset of the physical chemists, known as "colloid chemists" and led initially by Theodor Svedberg, denied that molecules this large could exist. They viewed proteins as amorphous conglomerations of smaller polypeptide strands. But by the mid-1920s, Svedberg had convinced himself from work with his ultracentrifuge that proteins were macromolecules with very high but specific molecular weights.

Pauling believed that Fischer's hypothesis was correct in its essence—that proteins consisted of linear chains of amino acids connected by peptide bonds. Pauling extended Fischer's ideas into three dimensions, asserting that a protein was held in a particular three-dimensional conformation by hydrogen bonds forming between different parts of the protein chain, giving the protein its specific physiological activity. In a 1936 paper with Alfred Mirsky, Pauling wrote, "Our conception of a native protein molecule (showing specific properties) is the following: The molecule consists of one polypeptide chain which continues without interruption throughout the molecule (or, in certain cases, of two or more such chains); this chain is folded into a uniquely defined configuration, in which it is held by hydrogen bonds."[20] This was only a hypothesis, and Mirsky and Pauling clearly identified it as such—they provided no evidence of such a structure in a particular protein.

In the mid-1930s, the most important group involved in the study of the structure of proteins was led by John Desmond Bernal, a Cambridge X-ray crystallographer. In 1934, he and his student Dorothy Crowfoot (later Dorothy Hodgkin) took the first X-ray photographs of hydrated pepsin protein crystals, and the photographs of pepsin as well as those of the next few proteins they examined showed surprising symmetry. Like Pauling, Bernal believed that the hydrogen bond was important in determining protein structure.[21] While Pauling and Bernal were scientific competitors, they were roughly aligned in their approach to determining protein structure. Both believed in the importance of experimental verification and were cautious in their claims. But there were differences in emphasis between the two: Bernal relied more on the use of X-ray diffraction data on proteins. Although Pauling had a strong background in X-ray diffraction, he tried to extrapolate from the structure of smaller molecules, believing that the same bond types, lengths, angles, strengths, and resonance stabilization would be found in proteins as in other molecules. He believed that proteins were too complicated for X-ray diffraction to be useful as the primary tool in determining structure, although he thought that it would be useful in choosing among possible structures.

Dorothy Wrinch and the Cyclol Theory

Weaver funded Rockefeller Foundation grants for both Pauling and Bernal, and in 1935 he approved a grant for Dorothy Wrinch, an English mathematician who hoped to fit into Weaver's plans to quantify biology. Wrinch had become interested in applying mathematical theory to the structure of proteins. Born in 1894 in Argentina to an English hydraulic engineer and his wife, she grew up in London and attended a public day school. Her father, an amateur mathematician, encouraged her to take up mathematics. She entered Girton College at Cambridge, where she graduated with honors and proceeded to graduate study under English mathematician G.F. Hardy. She received a doctorate from London University in 1921 and was an associate of both Bernal and philosopher-mathematician Bertrand Russell. Beginning in 1931, she spent time in Europe, learning what she felt she needed to know about chemistry and biology. She was a member of the Oxford Biotheoretical Gathering in the early 1930s, an informal group of scientists interested in the physical, chemical, and mathematical foundations of biology. In 1935, she published a paper, especially important when viewed retrospectively, whose abstract begins, "The genetic identity of the chromosome resides in its protein pattern and is expressible in its one-dimensional sequence."[22] This may be the first published association of the genetic code with the amino acid sequence of proteins.

Wrinch's 1922 first marriage to John Nicholson ended in separation in 1930 and eventual divorce in 1937, leaving her with a small daughter, Pamela; Nicholson was also a mathematician and was important in the development of quantum mechanics, but was permanently institutionalized for alcoholism and lunacy in 1930. Wrinch faced the problems of being both a single parent and a scholar, and she pseudonymously published a book, *Retreat from Parenthood*, on the problems that professional women faced in combining parenthood and careers. She applied to both the Rhodes Trust and the Rockefeller Foundation for sociology fellowships, and applied for a Rockefeller mathematics grant at the same time. She quickly received another lesson in the difficulties she would face as a woman scientist. All her grant applications were turned down. One of the judges for the Rhodes fellowships refused to approve any women,[23] and the Rockefeller Foundation denied her mathematics grant on the grounds that her simultaneous application for a sociology grant showed a lack of dedication to mathematics.[24]

But by 1935, Wrinch seemed to Weaver to be exactly the sort of person he wanted—a mathematician interested in his new molecular biology. Weaver offered her a five-year grant with the proviso that she must remain affiliated with Oxford. Since Wrinch had no laboratory of her own, Weaver expected that she would collaborate with chemists and biologists, taking their experimental data and fitting it into her mathematical theories, which were yet to be developed.

Wrinch had been greatly influenced by the ideas of D'Arcy Thompson, who is sometimes called the first biomathematician. Thompson, a Greek scholar, mathematician, and naturalist, had published *On Growth and Form* in 1917.[25] Thompson focused on symmetry in nature, using the hexagonal honeycomb of the bee and the spiral nautilus shell as examples. Wrinch was to approach protein structure in the same way, looking for symmetry to explain physiological specificity. In 1936 she heard a comment by Bernal, who said that F.C. Frank had suggested that polypeptides might rearrange themselves in another fashion in addition to the peptide bond. This hypothetical reaction, called lactim–lactam tautomerism, would become the basis of what Wrinch would call her "cyclol" theory of protein structure.

In 1936 and 1937, Wrinch published two papers on possible theories of protein structures. The first of these, published jointly with Jordan Lloyd, was a letter in response to the Mirsky-Pauling paper.[26] Wrinch and Lloyd noted that Lloyd had already proposed hydrogen bonds as important in protein structure, and asserted that hydrogen bonds are closely related to the lactim–lactam tautomerism that Frank and Wrinch had already discussed. The first comment was almost certain to antagonize Pauling, and the second was not true, at least if hydrogen bonds are understood in the way that Pauling had expressed them. His hydrogen bonds did not involve the breaking and making of other covalent bonds, as did the hypothetical lactim–lactam tautomerism. Wrinch's second paper, titled "On the Pattern of Proteins,"[27] was more innovative. She suggested that proteins might organize themselves through covalent bonds in two dimensions in addition to the one dimension that Fischer had first proposed and that Pauling espoused, and lactim–lactam tautomerism was at the heart of this idea. Linear polypeptides would cyclize end to end, forming a peptide bond between the ends of the chain—like a snake grabbing its own tail. Once that was done, the lactim–lactam reaction might tighten up the cyclized polypeptides and form a network of six-membered rings in a protein sheet.

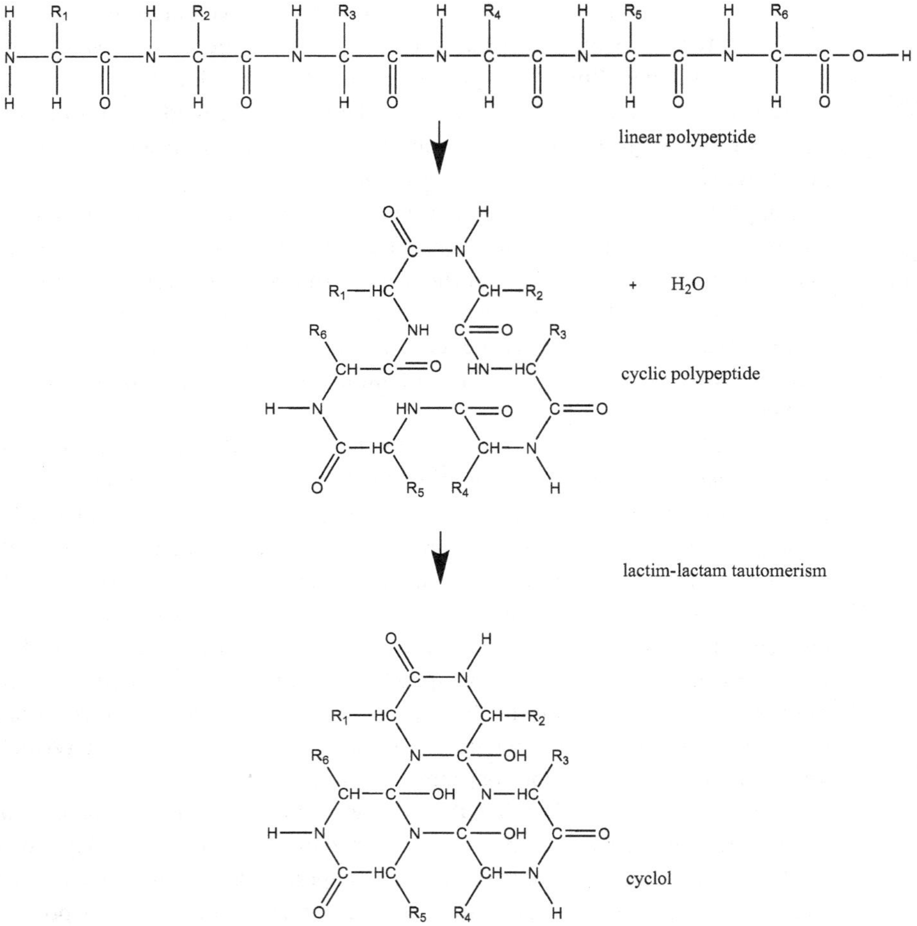

linear polypeptide

+ H_2O

cyclic polypeptide

lactim-lactam tautomerism

cyclol

Figure 9-3. Dorothy Wrinch's hypothetical cyclol structure for proteins. Linear polypeptides form cyclic polypeptides, which then form a network of six-membered rings via lactim–lactam tautomerism. The diagram shows formation of a "cyclol 6" in her terminology starting from a polypeptide with six amino acid residues.

All the side groups in a cyclol would be on one side of the protein sheet, and the two-dimensional array of different side groups would then provide the biological specificity shown by proteins. In addition, the sheets could then arrange themselves by stacking or folding, which would account for the observed three-dimensional globular nature of proteins. Wrinch elaborated this idea in a later paper, showing how the sheets could fold into symmetrical three-dimensional solids that matched the symmetry shown by X-ray studies of protein crystals.[28] She wrote, "We therefore suggest that the cyclol pattern...is the basis of the structure of the unimolecular film, when it is approximately one residue [molecule] thick."[29] Once again, a chemical surface one molecule thick—the continuing theme in Langmuir's work! Langmuir started to collaborate with Wrinch soon after her initial cyclol publication. From his point of view, it must have seemed that his ideas on surface chemistry, conceived a quarter of a century earlier and arising from his work on the electric lamp, might become the basis for the "secret of life." Although he knew little of biology or of X-ray crystallography, he was always game to attack a new problem, and he was already publishing on protein films even before his collaboration with Wrinch. Langmuir believed that simple protein films consisted of polypeptide chains, but that globular proteins were likely as described by Wrinch's cyclol hypothesis.[30]

This is another instance of a recurring pattern—scientists attempting to replicate their success in physical chemistry in unrelated fields. Arrhenius had tried to apply the mass action law to immunochemistry (chapter 1); Lewis had applied his reasoning on chemical equilibrium and his concept of activity to theoretical physics (chapter 7); and we now see Langmuir attempting to turn biology into a special case of surface chemistry.

Wrinch presented her ideas at a number of important scientific meetings in 1937. They were received as interesting but unproven hypotheses. Even Pauling showed reserved interest to Weaver in 1937:

> Some fifteen years ago a number of people indulged in extensive speculations regarding the structure of crystals, using the self-consistency of their systems as criteria rather than test by experimental methods. Despite the nicely symmetrical structures which they proposed, these speculations have turned out to be wrong. I feel that Dr. Wrinch's work suffers a little bit from being similarly too speculative and from being based too largely on the assumption that nicely symmetrical structures are the right ones. On the other hand, she seems to be quite conversant with what facts there are, and it is quite possible that her attempts to coordinate them with structural ideas will ultimately be of value in the solution of the great problem of protein structure.[31]

Pauling's quote shows his view of the difference between his and Wrinch's approaches. He was a chemist and believed that any new theory could stand or fall only by exhaustive experimental test. She was a mathematician and relied on the internal symmetry and consistency of her theory.

Since 1936, Wrinch had been attempting to persuade Weaver and the Rockefeller Foundation to provide her an assistant so that she could test her cyclol hypothesis. Weaver and his colleague W.E. Tisdale, who was Wrinch's European contact and ran the foundation's Paris office, denied all such requests. The reasons for this are understandable: she was a mathematician and had no experience in directing chemical research; furthermore, she had no laboratory, and the Rockefeller Foundation did not want to be responsible for setting one up. Weaver had made it clear to her that she needed to enlist interested scientists to test her work. But the correspondence between Tisdale and Weaver shows extraordinary sexism, even by the standards of the 1930s. From a letter from Tisdale to Weaver:

> I thought it would not only be foolish to think of attaching a protein chemist to her but I felt that I would overlook a bet if I did not seize the opportunity to head off similar proposals for a genetics assistant and so forth which as you will observe I did or at least think I did. My reason for expressing doubt is that as you well know my experience in managing the gentler sex is rather limited but I have found that the best any man can hope to get in argument with anyone of them is a half, to introduce a golfing term.[32]

And from Weaver's diary: "She continually raises the question of direct assistance under her supervision—'oh, Dr. Weaver: If only I had a handsome young man to twiddle the test-tubes for me!'—but repeated discouragement of this idea from WET [Tisdale] and WW [Weaver] should convince her that we are not prepared to furnish funds for this purpose."[33] Wrinch did attempt to have others test her ideas, but the way she approached collaborations led to unhappy results. John Jones, a peptide chemist and archivist at Balliol College, said that Wrinch asked Robert Robinson, one of England's most prominent organic chemists, to do experiments to test her cyclol hypothesis. Robinson did the experiments, and his results failed to support her ideas. Wrinch then asked a Swiss chemist to do the same experiments, without telling him that he would be duplicating Robinson's work. The Swiss chemist got in touch with Robinson, and the game was up. As Jones said, "She was very persistent."[34]

Harold Urey and Harry Sobotka, a protein chemist who became Wrinch's friend, became interested in the cyclol theory and agreed to perform some tests at Mt. Sinai Hospital in New York, as did William Astbury at the University of Leeds. Urey was tremendously impressed by Wrinch's theory (although his knowledge of protein chemistry was even less developed than Langmuir's), but Wrinch infuriated some other scientists. Weaver's diary records the following:

> I was calling on Linus Pauling at Ithaca.... He was in bed with a cold, and a Commonwealth fellow from Oxford was calling on him. When Dorothy Wrinch's name was mentioned he blushed furiously and had to draw on the deepest reserves of the English character to keep from being profane in Mrs. Pauling's presence concerning what the Oxford crowd thinks of her.... On the other hand I was talking to Urey..., and in commenting on a woman [probably Mildred Cohn] in his laboratory he said that she was exceedingly good but of course not a Dorothy Wrinch. When I asked him what he meant by that, he said: "Well I mean of course that she isn't an outstanding genius."[35]

Figure 9-4. Langmuir and Wrinch with a cyclol model, Dorothy Wrinch papers, Sophie Smith Collection, Smith College, Northampton, Mass.

Wrinch was beginning to be noticed in the popular press, which alarmed Weaver. In September 1937, he noted in his diary that "P. [Pauling] thinks W. [Wrinch] has had altogether too much publicity for ideas which are vague, speculative, and perhaps not very important if true [!].He refers to a recent picture in 'Time'[36] in which W. is fondly holding her elaborate model of a globular protein."[37] Weaver decided to arrange a meeting between Pauling and Wrinch in order to get Pauling's evaluation of her work. The propriety of having one of two competing researchers report on the other seems not to have bothered Weaver. Pauling and Wrinch met in Ithaca, New York, in January 1938, but even before they met, Pauling had already made up his mind that her work was worthless.[38] In his report to Weaver, Pauling wrote:

> I am sympathetic to the application of physical mathematical methods to chemical biological problems, and I began the interview with Dr. Wrinch

> with the hope that the unsatisfactory aspects of her published work would be removed in discussion. I found however, that her methods and results are still less scientific than they had appeared to be from her papers.
>
> I doubt that her attack on the problem of protein structure will lead directly to any valuable results.[39]

Weaver replied with a letter addressed for the first time to "Dear Linus" (all previous letters had begun "Dear Dr. Pauling"), saying, "I am certainly indebted to you.... I am sure all your comments will be extremely useful to us."[40]

Wrinch did in fact get a great deal of attention in the popular press. In addition to the *Time* article, she was featured prominently in four different *New York Times* articles and received secondary mention in several others, where her speculations about what her cyclol model might mean were quoted as assertions. It is likely that this transition from speculation to assertion was the work of the newspaper reporters, but the number of such occurrences indicates that she was sometimes indiscreet. For example, one newspaper reported:

> Offering an explanation of how viruses reproduce, she said that a protein unit gave birth to another by having a second layer form on its surface in exactly the same pattern, with the first splitting and flattening.
>
> Although protein cages are "hollow" they are filled with substances, the nature of which, it was stated, might determine differences between proteins which are similar in their molecular structure.[41]

She was the subject of a long article in the *New York Post*, accompanied by a large photo showing her holding a cyclol model with the headline "Woman Einstein—Dr. Wrinch," which included:

> Usually such solutions come from chemists and biologists who work with the materials in test tubes. But chemists, who have worked with proteins, never have been sure that they had the whole story because proteins change their structures so swiftly. The novel point about Dr. Wrinch's work is this: She studied the available knowledge on proteins. Then she took a pad, a pencil, and her own mathematical brilliance and built the facts into a consistent picture.[42]

A woman scientist was naturally a magnet for newspaper reporters of the time, but such publicity irritated not only Pauling but other chemists and biologists as well, almost all of whom were men.

"Proof" of the Cyclol Theory

Wrinch then found what she regarded as a striking confirmation of the cyclol hypothesis. Max Bergmann, who had moved to a position at the Rockefeller

Institute after leaving Nazi Germany, had published a paper with Carl Niemann suggesting that proteins typically occurred such that the number of amino acid residues could be expressed by the formula $2^m \times 3^n$, where m and n are integers.[43] This was based on Svedberg's suggestion that proteins seemed to show discrete patterns in their molecular weights and on Bergmann and Niemann's findings that egg albumin has 288 residues, a special case of their formula where $m=5$ and $n=2$. The formula seems only so much numerology today, but it was taken seriously at the time. Wrinch had become interested in a particular type of cyclol model, which she called a truncated tetrahedron; this required cyclols where the number of amino acid residues fit the formula $72n^2$, which also matched the number 288 with $n=2$. Wrinch took this numerical coincidence as strong evidence for what by 1937 she called her "cyclol theory"[44]—a promotion from her earlier references to it as a working hypothesis. Wrinch sent Bergmann a letter saying, "I have only just seen your wonderful paper.... It thrills me greatly as the numbers fit so well with my theory. You will remember I brought you my model of the globular protein containing exactly 288 residues when I first came to see you on October last. It gives me great pleasure that we have a definite point of contact between your work and my fantasies!"[45]

Bergmann's numerological ideas were viewed with skepticism by many, and he wanted to keep them as far as possible from Wrinch's similarly viewed cyclol hypothesis; he replied to Wrinch, saying, "What I do not understand is how the figure 288, which in itself must be compatible with any kind of protein structure, could be regarded as proof of the cyclol hypothesis."[46] A few months later he wrote a hostile note to Wrinch, saying:

> Thank you for your reprints. One of them contains a postscript which reads: "The Referees of the above communication have called the attention of the author to recent work of Bergmann and his collaborators, which, in their opinion, strongly confirms the conclusions there reached." A similar postscript is contained in your note in *Nature*, June 1937. I wonder what may have been your reason for concealing so eagerly the fact that already in the fall of 1936 you repeatedly visited our laboratory and became well acquainted with our recent work and our numerical findings on proteins.[47]

There are two more letters in the Bergmann-Wrinch correspondence; each accuses the other of concealing information in their earlier meetings.[48]

Wrinch had obtained X-ray data on the insulin protein from her friend Dorothy Crowfoot, which she believed provided proof for her cyclol structure. Wrinch took the internal symmetry that Crowfoot's X-ray data exhibited for insulin, pepsin, and other protein crystals to mean that the protein molecules themselves must also be highly symmetrical. She published a letter under her own name[49] and a longer paper coauthored with Langmuir, saying, "[A]ll the prominent features of the Crowfoot diagram are deducible from the [cyclol structure]," and "[W]e repeatedly found that as we introduced one by one the more delicate features the more perfect became the concordance between the Patterson-Harker diagrams [deduced from the cyclol structure] and the Crowfoot pictures. We feel that these X-ray data, in giving so perfect a picture of the [cyclol] structure, provide the experimental basis for the cyclol theory."[50] The Patterson-Harker maps to which Wrinch referred are derived from the X-ray data and give information about the distances between the atoms in the structure. David Harker, who had been Pauling's student at CalTech and was then at Johns Hopkins, had been a codeveloper of the idea of the maps and was interested in Wrinch's interpretation of them. There was a controversy at the time about the significance of the maps, because atoms could not be considered to be precisely defined points because of their vibration within the crystal. Many crystallographers believed that a single map could correspond to an infinite number of structures. Wrinch reduced this to a geometrical problem, showed how to solve the problem, and showed that there were at most a few possible solutions, which made a correspondence between a map and a possible structure much stronger evidence. Lipson and Cochran later wrote in *The Determination of Crystal Structures*, "In the ensuing controversy [over cyclols] Wrinch's important contribution to the theory of the interpretation of the Patterson function was lost sight of and little attention was paid to it until parallel results were found ten years later."[51]

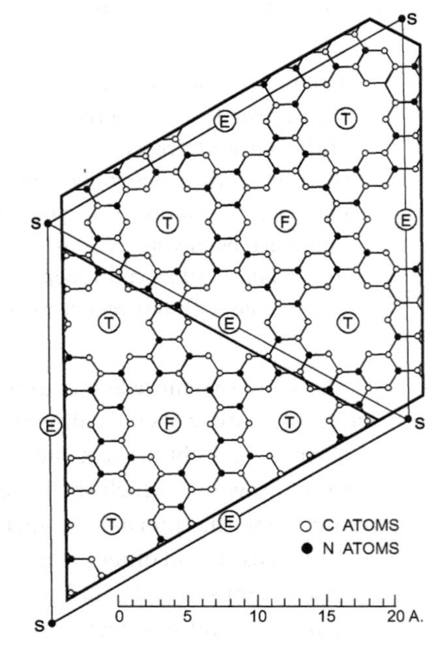

Figure 9-5. A cyclol model of a protein sheet. Dorothy Wrinch and Irving Langmuir. "The Structure of the Insulin Molecule." *Journal of the American Chemical Society* 60 (1938): 2247–55, on 2247.

Pauling's Attack

Wrinch's claims that Crowfoot's X-ray data proved the cyclol theory were particularly irritating to Bernal and his school. Neither Wrinch nor Langmuir had much experience in interpreting X-ray data, and Bernal attacked their X-ray analysis in an article in *Nature*.[52] The article shows that, although he did not understand Wrinch's new approach to the Patterson-Harker maps, he correctly saw that Wrinch and Langmuir were overinterpreting the data and coming to unwarranted conclusions on the structure of proteins. Pauling, after his 1938 meeting with Wrinch, decided that it was time to discredit the cyclol theory. Pauling's and Wrinch's ideas about proteins could not have been more different: for Pauling, a protein was a long coiling strand with no symmetry, lightly adhering to itself where two suitable hydrogen bonding sites were able to lock together; for Wrinch, a protein was a sheet of precise hexagonal symmetry, held together by covalent bonds, and perhaps folded into symmetrical three-dimensional polyhedra.

In an article with Niemann, Pauling laid out a methodical attack. He began by noting that the original idea for lactim–lactam tautomerism was due to Frank and had been developed by Wrinch. He wrote:

> It has been recognized by workers in the field [Pauling did not identify the workers, but presumably meant himself] that the lack of conformity of the cyclol structure with the rules seen to hold for simple molecules makes it very improbable that any protein molecules contain structural elements of this type. Until recently, no evidence worthy of consideration had been adduced in favor of the cyclol hypothesis [ignoring Wrinch's earlier claims that Bergmann's ideas supported her]. Now, however, there has been published an interpretation of Crowfoot's valuable X-ray data on crystalline insulin which is considered by the authors to provide proof that the insulin molecule actually has the structure of the space-enclosing cyclol.[53]

He then attacked the cyclol theory on multiple points:

- There was strong X-ray evidence that fibrous proteins were extended polypeptide chains.
- Crowfoot had said that her data could not be used to measure the positions of individual atoms and that it provided no support for either a polypeptide structure or a cyclol structure for insulin.
- Wrinch and Langmuir had made unwarranted assumptions in analyzing Crowfoot's data and had introduced seven adjustable parameters that could be used to fit almost any structure to the data.
- Bernal's analysis of the same data gave completely different results.
- Langmuir himself had found evidence to support the idea that protein films were extended chains.

- The bond energies of the proposed cyclol structure were almost certainly thermodynamically unfavorable.
- No organic structure had ever been synthesized with the cyclol structure.
- The cyclol structure would require the distances between nonbonded atoms to be much smaller than had been observed in other molecules.
- The cyclol hypothesis required the existence of a great number of hydroxyl groups in proteins, which had not been found experimentally.
- The breaking of covalent bonds, which the cyclol theory required to explain the denaturing of proteins, would be much slower than was observed experimentally.

Pauling concluded the paper with a short restatement of his own ideas on protein structure, which he admitted were not yet confirmed by experiment. In addition to what he had said in his 1936 paper with Alfred Mirsky (that proteins are long polypeptide chains held in place by hydrogen bonds), he said that he believed that the strands of polypeptides may be linked together in proteins in other ways than by simple linear peptide linkages, and that such linkages "may arise in part from peptide bonds between side-chain amino and carboxyl groups or from side-chain ester bonds or S–S [sulfur–sulfur] bonds."

Pauling and Niemann's paper was extraordinary. Although its title was "The Structure of Proteins," the bulk of the paper was about what Pauling thought the structure of proteins was *not*—cyclols. It is not common to dedicate a paper to refuting the ideas and claims of another scientist without advancing experimental evidence for one's own explanation. This paper shows the difference in approach between the chemist, Pauling, and the mathematician, Wrinch. They did not share a common understanding of even the scientific method. While Pauling had no strong evidence for his own ideas of protein structure, he saw overwhelming experimental evidence that her cyclol theory was wrong. His understanding of the scientific method was that hypotheses that are contradicted by experiment should be rejected. Wrinch thought that no other satisfactory theory had been advanced, and that the beauty and symmetry of the cyclol theory meant there must be truth there somewhere, even if some of the details might be wrong.

Wrinch wrote a twenty-one-page vitriolic paper in reply to Pauling and Niemann, and submitted it for publication. But Arthur Lamb, who was editor of the *Journal of the American Chemical Society*, was interested only in new findings and did not want to sponsor an extended and repetitious debate. He sent Wrinch's paper to Pauling and Niemann for their comments to be attached when her paper went to referees (although he had apparently not sent the Pauling-Niemann paper to her for her comments, even though it had been a direct attack on her theory). Pauling and Niemann made comments about Wrinch that were as vicious as her comments about them. The referees and

Lamb were shocked by the language used by all parties. Langmuir was staunch in his support of Wrinch in letters to Pauling both before and after the publication of the Pauling-Niemann paper,[54] and Langmuir nominated Wrinch for the Nobel Prize for Chemistry in 1939.[55] He wrote Lamb, "I think that Pauling and Niemann's article was extremely unfair and that the J.A.C.S. [*Journal of the American Chemical Society*] should never have accepted the paper for publication in that form. . . . I think she states the case very mildly and is entitled to present her case in the J.A.C.S. without having to submit her paper to Pauling and Niemann for further attacks."[56] Lamb offered to give Wrinch space for a short reply. Langmuir counseled her to accept, and this appeared in 1941.[57] At this point, Langmuir left the protein field to take up war work, but he continued to support the cyclol hypothesis in print as late as 1942, writing, "I think in general, the best picture of this structure is Dr. Wrinch's theory of the cyclol structure."[58] Langmuir never suffered the personal attacks that Wrinch did over the cyclol theory. He was too prominent and too well liked, and so the opprobrium fell upon Wrinch.

By this time, much of the scientific world was arrayed against Wrinch. She was particularly bitter toward Pauling, writing to her friend Eric Neville:

> This new Pauling business gets me down. He is a most dangerous fellow. Even decent people hesitate to stand up to LP [Linus Pauling]. He is bright and quick and mercyless [*sic*] in repartee when he likes and I think people are just afraid of him. It takes poor delta [her favored signature nickname] to point out where he is wrong: truly none of them would under any [circumstances]. The big paper on bond strengths and lengths has come back from JCPhys [*Journal of Chemical Physics*] with reports from six referees. They all seize upon my comments which apply to P. [Pauling] and want them deleted. They are cowards.[59]

Her attempts to present her ideas at scientific meetings must have seemed to her to be like the flailing of a deer attacked by a pack of wolves. The same letter to Neville continues: "Did I tell you that a fellow . . . told a friend of mine . . . 'I was with this group of men several nights in succession at Gibson Island [the site of a scientific meeting] and their one topic of conversation was working out how they would down Wrinch when she gave her paper?'"

Weaver hated involving the Rockefeller Foundation in controversies and must have longed to end its association with Wrinch, because he certainly had no plans to let Pauling go. The problem for Weaver was that Wrinch had too much prominent support. While the chemists and biologists were largely opposed to her, Langmuir and Urey were her backers, although they were very much on the physical side of chemistry and knew little biology. To Weaver's and Tisdale's alarm, another Nobel laureate was lining up behind her, as well.

Tisdale's log intersperses plans to terminate the foundation's association with Wrinch with reports that Langmuir's close friend Niels Bohr, who also knew little of biology, had become interested in cyclols. He was building metal models of cyclol structures and sending them to Wrinch.[60]

Pauling used his considerable influence to discourage others from associating with Wrinch. When Harker, Pauling's former student, published a paper with her,[61] Pauling sent him a letter saying, "I have just noticed again the letter by you and Dorothy Wrinch in the June issue of the *Journal of Chemical Physics* and have decided that its publication shows that you are in need of some advice. . . . I think that you could be about better business and in better company."[62] Harker replied,

> I was rather amazed and shocked by your letter of July 8th. . . . I have heard rumors from time to time concerning your alleged unfair attitude toward Dr. Wrinch and her right to dispute her theories in print. I have invariably thought—and said—that such an attitude on your part was impossible. I have always considered these criticisms of you as due to misunderstandings of your statements and unworthy professional jealousy. I should be most unhappy to be forced to believe otherwise.[63]

Wrinch's Rockefeller grant was due to expire in 1940, and Weaver and Tisdale repeatedly refused to even consider renewing it, although they did offer her a terminal extension for two years to tide her over while she found a new position. By this time, Wrinch had moved to America with her daughter Pamela. World War II had begun, and she was unwilling to return to England in the middle of the blitz. She was attempting to find a position in America, but with Pauling's and Bernal's opposition, she was having little success until Otto Glaser, a friend whom she later married, helped her gain a temporary appointment as a joint visiting professor at Amherst, Smith, and Mount Holyoke colleges in western Massachusetts. She eventually received an appointment to the Smith College faculty and continued to publish on her cyclol theory into the 1960s, but few were listening.

There is no question that, in the end, Wrinch was almost entirely wrong about the structure of proteins and Pauling was mostly (but not entirely) right. Proteins did turn out to be chains of polypeptides, cross-linked by sulfur–sulfur bonds among the side chains as Pauling and Niemann had asserted, and proteins' three-dimensional conformations are determined by noncovalent forces, as Pauling had forecast, rather than by the covalent bonds that the Wrinch's cyclol theory demanded. But Pauling's hydrogen bonds would not prove to be a complete explanation of the three-dimensional conformation of proteins: in support of the cyclol theory, Langmuir had advanced an important idea that was directly related to his earlier theories of surface chemistry and the orientation of oil films on water (chapter 5)—that proteins were held together by interactions among the hydrophobic* regions of the molecule as they clumped together to get away from the surrounding water molecules. And where Pauling had emphasized formation of hydrogen bonds between different parts of the protein chain as the force stabilizing a given three-dimensional configuration, hydrogen bonds turned out to be important in another way: many of the stabilizing hydrogen bonds actually form between the protein and the surrounding water. In general, proteins form three-dimensional configurations with the hydrophobic parts inside, away from the water, and the hydrophilic parts outside where they can form hydrogen bonds with water. Bernal realized that although Langmuir had advanced his idea of the hydrophobic interaction in support of the cyclol theory, it was a more general concept: "Langmuir has used this picture as a justification of the cyclol cage hypothesis, but it is strictly quite independent of it."[64] Pauling was dismissive of the importance of the hydrophobic interaction, believing the energetic contributions involved to be too small to be important.[65]

But discussion of who was right and who was wrong is largely hindsight. In 1936, and even in 1941, there was still plenty of room for scientific dispute about the structure of proteins, and Wrinch was effectively silenced and cut off from both funding and the academic community. As mathematician and historian Marjorie Senechal has written:

> When the appeal to experiment does not produce a clear answer, then the choice among theories may be deferred, and the argument continued. Or a consensus may be reached among scientists on other grounds. For example, one theory may seem more plausible than another, or may be preferable because it involves fewer assumptions. In the minds of most scientists, the cyclol controversy was settled in favor of the chains, years before the techniques were available which could decide the matter.[66]

When Dorothy Crowfoot Hodgkin, Wrinch's longtime correspondent, was asked in 1984 where Wrinch had gone wrong, Hodgkin wrote:

> In retrospect, I think that she was unlucky that three scientific observations made on proteins at that time seemed to fit so well with the cyclol hypothesis. The first—that protein molecules were highly symmetrical—was only favored for the first few we picked up.... The second was the Bergmann-Niemann numbers.... The third was that the molecular weights estimated by combined chemical analysis and the centrifuge came to exactly the predicted number 288 required by the cyclol hypothesis for egg albumin. All these were deductions from too little knowledge and this was realized by most protein scientists by 1937.[67]

---

* Hydrophobic regions of a molecule are repelled by water, while hydrophilic regions are attracted (see chapter 5).

Senechal commented on the contrast between Dorothy Wrinch and Dorothy Hodgkin: "Hodgkin was pleasant, noncompetitive, loyal to her mentor Bernal, a threat to no one; and she always insisted that being a woman was no obstacle to her career. Everyone compared the two Dorothys, then and now."[68] Wrinch was arrogant, acerbic, and publicity-seeking—all traits that Pauling shared, as did a number of her other male colleagues. She was a mathematician and had no idea how to talk her audience of chemists and biologists. She treated them as math students, albeit somewhat dim-witted, and did not use the normal chemical notation but devised her own abbreviated symbols. Wrinch saw chemistry as topology and believed in the primacy of mathematics over chemistry. In response to objections concerning implausible interatomic distances and bond lengths in her cyclol models, she wrote: "If and when the actual values of these angles and distances are known, these structures will presumably lend themselves to modification by a uniform or systematic deformation which leaves the topology unchanged. As will be seen..., a considerable part of the argument is concerned wholly with topological considerations and so is independent of any metrical data."[69] This was neither language nor argument that would persuade chemists. On the basis of fifteen years of study of the chemical bond, Pauling believed that proteins would have the same bond properties as were found in other organic molecules. Wrinch, who had only the knowledge of chemistry picked up during her time in others' laboratories, believed that "[p]roteins are so different from other substances that it is surprising that there is a reluctance to accept for them a structure for which no analogue in organic chemistry can be found."[70]

Three hundred years earlier, Johannes Kepler had suggested that the radii of the planetary orbits were determined by the ratio of the sizes of nested regular polyhedra, an idea that was eventually discarded by the scientific community. Like Kepler, Wrinch tried to impose more mathematics on the universe than the universe was prepared to accept.[71] Wrinch subscribed to an approach taken more often by theoretical physicists than by biologists or chemists: Paul Dirac once wrote, "This result is too beautiful to be false; it is more important to have beauty in one's equations than to have them fit experiment....It seems that if one is working from the point of view of getting beauty in one's equations, and if one has really a sound insight, one is on a sure line of progress,"[72] and Hermann Weyl had said, "My work always tried to unite the true with the beautiful, but when I had to choose one or the other, I usually chose the beautiful."[73]

Pauling's Approach to Protein Structure

It would take almost fifteen years after Pauling and Mirsky's 1936 paper for the single-chain polypeptide model to be fully validated. Pauling was cautious—he did not want to suffer Wrinch's fate. He worked from a number of assumptions:[74]

1. *Before trying to determine the structure of proteins, it was important to know the structures of their constituents—amino acids.* In the 1939 Pauling-Niemann paper, Pauling wrote:

> A protein molecule, containing hundreds of amino acid residues, is immensely more complicated than a molecule of an amino acid.... Yet... no complete structure determination for any amino acid had been made until within the last year, when Albrecht and Corey succeeded... in accurately locating the atoms in crystalline glycine.... The investigation of crystals of relatively simple substances related to proteins is being continued in these laboratories.[75]

Robert Corey joined Pauling's research group in 1937. He was an expert in X-ray crystallography, a shy man with a bad limp from infantile paralysis. He went to work on the problem of determining the structure of amino acids and small polypeptides. Pauling's style in managing research was quite different from that of Lewis, who worked in the laboratory himself and usually collaborated with, at most, one or two research assistants. Pauling had a reputation for running his research laboratory as a factory with himself as the manager and the researchers as the workers operating under his direction; this had caused him difficulties with both Arthur Noyes and Robert Millikan.[76] But Pauling's collaboration with Corey seemed to be closer, and it was Pauling and Corey whose names would appear on the important papers in 1951.

2. *Proteins have the same bond angles, lengths, and energies as other organic molecules.* Rather than trying to determine protein structures directly from X-ray diffraction photos, Pauling would build models (much like Tinkertoys, with balls showing the atoms and sticks showing the bonds) and then use the X-ray data to accept or reject possible models. Bernal would later say:

> By 1940 it was clear that a successful attack on the complete protein structure could be made, but there were still many difficulties. Two modes of attack suggested themselves: the first was a straightforward x-ray crystallographic study of crystalline proteins, using all the techniques of an advanced crystal analysis. Computers were not available for this until much later, in the mid 1950s. The second was a model building method based on an exact knowledge of the structure of the amino acids and smaller peptides themselves and an attempt to build up the protein a priori and then check the structure by x-ray methods. I remember very well discussing the problem with Pauling just before the war. He was in favor of the second method, which I thought indirect and liable to take a very long time. Nevertheless, it was Pauling's ideas that were to have a decisive effect on the result.... [Pauling] knew his atoms and their various states and binding

> conditions so well that he was prepared to break with what after all are only conventions—such as the regularities of classical crystallography—if they could not be fitted into these regularities.[77]

To return to an earlier analogy, using X-ray diffraction photos to determine protein structure was as difficult as using only soundless shadow projections of a play to determine the plot. But using only shadow projections, it might still be possible to tell the difference between performances of *Oklahoma!* and *The Importance of Being Earnest*. This was Pauling's approach.

3. *The peptide O=C–N–H bond, stabilized by resonance energy, is planar, and there is no free rotation about the C–N bond.* Pauling had arrived at this conclusion from his idea of resonance energy. He believed that the electrons would delocalize, giving the C–N a partial double bond character, so that rotation would be impeded. He emphasized this point in his 1954 Nobel lecture,[78] referring to a paper he had published on single bond–double bond resonance in 1935.[79] By restricting the models that he considered to only those including planar peptide bonds, the problem became much simpler.

4. *Protein conformation, as well as immunological interactions between proteins, is determined by noncovalent interactions*—principally hydrogen bonds. Pauling had made it clear in his 1939 paper with Niemann that covalent bonds would form and break too slowly to explain the speed with which proteins change conformation. In that same paper, he argued, however, against the importance of the hydrophobic effect, which did turn out to be another important noncovalent interaction.

5. *Whatever the principles of protein conformation are, they apply to all amino acids.* Pauling was looking for a model that was focused on the peptide bond rather than on particular side chains. While a side chain might kink the peptide chain or cause additional hydrogen bonding or cross-linking, the peptide bond, which linked the amino acids in a chain, had to be at the heart of the solution to the problem.

Like Wrinch, Pauling was speculating without much experimental evidence: he suggested a "stack of pancakes" model for protein conformation in 1940,[80] wherein the peptide chain ran back and forth, forming a pancake held together by hydrogen bonds. Several pancakes would then stack together, also held together by interpancake hydrogen bonds. But protein structure was not at the top of Pauling's list of important topics during the 1940s. Part of this was due to war work, but he was more interested in the physiological function of proteins, particularly the antibody–antigen interaction (the same question that Paul Ehrlich and Svante Arrhenius had disputed forty years earlier), than he was in protein structure.[81] In 1948, while sick with the flu in a hotel room, Pauling began working on the problem of protein conformation again. He drew chemical structures of polypeptides (with planar peptide bonds) on pieces of paper and began rolling them into tubes, making the peptide bond chain into a spiral or helix, and looking for places where possible hydrogen bonds might form across the chain. He found that it was possible to form a hydrogen bond either between each third amino acid or between each fifth amino acid. He called the first case the α-helix and the second the γ-helix. The surprising thing was that for neither structure was there an integral number of amino acids in each turn of the spiral—for the α-helix, it was about 3.7 amino acids per turn, and for the γ-helix, it was about 5.1 amino acids per turn. He was not ready to publish these structures in 1948, because neither of them agreed with some experimental evidence in the X-ray spectra. But when Sir Lawrence Bragg's group at Cambridge seemed to be getting close and published a paper titled "Polypeptide Chain Configurations in Crystalline Proteins," Pauling decided to risk it and publish, because he was also feeling competition from his former student (and Wrinch coauthor) David Harker, who was setting up a laboratory for protein research with Langmuir's backing.[82] The structure proposed by Bragg's group also included a helix or spiral configuration, but Pauling knew that their structure was wrong—the structure's peptide bonds were not planar. Furthermore, Bragg had only considered structures with integral numbers of amino acids per turn of the spiral: "We have preferred structures in which the repeat (or pseudo-repeat) contains three amino acid residues.... We do not feel however, that we can entirely exclude the possibility that the repeat contains four residues."[83] Pauling and

Figure 9-6. Pauling lecturing on the structure of proteins and the planar peptide bond, 1957. Oregon State University, Ava Helen and Linus Pauling Papers.

Corey sent a note to the *Journal of the American Chemical Society* describing the essence of their structures—the nonintegral number of amino acids per turn and the planar peptide bonds—and presented the structures to CalTech's biologists. Pauling pulled the shroud off a tall molecular model of the helix at the climactic moment of the talk.[84]

Shortly thereafter, Pauling proposed his β-sheet structure for proteins (somewhat like his old "pancake" structure). Although his helices and his β-sheet were based on X-ray studies of amino acids and small polypeptides rather than proteins, they would eventually be shown to be fundamental subunits in a great many protein structures. Max Perutz, one of the coauthors of Bragg's paper, found that the α-helix was supported by the X-ray photos of hemoglobin and published accordingly, giving up on the Bragg structure.[85] Bragg had missed out. In a state of obvious agitation, he walked into the Oxford office of Alexander Todd, an organic chemist who was Pauling's friend, and demanded to know why the Pauling-Corey structure should be preferred to his own; when Todd explained that Pauling's structure showed a planar peptide bond whereas his did not and explained Pauling's ideas on resonance to him, Bragg was crestfallen. Pauling later crowed to Todd, "I judge that he did not read *The Nature of the Chemical Bond* carefully enough!"[86]

In the May 1951 issue of the *Proceedings of the National Academy of Sciences*, Pauling and Corey published seven papers back to back,[87] all on the detailed structures of proteins. They provided coordinates for the α-helix, γ-helix, and β-sheet subunits and suggested detailed models for polypeptides, for globular proteins, and for the proteins in feather rachis (the shaft or quill of the feather), collagen, and muscle.

Pauling was a chemist's chemist: following in Lewis's path in 1939 by publishing *The Nature of the Chemical Bond*, Pauling explained a difficult subject, quantum mechanics, in terms that were useful to ordinary chemists. Both chemistry and physics were now grounded in the same equations, and while the actual practices of the two disciplines were quite different, there was no argument between the two sciences on the underlying principles. With the publication of the Pauling-Corey models for protein structure, Pauling had become a leader in the similar union of chemistry and biology. Weaver's dream of "molecular biology" was becoming reality. Pauling received the Nobel Prize in Chemistry in 1954 "for his research into the nature of the chemical bond and its application to the elucidation of the structure of complex substances."

10

Pathological Science

Langmuir

On 25 January 1950, Irving Langmuir presented a paper to the American Meteorological Society that his diary called the "most important that I have ever given."[1] He stood before the U.S. experts on weather and climate and claimed that one person, releasing a small amount of silver nitrate once a week in New Mexico, had changed the weather patterns of the entire United States. His evidence was that the rains had followed the same periodicity as the silver nitrate release. Langmuir's claims were unconvincing to many in his audience, who thought that he had fallen into the kind of trap he had described many times with regard to the work of others, "where there is no dishonesty involved, but where people are tricked into false results by a lack of understanding about what human beings can do to themselves in the way of being led astray by subjective effects, wishful thinking, or threshold interactions."[2] The meteorologists may not have known his term for this sort of research, but if they had, they might have used it in the discussion period after his talk—"pathological science."

Almost four years later, when he was seventy-two years old, Langmuir would give a talk with the title "Pathological Science" at the Knolls Research Laboratory at General Electric. (His talk's examples did not include his rain-making research—he would believe wholeheartedly in its validity until his death in 1957.) In his talk, he said that he saw a pattern that showed up frequently in the scientific literature: respected scientists made astounding claims that caused a great deal of excitement but in the end amounted to nothing. But he meant more by "pathological" than simply "wrong," and he certainly did not mean "fraudulent." Pathological science was not willful misstatement of results, like Woo Suk Hwang's falsification of his data in stem cell research, revealed in Korea in 2006. Pathological science was based on self-delusion, on the desire by a researcher to reach a certain conclusion. It usually involved what Langmuir called "threshold events"—observations that were difficult to classify, and that the researcher unwittingly interpreted in a way that would lead to the desired result.

Langmuir's 1953 talk did not put an end to pathological science: In the mid-1960s Russian scientists claimed to have made small amounts of "polywater," a new polymerized form of water that had properties more like syrup than like normal water. By the late 1960s many scientists had accepted polywater's existence, and some feared that an uncontrolled chemical chain reaction might convert the oceans from normal water to polywater (a plot device used by Kurt Vonnegut in *Cat's Cradle** and that would also become the subject of a *Star Trek* episode). In 1970, Dennis Rousseau at Bell Labs experimented with his own sweat at Bell Labs after playing handball and found that it had the same properties as polywater.[3] Polywater had been due to human contamination of the apparatus involved in its preparation—it could not be produced once the glassware was thoroughly cleaned.

The most famous example of what most would consider pathological science is "cold fusion." In 1989, Stanley Pons and Martin Fleischmann, both chemists, announced at a press conference that they had achieved controlled nuclear fusion—something that the physicists had been able to do only in apparatus operating at extremely high temperatures and pressures, backed by billions in research funding—at room temperature on a laboratory bench top. Pons and Fleischmann's claims ran contrary to every accepted physical theory of nuclear behavior, and attempts to reproduce their experimental results have been, at best, inconsistent. A community of cold-fusion believers remains,[4] and there is some continuing research in the area, but it is viewed as "pariah science" by the mainstream physics community.[5]

I have reproduced part of Langmuir's 1953 talk on pathological science below, as transcribed by R.N. Hall from an audio recording in the Library of Congress (the full text is available online at www.cs.princeton.edu/~ken/Langmuir/langmuir.htm). It was published in a GE technical report[6] and later in *Physics Today*.[7] There is also a 1953 folder labeled "Pathological Science" in Langmuir's papers at the Library of Congress that contains a draft of the talk. I have made what I consider to be improvements to some of Hall's punctuation and added some footnotes in order to clarify some of the Langmuir's points.

---

* Kurt Vonnegut was the younger brother of Bernard Vonnegut, a collaborator of Langmuir's who appears later in this chapter. Kurt Vonnegut was hired as a publicist at GE and knew Langmuir. The idea for "ice-nine," the name for the fictional polywater in Vonnegut's novel *Cat's Cradle*, was in fact originated by Langmuir, who had invented it as a suggested plot device for H.G. Wells when he visited GE. When Wells did not use the idea, Kurt Vonnegut did so after Langmuir's death. Vonnegut modeled the novel's eccentric scientist Felix Hoenikker on Langmuir. [Interview with Kurt Vonnegut in Roger Summerhayes, *Langmuir's World* (CustomFlix, 1996).]

The figures included are taken from the papers in which the "pathological" theories were published, which are likely what Langmuir displayed on the slide projector. Where Langmuir seems to be pointing to the blackboard or a slide, I have indicated that in brackets.

I have included the entire text of one section, and some readers may find the technical discussion a bit overwhelming—the talk was meant for GE's scientists and engineers. My advice is to skip over whatever parts of the talk might seem incomprehensible and concentrate on Langmuir's analysis and presentation. I have left the talk exactly as Langmuir delivered it—extemporaneous and somewhat ungrammatical. The talk captures Langmuir's personality and shows why he was considered an outstanding speaker; I have found that reading it aloud enhances its effect. The talk included six examples of what Langmuir called pathological science: the Davis-Barnes effect, which is reproduced below; N-rays; mitogenetic rays; the Allison effect, of which parts are quoted below; J.B. Rhine's ESP experiments at Duke; and flying saucers. Here is Langmuir, speaking on the Davis-Barnes effect at GE in 1953:

> The thing started this way. On April 23, 1928, Professor Bergen Davis from Columbia University came up and gave a colloquium in this laboratory [at GE], and it was very interesting. He told Dr. Whitney, and myself, and a few others something about what he was going to talk about beforehand and he was very enthusiastic about it, and he got us very interested in it, and well, I'll show you right away on this diagram what kind of thing happened.

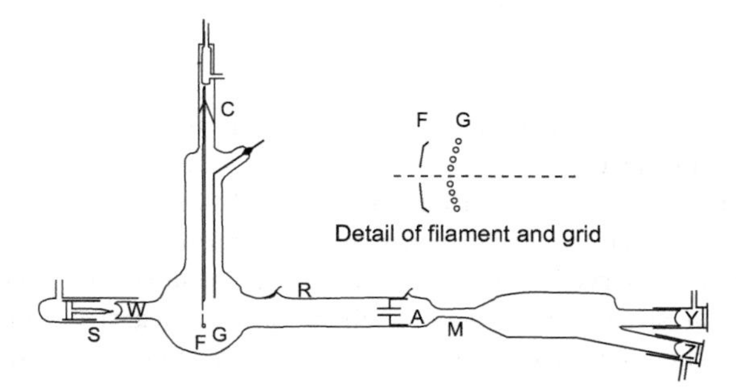

S, radioactive source; *W*, thin glass window; *F*, filament; *G*, grid; *R*, lead to silvered surface; *A*, second anode; *M*, magnetic field; *C*, copper seals; *Y* and *Z*, zinc sulphide screens.

Figure 10-1. The apparatus for the Davis-Barnes effect. Reprinted with permission from Arthur Barnes, "The Capture of Electrons by Alpha-Particles." *Physical Review* 36 (1930): 217–228, on 218. Copyright 1930 by the American Physical Society.

> He produced a beam of alpha rays from polonium in a vacuum tube [starting from the radioactive source S, fig. 10-1]. He had a parabolic hot cathode electron emitter with a hole in the middle [F], and the alpha rays came

through it and could be counted by scintillations on a zinc sulfide screen with a microscope over here [Y]. The electrons were focused on this plate, so that for a distance there was a stream of electrons moving along with the alpha particles. Now you could accelerate the electrons and get up to the velocity of the alpha particles. To get an electron to move with that velocity takes about 590 volts; so if you put 590 volts here [between F and A], accelerating the electrons, the electrons would travel along with the alpha particles and the idea of the experiment was that if they moved along together at the same velocity, they might recombine so that the alpha particle would lose one of its charges, would pick up an electron, so that instead of being a helium atom with two positive charges, it would only have one charge. Well, if an alpha particle with a double charge had one electron, it's like the Bohr theory of the hydrogen atom, and you know its energy levels. It's just like a hydrogen atom, with a Balmer series, and you can calculate the energy necessary to knock off this electron and so on.*

Well what they found, Davis and Barnes, was that if this velocity was made to be the same as that of the alpha particles, there was a loss in the number of the deflected particles. If there were no electrons, for example, and no magnetic field, all the alpha particles would be collected over here [Y] and they had something of the order of 50 per minute which they counted over here. Now if you put on a magnetic field you could deflect the alpha particles so they go down here [Z]. But if they picked up an electron then they would only have half the charge and therefore they would only be deflected half as much and they would not strike the screen.**

---

* Davis and Barnes's apparatus produced a stream of alpha particles, which are helium nuclei with no electrons attached—they have a charge of +2. The alpha particles are traveling from the source S toward Y, which is a screen. If they hit the screen, they will cause a scintillation—a bright spot—to appear, which can be observed with a microscope. At the same time, Davis and Barnes had added a stream of electrons, which have a negative charge of −1. The velocity of the electrons is determined by the voltage between F and A, and the voltage was tuned so that the electrons were traveling the same path at the same speed as the alpha particles. Davis and Barnes hoped that the positive alpha particles (with charge +2) would combine with the negative electrons (with charge −1) to form helium ions with a net charge of −1. The energy levels—the velocity of the electrons in their orbits—of helium ions with one electron were known exactly according to Bohr's theory.

** Davis and Barnes applied a magnetic field to deflect the stream of moving alpha particles from their ordinary path toward Y. For an alpha particle that had not picked up an electron—with charge +2—the apparatus was arranged so that the particle would strike the screen Z, where its scintillation could be observed. But if the alpha particle had picked up an electron and had become a helium ion with

Now the results that they got, or said they got at the time, were very extraordinary. They found that not only did these electrons combine with the alpha particles when the electron velocity was 590 volts, but also at a series of discrete differences of voltage. When the velocity of the electrons was less or more by perfectly discrete amounts, then they could also combine. All the results seemed to show that about 80% of them combined. In other words, there was about an 80% change in the current when the conditions were right. Then they found that the velocity differences had to be exactly the velocities that you can calculate from the Bohr theory. In other words, if an electron coming along here happened to be going with a velocity equal to the velocity it would have if it was in a Bohr orbit, then it will be captured.*

Well that makes a difficulty right away because in the Bohr theory, when there is an electron coming in from infinity it has give up half its energy to settle into a Bohr orbit. Since it must conserve energy, it has to radiate out, and it radiates out an amount equal to the energy that it has left in the orbit. So, if the electron comes in with an amount of energy equal to the amount you are going to end up with, they have to radiate an amount of energy equal to twice that, which nobody had any evidence for. So there was a little difficulty which was never quite resolved, although there were two or three people, including some in Germany, who worked up theories to account for how that might be. Sommerfeld, for example, in Germany. He worked up a theory to account for how the electron could be captured if it had a velocity equal to what it was going to have after it settled down into the orbit.**

---

charge +1, it would not deflect as far, would miss the screen, and would not be observed.

* The Bohr theory, which reproduces the spectra of a helium ion with a charge of +1 very well, holds that only certain energy levels, corresponding to certain electron velocities, are possible. Barnes and Davis claimed that if the electrons were tuned to have exactly those velocities, 80% of them were captured—the number of scintillations counted at Z dropped by 80%.

** There are two components to the energy of an electron in a Bohr orbit. One is the kinetic energy, which is due to the electron's velocity, which Davis and Barnes were trying to match by accelerating the electron at a particular voltage. The other is the potential energy, which is due to the attraction between the negative electron and the positive alpha particle. The Bohr theory predicts (as does modern quantum theory) that the potential energy has twice the magnitude of the kinetic energy but an opposite sign—the kinetic energy is positive, while the potential energy is negative. When an electron was hypothetically trapped by the alpha particle, if energy were to be conserved, there would need to be a corresponding radiation of the same magnitude as the potential energy, which had not been observed. Arnold Sommerfeld was the German physicist who extended the Bohr theory in the 1920s and with whom Pauling studied after he received his Ph.D.

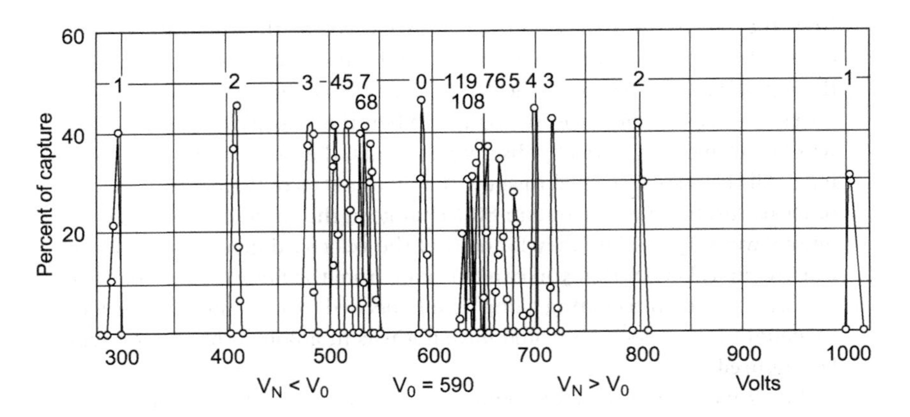

Figure 10-2. The Davis-Barnes capture of electrons as a function of voltage. Reprinted with permission from Arthur Barnes, "The Capture of Electrons by Alpha-Particles." *Physical Review* 36 (1930): 217–228, on 219. Copyright 1930 by the American Physical Society.

Well, there were these discrete peaks, each one corresponding to one of the energy levels of the Bohr theory of the helium atom, and nothing else. These were the only things they recorded. So you had these discrete peaks. Well how wide were they? Well, they were one-hundredth of a volt wide. In other words, you had to have 590 volts. That would give you equal velocities, but there were other peaks, and I think the next velocity would be about 325.1 volts. If you had that voltage, then you got beautiful capture. If you didn't, if you changed it by one-hundredth of a volt—nothing. It would go right from 80% down to nothing. It was sharp. They were only able to measure to a hundredth of a volt, so it was an all or nothing effect. Well, besides the peak at this point, there were ten or twelve different lines in the Balmer series, all of which could be detected, and all of which had an 80% efficiency. They almost completely captured all the electrons when you got exactly on the peak.

Well, in the discussion we questioned how, experimentally, you could examine the whole spectrum, because each count, you see, takes a long time. There was a long series of alpha particle counts, that took two minutes at a time, and you had to do it ten or fifteen times, and you had to adjust the voltage to a hundredth of a volt. If you have to go through all the steps at a hundredth of a volt each and to cover all the range from 330 up to 900 volts, you'd have quite a job. (Laughter.) Well, they said that they didn't do it quite that way. They had found by some preliminary work that they did check with the Bohr orbit velocities so they knew where to look for them. They found them sometimes not exactly where they expected them, but they explored around in that neighborhood, and the result was that they got them with extraordinary precision. So high, in fact, that they were sure they'd be able to check the Rydberg constant more accurately than it can be done by studying the hydrogen spectrum, which is something like one in 10^8.* At any rate, they had no inhibitions at all as to the accuracy which could be obtained by this method, especially since they were measuring these voltages within a hundredth of a volt.

Anybody who looks at the setup would be a little doubtful whether the electrons had velocities that were definite within a hundredth of a volt, because this is not exactly a homogeneous field. The distance was only about 5 mm in which they [the alpha particles and the electrons] were moving along together.

Well, in this talk, a few other things came out that were very interesting. One was that the percentage of capture was always around 80%. The curve would come along like this as a function of voltage.** The curve would come along at about 80%, and there would be a sharp peak up here and another sharp peak here, and, well, all the peaks were about the same height.

Well, we asked, "How did this depend upon current density?"†

"That's very interesting," he said, "it doesn't depend at all upon current density."

We asked, "How much could you change the temperature of the cathode here?"

"Well," he said, "that's the queer thing about it. You can change it all the way down to room temperature." (Laughter.)

"Well," I said, "then you wouldn't have any electrons."

"Oh yes," he said, "if you check the Richardson equation and calculate, you'll find that you get electrons even at room temperature, and those are the ones that are captured."‡

"Well," I said, "there wouldn't be enough to combine with all the alpha particles, and, besides that, the alpha particles are only there for a short time as they pass through and the electrons are a long way apart at such low current densities, at 10^{-20} amperes or so." (Laughter.)

He said, "That seemed like quite a great difficulty. But," he said, "you see it isn't so bad because we now know that the electrons are waves. So the electron doesn't have to there at all in order to combine with something. Only the waves have to be there, and they can be of low intensity, and the

---

* The Rydberg constant is determined by the spacing of the lines in the atomic spectrum.

** Fig. 10-2 shows the voltage drop at about 40%, not 80%. This figure was taken from Davis and Barnes's paper, published about a year after their talk at GE, which may explain the discrepancy.

† The current density is determined by the number of electrons.

‡ The Richardson equation is for thermionic emission, in which Langmuir had done extensive work (see chapter 5).

quantum theory causes all the electrons to pile in at just the right place where they are needed." So he saw no difficulty. And so it went.

Well Dr. Whitney likes the experimental method, and these were experiments, careful experiments, described in great detail, and the results seemed to be very interesting from a theoretical point of view. So Dr. Whitney suggested that he would like to see these experiments repeated with a Geiger counter instead of counting scintillations, and C. W. Hewlett, who was here working on Geiger counters, had a setup, and it was proposed that we would give him [Davis] one of these, maybe at a cost of several thousand dollars for the whole equipment, so that he could get better data. But I was a little more cautious. I said to Dr. Whitney that before we actually give it to him and turn it over to him, it would be well to go down and take a look at these experiments and see what they really mean. Well, Hewlett was very much interested, and I was interested, so only about two days later, after this colloquium, we all went down to New York. We went to Davis's laboratory at Columbia University, and we found that they were very glad to see us, very proud to show us all their results, so we started early in the morning.

We sat in the dark room for half an hour to get our eyes adapted to the darkness so that we could count scintillations. I said, first I would like to see these scintillations with the field on and with the field off.* So I looked in and I counted about 50 or 60 [counts in two minutes]. Hewlett counted 70, and I counted somewhat lower. On the other hand, we both agreed substantially. What we found was this: These scintillations were quite bright with your eyes adapted, and there was no trouble at all about counting them when these alpha particles struck the screen. They came along at the rate of about 1 per second. When you put on a magnetic field and deflected them, the count came down to about 17, which was a pretty high percentage, about 25% background. Barnes was sitting with us, and he said, "That's probably radioactive contamination of the screen." Then Barnes counted, and he got 230 on the first count and about 200 on the next, and when he put on the field it went down to about 25. Well Hewlett and I didn't know what that meant, but we couldn't see 230. Later we understood the reason.

I had seen, and we discussed a little at this point, that the eyepiece was such that, as you looked through, you got some flashes of light which I took to be flashes just outside the field of view that would give a diffuse glow that would be perceptible. And you could count them if you wanted. Well,

* Langmuir was observing the screen at Z in fig. 10-1. The magnetic field deflects the alpha particles to screen Z. With the magnetic field on, all the alpha particles should strike the screen at Z; with the field off, none should go to Z, and any scintillations still observed at Z should be background scintillations from something other than the alpha particle stream. This number would be used to correct the experimental observations.

Hewlett counted those, and I didn't. That accounted for some difference. Well, we didn't bother to check into this, and we went on.

Well, I don't want to spend too much time on this experiment. I have a 22-page letter that I wrote about these things and I have a lot of notes. There was a long table at which Barnes was sitting, and he had another table over here where he had an assistant of his named Hull who sat there looking at a big scale voltmeter, or potentiometer really, but it had a scale that went from one to a thousand volts, and on that scale that went from one to a thousand, he read hundredths of a volt. (Laughter.) He thought he might be able to do a little better than that. At any rate, you could interpolate and put down figures, you know. Now the room was dark except for a little light here on which you could read the scale on your meter. And it was dark except for the dial of a clock, and he counted scintillations for two minutes. He said he always counted for two minutes. Actually, I had a stop watch and I checked him up. They sometimes were as low as one minute and ten seconds, and sometimes one minute and fifty-five seconds, but he counted them all as two minutes, and yet the results were of high accuracy!

Well, we made various suggestions. One was to turn the voltage off entirely. Well, then Barnes got some low values around 20 or 30, sometimes as high as 50. Then to get conditions on a peak, he adjusted the voltage down to two hundred and—well, some of these readings are interesting: 325.01. That's the figure I put down, and there he got a reading of only 52, whereas before, when he was on the peak, he got about 80. He didn't like that very much, he tried changing this to .02, a change of one hundredth of a volt. And then he got 48. Then he went in between. (Laughter.) They fell off, you see, so he tried 325.015, and then got 107. So that was a peak.

Well, a little later, I whispered to Hull who was over here adjusting the voltage, holding it constant, I suggested to him to make it one tenth of a volt different. Barnes didn't know this, and he got 96. Well, when I suggested this change to Hull, you could see immediately that he was amazed. He said, "Why, that's too big a change. That will put it way off the peak." That was almost one tenth of a volt, you see. Later, I suggested taking a whole volt. (Laughter.)

Then we had lunch. We sat for half an hour in that dark room so as not to spoil our eyes, and then we had some readings at zero volts, and then we went back to 325.03. We changed by one-hundredth of a volt, and there he got 110. And now he got two or three readings at 110.

Then I played a dirty trick. I wrote out on a card of paper ten different sequences of V and 0. I meant to put on a certain voltage and then take it off again. Later, I realized that this wasn't quite right, because when Hull took off the voltage, he sat back in his chair—there was nothing to regulate, so he didn't. Well, of course Barnes saw him whenever he sat back in his chair. Although the light wasn't very bright, he could see whether he was sitting back in his chair or not, so he knew the voltage wasn't on, and the result was that he got a corresponding result. So later I whispered,

"Don't let him know that you're not reading," and I asked him to change the voltage from 325 down to 320 so he'd have something to regulate, and I said, "Regulate it just as carefully as if you were sitting on a peak." So he played the part from that time on, and from that time on Barnes' readings had nothing whatever to do with the voltages that were applied; whether the voltage was at one value or another didn't make the slightest difference. After that, he took twelve readings, of which about half of them were right and the other half were wrong, which was about what you would expect out of twelve sets of values.

I said, "You're through. You're not measuring anything at all. You have never measured anything at all."

"Well," he said, "the tube was gassy." (Laughter.) "The temperature has changed and therefore the nickel plates must have deformed themselves so that the electrodes are no longer lined up properly."

"Well," I said, "isn't this the tube in which Davis said he got the same results when the filament was turned off completely?"

"Oh yes," he said, "but we always made blanks to check ourselves, with and without the voltage on." He immediately—without giving any thought to it—he immediately had an excuse. He had a reason for not paying any attention to any wrong results. It was just built into him. He just had worked that way all along and always would. There was no question but that he was honest; he believed these things absolutely.

Hewlett stayed there and continued to work with him for quite a while, and I went in and talked it over with Davis, and he was simply dumbfounded. He couldn't believe a word of it. He said, "It absolutely can't be." He said, "Look at the way we found those peaks before we knew anything about the Bohr theory. We took those values and calculated them up and they checked exactly. Later on, after we got confirmation, in order to see whether the peaks were there, we would calculate ahead of time." He was so sure from the whole history of the thing that it was utterly impossible that there never had been any measurements at all, that he just wouldn't believe it.

Well, he had just read a paper before the research laboratory at Schenectady, and he was going to read the paper the following Saturday before the National Academy of Sciences, which he did, and gave the whole paper. And he wrote me that he was going to do so on the 24th. I wrote to him on the day after I got back. Our letters crossed in the mails, and he said that he had been thinking over the various things that I had told him, and his confidence wasn't shaken, so he went ahead and presented the paper before the National Academy of Sciences. Then I wrote him a 22-page letter, giving all our data and showing really that the whole approach to the thing was wrong: that he was counting hallucinations, which I found is common among people who work with scintillations if they count for too long. Barnes counted for six hours a day and it never fatigued him. Of course it didn't fatigue him, because it was all made up out of his head. (Laughter.) He told us that you mustn't count the bright particles. He had a beautiful reason for why you mustn't pay any attention to the bright flashes. When Hewlett tried to check his data, he said, "Why, you must be counting those bright flashes. Those things are only due to radioactive contamination or something else." He had reason for rejecting the very essence of the thing that was important. So I wrote all this down in this letter and I got no response, no encouragement. For a long time, Davis wouldn't have anything to do with it. He went to Europe for a six month leave of absence, came back later, and I took up the matter with him again.*

In the meantime, I sent a copy of the letter that I had written to Davis to Bohr asking him to hold it confidential but to pass it on to various people who would be trying to repeat these experiments—to Professor Sommerfeld and other people—and it headed off a lot of experimental work that would have gone on. And from that time on, nobody ever made another experiment except one man in England who didn't know about the letter that I had written to Bohr, and he was not able to confirm any of it. Well, a year and a half later, in 1931, there was just a short little article in the *Physical Review* in which they say that they haven't been able to reproduce the effect: "The results reported in the earlier report depended upon observations made by counting scintillations visually. The scintillations produced by alpha particles on a zinc sulfide screen are a threshold phenomenon. It is possible that the number of counts may be influenced by external suggestion or auto-suggestion to the observer," and later in the paper they said that they had not been able to check any of the older data. And they didn't even say that the tube was gassy. (Laughter.) To me the thing that is extremely interesting, that men, perfectly honest, enthusiastic over their work, can so completely fool themselves....

These are cases where there is no dishonesty involved, but where people are tricked into false results by a lack of understanding about what human beings can do to themselves in the way of being led astray by subjective effects, wishful thinking, or threshold interactions. These are examples of pathological science. These are things that attracted a great deal of attention. Usually hundreds of papers have been published upon them. Sometimes they have lasted for fifteen or twenty years and then they gradually die away. Now the characteristic rules are these:

Symptoms of Pathological Science

1. *The maximum effect that is observed is produced by a causative agent of barely detectable intensity, and the magnitude of the effect is substantially independent of the intensity of the cause....* Davis-Barnes worked just as well when the filament was turned off.

* Eight months later, Davis and Barnes submitted a paper to *Physical Review*, which was accepted.

2. *The effect is of a magnitude that remains close to the limit of detectability; or, many measurements are necessary because of the low statistical significance of the results*.... There is a habit among most people, that when measurements of low significance are taken, they find means of rejecting data. They are right at the threshold value, and there are many reasons why you can discard data. Davis and Barnes were doing that right along. If things were doubtful at all, why, they would discard them or not discard them depending on whether they fit the theory. They didn't know that, but that's the way it worked out.
3. *There are claims of great accuracy*. Barnes was going to get the Rydberg constant more accurately than the spectroscopists could....
4. *Fantastic theories contrary to experience*. In the Bohr theory, the whole idea of an electron being captured by an alpha particle when the [electrons] aren't there just because the waves are there doesn't make a very sensible theory.
5. *Criticisms are met by ad hoc excuses thought up on the spur of the moment*. They always had an answer—always.
6. *Ratio of supporters to critics rises up to somewhere near 50% and then falls gradually to oblivion*. The critics can't reproduce the effects. Only the supporters could do that. In the end, nothing was salvaged. Why should there be? There isn't anything there. There never was. That's characteristic of the effect.

The Allison Effect

Langmuir's talk on pathological science also included the Allison effect, named after Fred Allison, a professor of physical chemistry at the University of Alabama. Langmuir said, "It started in 1927. There were hundreds of papers published.... Why, they discovered five or six different elements... Alabamine, Virginium, a whole series of elements and isotopes were discovered by Allison." The Allison effect was based on the Faraday effect, in which a beam of polarized light moving through a liquid is rotated if a magnetic field is applied in the direction of the light beam. Allison investigated the time lag between applying the field and observing the rotation of the light beam. He found that the time lag varied depending on the liquid used—carbon disulfide gave a different time lag than carbon tetrachloride, for example. If a metallic salt, copper sulfate, for example, were added to the liquid, a time lag due to the salt could be observed, and if multiple salts were added, their effects were additive. But there were some odd things, according to Langmuir:

> They found that only the metal elements counted, but they didn't act as an ion. That is, all potassium ions weren't the same, but potassium nitrate

and potassium chloride and potassium sulfate all had quite characteristic different points that were characteristic of the compound. It was only the positive ions that counted, and yet the negative ions had a modifying effect. But you couldn't see the negative ions directly.

Now they began to see how sensitive it was. Well they found that any intensity more than about 10^{-8} molar would always show the maximum effect, and you'd think that would be kind of discouraging from the analytical point of view, but no, not at all. And you could make quantitative measurements down to about three significant figures by diluting the solutions down to the point where the effect disappeared. . . .

They then went on to find that all isotopes stuck out like sore thumbs with great regularity. In the case of lead, they found 16 isotopes. . . . Well, it became important as a means of detecting elements that hadn't yet been discovered, like alabamine and elements that are now known, and filling out the periodic table. All the elements in the periodic table were filled out that way and published.

But a little later, in 1945 or 1946, I was at the University of California. . . .

Wendell Latimer had a bet with G. N. Lewis in 1932. He [Latimer] said, "There's something funny about this Allison effect, how they can detect isotopes. . . . I think I'll go down and see Allison, to Alabama, and see what there is in it."

Now people had begun to talk about spectroscopic evidence that there might be traces of hydrogen of atomic weight 3.* . . . Latimer said, "Well this might be a way of finding it, I'd like to find it." So he went and spent three weeks at Alabama with Allison, and before he went he talked it over with G. N. Lewis about what he thought the prospects were, and Lewis said, "I'll bet you ten dollars you'll find that there's nothing in it." And so they had this bet on. He went down there and he came back. He set up the apparatus, and he made it work so well that G. N. Lewis paid him his ten dollars. (Laughter.) He then discovered tritium and he published an article in the *Physical Review*. . . . I saw him then, seven or eight years after that . . . I told him . . . how the Allison effect had all these characteristics [of pathological science]. . . . Anyway, Latimer said, "You know, I don't know what was wrong with me at that time." He said, "After I published that paper, I never could repeat the experiments again. I haven't the least idea why. But . . . those results were wonderful. I showed them to G. N. Lewis, and we both agreed that it was all right. They were clean cut. I checked myself every way I knew how to. I don't know what else I could have done, but later on I just couldn't ever do it again."

During the talk's question period, Langmuir said, "I understand that Lewis got his ten dollars back." All the elements and isotopes that had been announced based on the Allison effect had to be retracted, and when the

* Now known as tritium.

elements were discovered later, they were renamed by their new discoverers. Alabamine is now astatine, and virginium is now francium. Luis Alvarez and Robert Cornog in 1938, not Latimer in 1933, discovered tritium. There is a telling comment in a paper by Allison and Edgar Murphey that fits Langmuir's symptoms of pathological science exactly: "[The method's] operation, however requires the attainment of a considerable amount of experience and technique on the part of the observer. It is not to be expected, in the present stage of development, that the method will yield dependable results in the hand of the average observer."[8]

Langmuir's Pathological Science?

Langmuir's talk touches directly at one of the recurring themes of this book—how scientific theories are developed, tested, and then accepted or rejected. It also shows Langmuir's gripping speaking style, his incisive mind, and his somewhat tactless approach to correcting those that he considered to be in error. Langmuir's biographer Albert Rosenfeld says that "Langmuir was always 'pointing out errors' in other men's work. And occasionally the erring soul would get angry, and Langmuir would be surprised and puzzled.... 'Having Langmuir point out an error,' one of his colleagues recalls, 'was like having a ton of bricks land on you.'"[9]

Langmuir was, of course, not alone in enjoying pointing out others' errors. Walther Nernst had made a point of picking apart Svante Arrhenius's immunochemistry, Gilbert Lewis had published every one of Nernst's errors he could find, and Linus Pauling had dedicated an entire paper to dissecting Dorothy Wrinch's cyclol theory. But Langmuir, although somewhat rough at times, does not seem to have been vindictive or to have focused his attacks on those he regarded as personal enemies.

Now let's turn the table for a moment. Was Langmuir guilty of "pathological science"? His attempts to attack the problem of protein configuration with Wrinch, although unsuccessful, did not fit any of Langmuir's symptoms of pathological science, with the possible exception of the last—"Ratio of supporters to critics rises up to somewhere near 50% and then falls gradually to oblivion."[10] The cyclol theory was wrong, but it was eventually testable, and it did not involve threshold effects or self-delusion by the observer. But there is another area, cloud seeding, that became Langmuir's great passion at the end of his career, and it is much closer to his description of pathological science. In the same way that his work on the lightbulb eventually led to his theories of surface chemistry, his work on rainmaking was at the end of a chain of other investigations. The chain went like this: masks for protection against poisonous smoke → smoke generators → electrical interference in radio transmission → icing of airplanes → supercooled water vapor → cloud seeding with smoke generators. It all connects in the end, but it will take a few pages to track Langmuir's progression.

The Path to Cloud Seeding

In 1940, Langmuir's research assistant was retiring, and Langmuir decided to hire Vincent Schaefer, an apprentice machinist, as his replacement. Both Langmuir and his colleague Katherine Blodgett had cabins at Lake George, and Schaefer had become friendly with both of them through his conservation work there.[11] As World War II approached, Langmuir was again summoned to the defense effort. The U.S. Army was concerned that the Germans might develop poisonous smoke and that the gas masks that the army was using would be ineffective. A protective mask would need to include a filter that would block the smoke. Langmuir and Schaefer needed to be able to generate smoke of known particle sizes if they were to be able to test filters. The generated smoke had to be stable, so that the particle size would not change during the test, and it had to be of small particle size, so that it would be difficult to filter. After the war, Langmuir said that he and Schaefer began learning

> to make particles of uniform size, determine how to measure them, and to learn how much of the material went through the filter. That work lasted for about one year. We obtained fairly successful theoretical results and a better understanding of how to build a good filter. But notice what we did incidentally: we acquired a great deal of detailed knowledge as to how to make a smoke which would be nonvolatile, which would consist of very small particles, far smaller than that of ordinary smokes, and we learned much about their optical properties.[12]

By August 1941, Langmuir and Schaefer had largely solved the problem of developing filters and making smoke with small, uniform particles. The United States had not yet entered the war, but the Allied bomber pilots had noted that the Germans were using smoke screens effectively to obscure industrial targets against Allied bombers, and the War Department asked for ideas on how to do the same—to create artificial fog or smoke screens that could cover large areas. The optimum particle size for scattering light is of approximately the same size as the wavelength of the light, and for visible radiation, this meant a particle size of 0.6 micrometers; the smoke that Langmuir and Schaefer had generated for their filter tests was 0.5 micrometer smoke, coincidentally almost exactly the same size. But they had been generating smoke at only 0.01 gram per second, and a smoke-screen generator would require a great deal more, so the two of them split up—Schaefer to design a smoke generator by trial and error, and Langmuir to work out the theory. They met a week later with almost exactly the same plans for a device, which they tested in the Schoharie Valley, an area in New York State that they had both hiked extensively, which had a flat bottom and steep walls. The new test device was still much smaller than the eventual generator would be—it burned only ten gallons of oil per hour, about a tenth of what would be required—but it was enough. Preliminary tests showed that the new generator was 400 times as effective as the generator the army was then purchasing. Langmuir arranged a demonstration for the military for 24 June 1942. When he discovered that the brass expected to start for the valley at 9:00 A.M., Langmuir told them they needed to start the demonstration at sunrise, when there would be no wind, so they would all be getting wakeup calls at 2:30. The demonstration was a great success: the valley was completely filled with smoke in a short time, and the air cleared about an hour later when the wind picked up. The Allies used the new smoke generators to cover troop movements in Holland and North Africa, to screen the crossing of the Rhine and the invasion of Italy, and to protect navy ships in the Pacific against kamikaze attacks.[13]

Early in 1943, the Allies were planning to invade Japan through the Aleutians, the chain of islands that stretches from Alaska to Siberia, and were worried because airplanes flying through snowstorms were becoming electrically charged, causing corona discharges. They were losing radio contact, and the discharges also impeded directional navigation devices. When Langmuir and Schaefer began to attack this problem, they looked for a test environment and found one close at hand, Mount Washington in the White Mountains of New Hampshire. Mount Washington's weather is some of the world's most extreme: the world's highest surface wind speed, 231 miles per hour, was recorded there, and more than twenty feet of snow falls in a typical winter. Langmuir and Schaefer planned to take different materials with different coatings up to the summit and measure the electrical charges as snow struck the surfaces. They found was that this was impossible, because rime ice (a white ice that forms when fog droplets freeze) quickly coated everything they brought up. They moved the tests on electrical charging indoors, where they could better control the environment.

But rime icing on Mount Washington made it an excellent test laboratory for aircraft icing, and the War Department was interested in that, as well. Langmuir and Schaefer began investigating icing, using both Mount Washington and a B-17 as test laboratories. To reach his test equipment at the top of Mount Washington, Langmuir, then more than sixty years old, skied up and down the mountain, or climbed using ice cleats.[14]

When the war ended, GE got out of the meteorology business. Because of his status at GE, Langmuir had free rein to do whatever research he wished, and Schaefer, as his assistant, shared the privilege. The two of them were still interested in ice formation, and Schaefer continued his research using a home freezer, which he had lined with black velvet. He shone a beam of light into it and then breathed into the beam, where he could see a fog form. Although the

Figure 10-3. Langmuir in the Adirondacks. Picture by Vincent Schaefer. Courtesy of Roger Summerhayes.

temperature was –23°C inside the chest, no ice crystals formed, and the water in the air remained supercooled although well below its freezing point of 0°C. In an attempt to lower the temperature further, Schaefer threw a piece of dry ice (solid carbon dioxide) into the freezer, and the air began to fill with ice crystals. The effect did not depend on the size of the piece of dry ice—the smallest piece had the same effect. And it had nothing to do with the dry ice itself—a needle dipped in liquid air had the same effect. It seemed that any object with a temperature lower than –40°C would cause the water in the air to crystallize in a chain reaction, limited by the diffusion or mixing rate of the ice crystals. As ice crystals formed, they would act as nuclei for the formation of further ice crystals, eventually filling the chamber. Schaefer decided to try the experiment on a cloud. He took off in a small plane from Schenectady, carrying six pounds of granulated dry ice, which he dumped into a cumulus cloud over a four-mile stretch. *Time* magazine described the result: "Almost at once the cloud, which had been drifting along peacefully, began to writhe as in torment. White pustules rose from its surface. In five minutes the whole cloud melted away, leaving a thin wraith of snow. None of the ice reached the ground (it evaporated on the way down), but the dry ice treatment had successfully broken up the cloud."[15]

Schaefer and Langmuir continued seeding clouds—Langmuir's bargain with GE was that he could work on what he wanted. "In one case," Langmuir said, "I happened to be driving an automobile under a cloud which was being seeded. I never saw such heavy rain in all my life. It came down in torrents for about fifteen minutes.... It then let up a little, so I drove on, less than half a mile. The rain stopped. I got out and looked, and the road was dry. It had not rained there at all."[16] In the winter of 1946, they seeded clouds over Schenectady, and a blizzard occurred shortly thereafter.

The legal department at GE was becoming worried. Was the company going to be liable for storm damage or for stealing rain that would have fallen elsewhere? The company decided to release all its patents in this area for public use, and the U.S. government took over the rainmaking project, now named "Project Cirrus," although both Langmuir and Schaefer remained involved and were employees of GE. In October 1947, the two of them seeded a hurricane over the Atlantic, which was heading northeast away from the Florida coast. Shortly thereafter, the hurricane made a sudden turn to the west, slammed into Georgia, and further panicked the GE lawyers when some who had suffered damages from the hurricane filed lawsuits against GE.

Langmuir began working out a mathematical theory of cloud seeding, which he published in 1948. He predicted that chain formation of droplets, analogous to a nuclear reaction, could occur in a cloud and that an entire cloud might be brought down by seeding with even a single drop of water if the conditions were appropriate.[17]

Another of Langmuir's assistants, Bernard Vonnegut, found that seeding with silver iodide was even more effective than with dry ice.[18] Dry ice initiated the condensation process by cooling water below the temperature where it could remain supercooled at atmospheric pressure, but in a warm cloud, the dry ice would quickly evaporate to gaseous carbon dioxide. Silver iodide worked differently—it mimicked the shape of ice crystals. Vonnegut had searched the scientific literature to find a compound that had the symmetry and dimensions that matched those of ice crystals, and came up with several, silver iodide being the most promising. Since it was crystalline even at room temperature, he hoped that it could repeatedly nucleate crystallization of water. Furthermore, it could be generated on the ground as a smoke, allowing it to drift into a suitable cloud. Langmuir, Schaefer, and Vonnegut adapted the design of the smoke generator that they had developed for the military into a silver iodide smoke generator, burning fibers that had been coated with silver iodide and forming finely granulated smoke.

Larger Claims

So far, so good—no pathological science yet. Rainmaking before 1946 was mostly a story of charlatans and the deluded, not much advanced from primitive rain dances, although the basic meteorological principles of condensation had been understood since the 1870s.[19] Langmuir and Schaefer's work was the first to attack the problem both methodically and practically. What brought

Figure 10-4. *Time* magazine cover, 28 August 1950, © Time, Inc.

accusations of "pathological science" was not the work that they did or the theories that they developed, but their claims for its practical significance and their appeals to the mass media. *Time* magazine, for example, reported:

> Near Albuquerque in July, 1949, Langmuir performed an experiment that is still debated heatedly in meteorological circles. He started his silver iodide generator early on a morning when the Weather Bureau had predicted no substantial rain. Then he watched the developments with a radar.
>
> At 8:30 AM, a cloud starting growing 25 miles away downwind. When the cloud reached 26 miles, it suddenly spurted, bulging upward at 15 mph. Soon a radar echo showed that the cloud was full of rain or snow. Heavy rain fell near the Manzano Mountains. A short while later, a second cloud showed a similar convulsion and also produced a heavy rain.
>
> Langmuir insisted that both these thunderstorms formed in the trajectory of his silver iodide particles and at about the time when the particles must have been entering their bases. He therefore took credit for the rain that dropped as well as for other rain from later storms.[20]

Langmuir published these data with a statistical analysis, saying that it seemed to "prove conclusively that silver iodide seeding produced almost all of the rain in the State of New Mexico on these two days."[21] There was no scientific doubt that silver iodide could nucleate the condensation of supercooled water—Vonnegut had clearly demonstrated this—but its practical efficacy in rainmaking was in question. There was also no doubt that seeding high, supercooled stratus clouds with dry ice could cause snow to fall from their bases. Langmuir and Schaefer had flown different patterns over such clouds, dumping dry ice as they went, and had photographed the clouds with time-lapse photography, showing the cloud breaking up in the same pattern. What *was* in question was whether rain could be reliably stimulated in much warmer cumulus clouds, which are the primary source of summer rainfall.

There is a problem in any particular cloud-seeding experiment of determining cause and effect. Would the rainfall have occurred even without the seeding? Was the observer biased, either in selecting or rejecting particular clouds for seeding, or in throwing out failures where the desired results were not achieved? Langmuir had said of pathological science, "These are cases where there is no dishonesty involved, but where people are tricked into false results by a lack of understanding about what human beings can do to themselves in the way of being led astray by subjective effects, wishful thinking, or threshold interactions." He should have been worried. Cloud-seeding experiments involved threshold interactions.

Skeptical Meteorologists

Meteorologists found Langmuir's statistical arguments unconvincing. Ferguson Hall, a meteorologist at the U.S. Weather Bureau, pointed out that Langmuir's claims to have produced large amounts of rain in New Mexico were based on an assumption that rainfall amounts in adjacent areas were statistically independent of one another, something that was certainly not the case.[22] In the same communication, five scientists and mathematicians from New York University and MIT concluded that "the periodicity exhibited in the present case is a likely occurrence by pure chance... [and] we cannot accord credence to the sweeping inference that the abnormal character of the basic weather pattern was the result of the silver iodide operations at Socorro [New Mexico]."

Langmuir presented his results to the meteorologists, as described at the beginning of this chapter, but was not well received. From his diary of 25 January 1950: "I read the most important paper that I have ever given at the afternoon session of the American Meteorological Society, about 1 1/2 hours. Reichfelder is Chairman. Wexler and Lewis out on a limb."[23] These three were meteorologists who did not believe Langmuir's theories, and whose own seeding experiments had showed negative results. In an effort to convince the experts, Langmuir had begun releasing a kilogram (2.2

pounds) of silver iodide at regular weekly intervals from Albuquerque in the summer of 1949, which he claimed, based on statistical analysis, caused large, periodic rainfall over the entire eastern United States for the summer and into the next year. He gave no physical explanation for how releasing a kilogram of silver iodide in New Mexico could cause massive rainfall in New England. Clyde Harris, a physicist, showed that silver iodide was destroyed by sunlight, and that it could not drift with the wind to cause storms in the eastern United States.

The meteorologists and statisticians were aware of the problems of experimental bias—the seeder unconsciously seeding only clouds that looked "ripe," for example—and they would accept only randomized experiments; Langmuir had not done these. Meteorologist James McDonald wrote,

> [T]he seeder must do something equivalent to *first* selecting the cloud or time interval as "seedable," and *after* announcing his decision, he must then effectively flip a coin (randomize) to decide whether he will actually seed that "seedable" situation. After many repetitions of this process, two sets of results will be obtained, those for the seedable-and-seeded and those for the seedable-and-not-seeded cases. These data, processed in any number of ways, fulfill the philosophic requirements for sampling theory.[24]

McDonald went on to point out that this is difficult to do in practice, because commercial seeders are not willing to let "seedable" clouds pass by.

It is perhaps unfair to blame Langmuir for pathological science. The experimental work on nucleation by Langmuir, Schaefer, and Vonnegut was accurate and important, and Langmuir's theories on condensation and cloud dynamics were sound. Where Langmuir failed, he did so for the same reasons that Lewis failed with his theories of radiation, and that Wrinch failed with her theories of protein structure. He entered a new field, meteorology, full of ideas. He did not bother to ground himself in the literature of that field, or to understand that meteorologists, who knew that they were dealing with threshold phenomena, would accept only statistical evidence that was protected against experimental bias. Like Wrinch, Langmuir addressed the scientists in the new field as if they were slightly dim, rederiving equations that they had used for years but expressing them in forms unfamiliar to them. And like Wrinch, Langmuir found that his substantial contributions to the field were overshadowed by his spurious claims.

The general consensus at present among meteorologists is that cloud seeding causes some increase in rainfall. The American Meteorological Society declared in its policy statements on weather modification:

> Operations that dissipate supercooled fog and low stratus (clouds containing water droplets at subfreezing temperatures) by seeding with ice-forming agents (e.g., dry ice, liquid nitrogen, compressed air, silver iodide, etc.) have

> become routine at some airports.... There is statistical evidence that precipitation from supercooled orographic clouds (clouds that develop over mountains) has been seasonally increased by about 10%. The physical cause-and-effect relationships, however, have not been fully documented.... Some experiments with warm-based convective clouds (bases about 10°C or warmer) involving heavy silver iodide seeding have suggested a positive effect on individual convective cells, but conclusive evidence that such seeding can increase rainfall from multicell storms has yet to be established. (1998: 2771)

> Hurricane modification experiments of the 1950s and 1960s were inconclusive. Although strong research interest continued into the 1970s, no organized research effort was undertaken, and few studies have been devoted to this subject for the past 20 years. (1992: 334, 1998: 2776)

> The use of untested weather modification techniques during severe droughts, as a means of increasing precipitation, is not recommended. Opportunities to increase precipitation are typically minimal during droughts and only well-tested techniques should be considered, realizing that only limited precipitation augmentation will probably result. (1992: 336; similar language at 1998: 2772, 1998: 2778)[25]

Langmuir's conclusions may have been overstated, but their essence has not been disproven—unlike the Davis-Barnes effects and the Allison effect, which Langmuir had cited as examples of pathological science—and has even found some support. Langmuir never admitted any possibility that he was wrong, holding firm until his death in 1957.

Langmuir called his paper on cloud seeding "the most important paper I have ever given" in his diary—more important than surface chemistry, more important than his electronic theories. He was a practical man and was always most interested in those problems that would affect the greatest number of people. Success in controlling the weather, had he achieved it, would have been as important as he thought.

Atomic Energy and a Trip to Russia

Langmuir was not directly involved in the Manhattan Project, but at the completion of the war he was one of the best-known American scientists and was asked for his views on atomic policy. Lewis, isolated in Berkeley, was not asked for his opinions.

Starting on 9 September 1945, only a month after the atomic bombs had been dropped on Japan and the war had ended, a conference on atomic energy control—addressing the question as to whether the United States should keep the atomic bomb as a national secret or turn it over to some sort of international control—was held at the University of Chicago. Forty participants were invited, including many of the nation's most prominent physicists, chemists, political scientists, economists, university presidents, and foundation heads; the chairman of the Federal Reserve; and a few businessmen. Among the scientists, Glenn Seaborg, Harold Urey, and Langmuir were invited. Seaborg's journal shows that the scientists were in general consensus that the secrets could not be kept, that other nations would have the bomb within a few years, and that some form of international control would be necessary. According to Seaborg, Langmuir was an active participant, suggesting "a statement that we are in favor of no secrets . . . , and to issue it with the consent of the War Department. He [Langmuir] is willing to discuss it with the State Department, President, etc."[26] Later in 1945, Langmuir testified to the same effect before the Senate hearings on atomic energy, saying that he believed a Russian feeling of insecurity was understandable in view of the fact that the United States was the only country possessing the atomic bomb.[27] There were attempts through late 1945 and early 1946 in the direction of international control of atomic energy, including the Truman-Attlee-Mackenzie declaration, the Acheson-Lilienthal report, and the Baruch plan, but by late 1947 all of this collapsed, and the nuclear arms race was on.

Langmuir's proposals for control of atomic energy were well intentioned and possibly sound, but his understanding of the situation in the Soviet Union was completely naive, so much so that it might have been called "pathological social science." In June 1945, before the United States used the bomb against Japan but after the end of the war in Europe, Langmuir and other prominent scientists were invited to the Soviet Union as guests of the Soviet Academy of Sciences. The War Department, worried about Soviet intelligence agencies gaining information about the atomic bomb, refused to allow any of those involved in the development of the bomb to travel and pressured the British to do the same, which Langmuir accurately noted in his Senate testimony must have been a dead giveaway to the Russians—by process of elimination, the missing scientists were those involved in the development of the bomb. But Langmuir, who was only marginally involved in the Manhattan Project, was permitted to go after some initial attempts to dissuade him from the trip. At first, the scientists were scheduled to fly on a Russian plane from LaGuardia Field in New York. When the travelers assembled at the Harvard Club in New York, they met two Americans who had just gotten off the same Russian plane, one coming from San Francisco and one from Chicago, and they learned what was in store for them. The plane was not a passenger plane, but a fully loaded cargo plane into which nine seats had been crammed. There were eighteen scientists traveling, so half of them would have to stand. The plane was unheated, and they were scheduled for a five or six day trip over the Arctic Circle. One of the two passengers to whom they spoke said that he had thought that he was going to die on the way to New York when it became so cold that the circulation in his legs had stopped. The plane's radios could not be used for communication or navigation within the United States—the frequencies were wrong—so while in American airspace, they would be flying without instruments and in complete radio silence. (When the Russian plane had landed without radios in bad weather in Denver on the way to New York, it had collided with an American plane on a taxiway.) Some of the scientists were still game to go, but when they learned that the Russian plane had run into a truck at LaGuardia but was being repaired, they began to lose heart. Finally, President Truman made one of his own planes available, and the trip was on.

Langmuir was tremendously impressed with his visit to the Soviet Union and seems to have believed everything he was told by the Russians on the trip. He was happy to share his new knowledge in further testimony before a Senate committee, saying that he found the treatment of leading scientists in the Soviet Union much better than in the United States:

> The [Soviet] government has deliberately chosen to adopt an incentive system which is at least as effective and as logical as that which we have inherited from our capitalist system before it was restrained by government control. Among the directors of the scientific institutes in the Russian Academy of Sciences I found striking evidence of this incentive system. Automobiles with chauffeurs, who could be called upon at 3 AM, were supplied at government expense to such men. One director told me that although he already has a summer home provided for him, the government has recently offered to build him another summer home in the mountains.... As illustrating the importance attached to science I quote from an article "Science Serves the People" in the *Moscow News* [an English-language Soviet newspaper that he likely found at breakfast in his hotel each morning]: "Never before has the scientist been accorded such attention by the state and such esteem by society as in the Soviet Union. The state provides the maximum amenities of life and facilities for research to the scientist and insures a comfortable life to his family after his death."[28]

In evaluating Soviet incentives for scientists, Langmuir might have done better had he asked about any of the hundreds of agricultural scientists and biologists who were then incarcerated in the gulag for opposing Trofim Lysenko's bizarre theories of genetics. The most prominent Soviet biologist, Nikolai Vavilov, had starved to death in prison in 1943. Admittedly, there were a great many American scientists on the left who were accepting of Soviet abuses, but Langmuir had no particular political bias. It seems that in politics and economics Langmuir was particularly naive, accepting political claims at face value in ways that he never would have done in the scientific arena.

11
Lewis's Last Days

After Gilbert Lewis finished working on acids and bases in 1938, his laboratory was full of the organic dyes that had been used as colorimetric indicators of acidity—many dyes have one color in an acidic solution and another in a basic solution. Lewis had been fascinated with the idea of color for years. His first publication on color had been in 1906,[1] and in his 1916 paper "The Atom and the Molecule," he had portrayed the electron pair as central to the color of chemical compounds. For Lewis, color was the result of the interaction of light with chemicals, shifting a molecule from one electronic state to another, with light absorbed or emitted at a frequency that was equivalent to the difference between the energy levels of the starting and ending states—this became known as "photochemistry." In 1939 Lewis and Melvin Calvin, then a junior faculty member at Berkeley, published a theory of color.[2] Robert Connick remembers the two of them working together: "I remember that Lewis and Calvin were working on color, and there was a seminar room in Gilman Hall with a long table in the back of it. I walked in and he and Calvin had thirty books on the table, all open, and Lewis was walking around, moving from one to another, stopping to look at one article and then the next. He was trying to get into his mind where the color was coming from."[3] Because it was impossible to perform detailed quantum chemical calculations before the age of computers, Lewis and Calvin were forced to approximate. For each molecule, they used experimental data to assign the different energy levels to different electronic states. The transitions between these levels explained the colors of substances. Lewis saw the rearrangement of the electron pairs as fundamental to the color of molecules, particularly for molecules containing several staggered double bonds that were often strongly colored. These included his colorimetric indicators and most organic dyes. For example, he saw the color of molecules such as (where the "R" groups represent arbitrary organic substituents) as due to the polarization of the double bonds. In terms of Lewis electron dot structures, this molecule could be drawn as

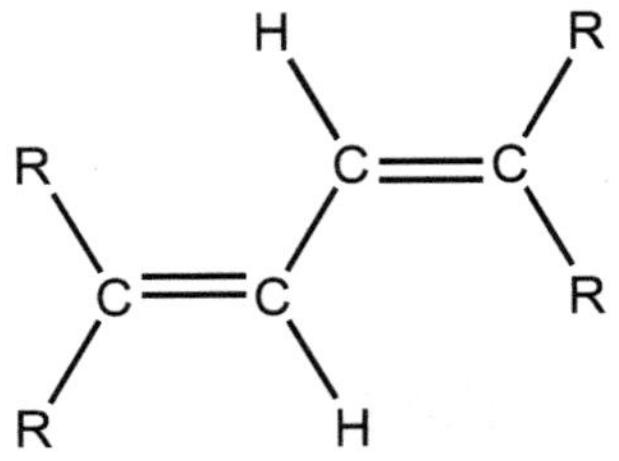

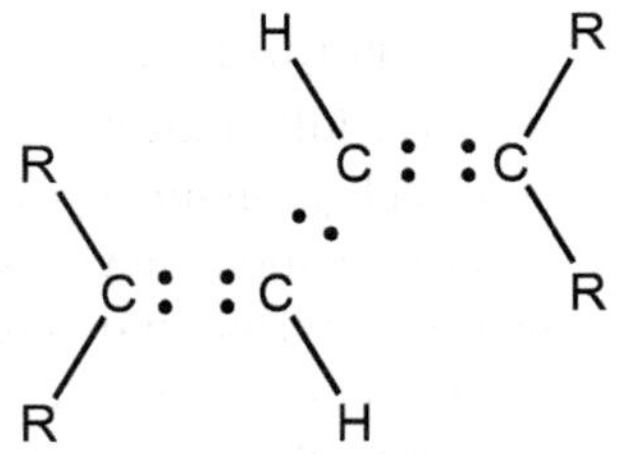

and typical polarized structures as

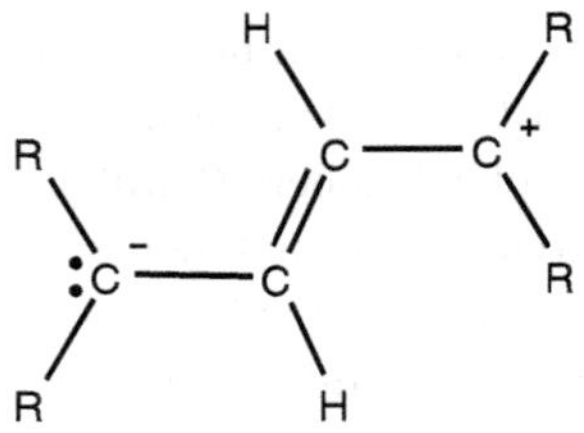

and

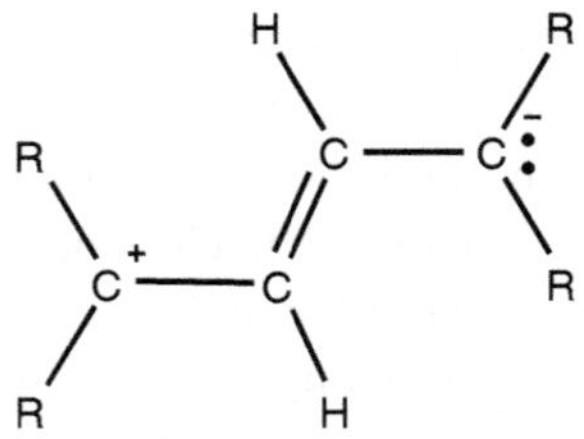

where the electron pairs are relatively free to move around the molecule. (By 1940, Lewis did not draw his dot pairs this explicitly, because both the concept and the notation were by this time well understood. I have drawn them in this fashion to show the continuity with his earlier work.)

Looking *Inside* a Chemical Reaction

Lewis began using the leftover organic dyes in his laboratory in experiments. It was not an absolute rule that molecules consisted only of electron pairs. The exceptions were what Lewis called "odd molecules," his term for what are now called free radicals. In molecules with several bulky substituents—large chemical groups that crowded each other—it was sometimes energetically unfavorable for a molecule to preserve its normal valence structure. For example, a carbon atom with four large substituents might be able to relieve some strain by the loss of one of the substituents. Such molecules were often particularly susceptible to photochemical reactions, where light could strike the molecule and rupture an electron-pair bond, leaving two free radicals. For the molecule

$$\begin{array}{ccccccc} & & R & & R & & \\ & & | & & | & & \\ R & — & C & — & C & — & R \\ & & | & & | & & \\ & & R & & R & & \end{array}$$

where the "R" symbols represent bulky substituents, such a reaction can be represented in terms of the Lewis electron pair as

$$\begin{array}{ccccccccccccccccc} & & R & & R & & & & & & & R & & & & R & \\ & & | & & | & & & & & & & | & & & & | & \\ R & — & C & : & C & — & R & + & \text{light photon} & \rightarrow & R & — & C\bullet & & + & \bullet C & — R \\ & & | & & | & & & & & & & | & & & & | & \\ & & R & & R & & & & & & & R & & & & R & \end{array}$$

Such free radical products were usually reactive. One or both of the free radicals might quickly lose or gain an electron, resulting in positive or negative ions, respectively, with even numbers of electrons, which the Lewis theory predicted would be more stable than the free radicals themselves. Either the free radicals or the resulting ions would usually combine with other molecules in a matter of microseconds, but such free radicals and ions were believed to form important intermediate compounds in a number of reactions. Because of their instability, the intermediate compounds were impossible to isolate and observe under normal conditions. Lewis and his coworkers developed methods to trap both free radical and ionic intermediates in what he called "rigid media." He would chill a dilute solution of a compound to a glasslike consistency and then irradiate the solution with light, forming the intermediates. Because the rigid solvent trapped them, the intermediates were unable to move to combine in bimolecular reactions. Lewis chose molecules that were colored (often the same colorimetric acid–base indicators that he had used in developing his theory of acids and bases) and that could be evaluated spectroscopically (with instruments designed to measure light frequencies and intensities quantitatively).

As an example, Lewis, Theodore Magel, and David Lipkin investigated the crystal violet cation (a positively charged ion),[4] with the chemical formula

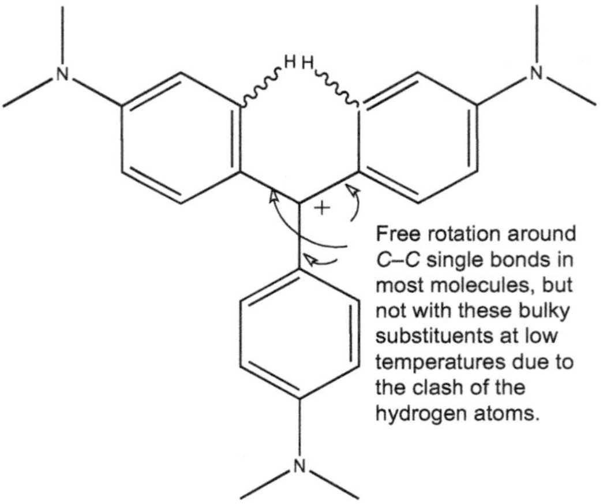

At normal temperatures, molecular models suggest that two of the hydrogen atoms (shown by the wavy bonds at one of the three equivalent sites in the molecule) collide when the structure swivels around one of the three marked C–C single bonds. Using spectroscopic techniques, which measure color quantitatively, Lewis observed two bands (two slightly different colors) for the molecule in rigid media, indicating the existence of two separate forms at low temperatures—the molecule did not have the energy to allow rotation around the marked bonds, and the structure twisted so that the rings do not all lie in a single plane, thereby relieving the strain caused by the interaction of the hydrogen atoms. Lewis assigned these bands to the two rotational isomers shown below, where the bonds drawn in bold should be viewed as extending out from the plane of the paper:

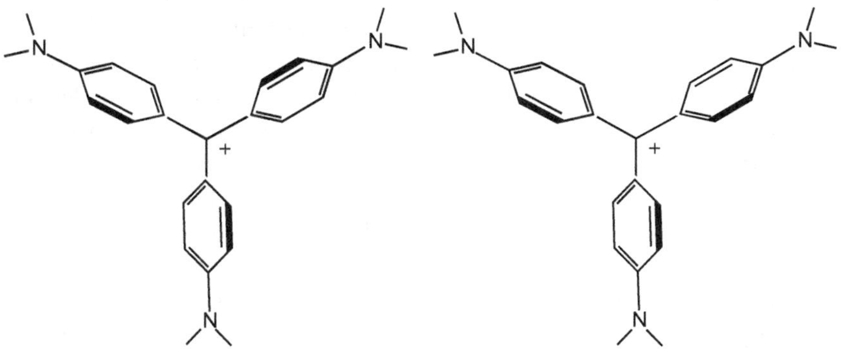

Lewis said that one isomer was like an airplane propeller with its blades twisted (the structure on the left), and the other was like a propeller with one of the blades turned the wrong way (the structure on the right). At low temperatures, a molecule in one of these forms did not have enough energy to push past the colliding hydrogen atoms and transform itself into the other. One could watch the intensity of the spectroscopic bands change as a function of temperature, indicating that the concentration of the two isomers was changing, as well. As the temperature increased, the molecule finally had enough thermal energy that the colliding hydrogen atoms were knocked out of the way during rotation, and the two isomers then interconverted so rapidly that only a single band was observed. This was astounding work for 1941. Chemists were able to look *inside* a reaction, examining short-lived intermediates.

Lewis's interest in color and in electronic transitions was behind his interest in *fluorescence* and *phosphorescence*. When a molecule fluoresces, it absorbs radiation at some frequency, is excited to a higher electronic energy level, and then emits radiation at a different frequency than that absorbed as it relaxes to some third energy level. All of this happens very quickly, in fact, instantaneously as far as the human eye can see. Fluorescent paint is an example of this effect, where the light that is emitted makes the paint intensely colored at the emission frequency. When a molecule phosphoresces, the process is similar—radiation is absorbed and emitted—but the emission is delayed. Common examples are glow from algae and from the strontium oxide aluminate paint used on exit signs. In 1941, in the first of his papers on rigid media,[5] Lewis postulated that phosphorescence is due to a transition from a molecule's triplet state (a state where two electrons have the same, or unpaired magnetic moments) to a lower energy singlet state (where two electrons have different, or paired magnetic moments).

Lewis had first discussed the pairing of the magnetic moments of electrons in his 1916 paper "The Atom and the Molecule." He had apparently confirmed his opinion on phosphorescence by 1942, when after a visit to Berkeley by the chemical physicist Robert Mulliken, he commented to his student Jacob Bigeleisen, "You know, those triplet states that Mulliken was talking about in ethylene are probably the phosphorescent state."[6] Lewis's statement was controversial because the transition from a triplet state to a singlet state is "forbidden" by the rules of quantum mechanics, and Lewis's idea was opposed by Edward Teller, James Franck, and others. The controversy on the triplet state has been described by Lewis's former student Michael Kasha.[7] Lewis was proved right only after his death. Franck said in an interview, "I did him injustice at the very end of his life, and I still regret it. Because I always said, I cannot understand this idea of the triplet state responsible for . . . phosphorescence. . . . And he had such a state."[8] The forbidden nature of the triplet-singlet transition was not absolute, but was the reason why the phosphorescence lingered for such a long time—for any particular molecule, the transition was unlikely but possible, so it did not occur very often. Now well into his sixties, Lewis had again returned to his concept of the importance of the electron pair, applying it to the study of color and the interaction between light and chemistry—photochemistry.

Lewis and the Nobel Prize (Again)

In 1940, when Lewis was sixty-five, the Nobel Committee for Chemistry asked for one more report on his theory of the electron pair and the chemical bond. This time they assigned the task to Ludwig Ramberg, who was a member of the chemistry committee and a professor of organic chemistry at Uppsala University, although he had a strong leaning toward physical chemistry. It was unlikely that Ramberg's opinion of Lewis's theory would be positive. Quoting from Göran Bergson's unpublished history of the Uppsala chemistry department:

> In certain ways, Ramberg was ahead of his time. He strongly disliked the so-called electronic theory of organic chemistry [the Lewis-Langmuir theory and related pre-quantum mechanical ideas] which was very popular from the 1920's to the 1950's. Already in his inaugural lecture in 1919, Ramberg expressed his conviction that quantum theory would become the only and final basis for understanding of chemical phenomena at the molecular level.[9]

When Ramberg was asked to prepare a report on Lewis's valence theory, no one (including Theodor Svedberg) remembered that Svedberg had submitted a report on the subject eight years earlier. In the letter submitting his 1940 report to the chemistry committee, Ramberg wrote: "I submit my report on Lewis's valence theory. It has become a long story, but this is mainly because I discovered Svedberg's report of 1932, which both he and I had forgotten about. It contained everything that could be stated briefly, and therefore I had nothing to do but to put into writing that which could *not* be stated briefly." Ramberg's 1940 report was dismissive. The best that he could find to say about Lewis's theory was, "From a pedagogical point of view, Lewis's valence theory undeniably holds quite a few advantages, perhaps mostly on an elementary level."[10]

That was the last word the committee had to say on Lewis's theory of the chemical bond. For his theory of acids and bases, Lewis received only a single nomination from Norris Hall in 1944, and the committee never commissioned a special report on the subject. Lewis's theory of acids and bases was not widely adopted until after his death, but it has since become an important concept in physical organic chemistry.

In 1944, Lewis was nominated for his work on photochemistry. This was to be the Nobel committee's last look at Lewis. By this time, Ramberg and Wilhelm Palmaer had left the committee, and if the evaluation had been broadened to include all of Lewis's related work—the covalent bond, acid–base theory, and photochemistry—it is possible that Lewis might have had a chance. In Lewis's mind, all of this research and these ideas were tied together by one concept—the importance of the electron pair. While he was nominated for the Nobel Prize many times for his work in separate areas involving the electron pair—the theory of chemical bonding, acid–base theory, and photochemistry— the Nobel committee treated these as separate topics rather than as components of a single theory. The committee's attempt to segment his work—looking for a single "discovery," as the Nobel regulations stipulated—obscured the importance of the general concept.

The author of the 1944 Nobel report was Arne Fredga, one of Ramberg's students who had joined the committee, who began as follows: "This year two nominations have been submitted which refer to Lewis's collective scientific work, and a third one that refers to his publications during the past two years on the light absorbance of organic molecules. This consists of nine publications in *J. Amer. Chem. Soc.* during 1942–1943."[11] But Fredga's evaluation was to be restricted to photochemistry and was not to cover his "collective scientific work." Fredga continued:

> Several of these treat or are connected to photochemical experiments in "rigid solvents," that is, solvents that have been chilled until the consistency of the solvent becomes glass-like. . . . Lewis probably got the idea during his fluorescence research. It has been known for a long time that fluorescent compounds in a solvent that has been chilled to a glass-like consistency can show phosphorescence, which can at least partially be regarded as a delayed fluorescence. By illuminating the dissolved material, various reactions can

occur: dissociation into radicals, dissociation into ions, or creation of ions by emission of electrons. The primary products in many cases are those that under normal circumstances have a very short lifespan. In rigid solvents, where bimolecular reactions are excluded and even emitted electrons are caught and retained by the solvent medium, the primary products have a considerable lifespan and can however be studied spectroscopically. . . .

It is clear that from this basis Lewis has attacked the problem of the connection between color and molecular structure on a broad front. Definite and decisive results do not seem to me to be won yet, but his work even without this is of definite interest. His chemical explanations of optical phenomena are ingenious and seem plausible, although other explanations in certain cases may not be excluded. Sometimes present irregularities are attributed to association, solvation, etc., conditions that in some cases have been objects of certain studies. The most important in my opinion are the investigations into photochemical effects in rigid media, which have given the most interesting results, and which could show themselves to be of primary importance, but a closer examination seems to be wished for.[12]

The committee accepted Fredga's report, judged the work of significance, but concluded its report to the Royal Swedish Academy of Sciences by saying, "The committee however wishes to wait for further development in this area and does not consider itself ready to award the prize to Lewis."[13]

Despite four evaluations (by Arrhenius in 1924, Svedberg in 1932, Ramberg in 1940, and Fredga in 1944), the committee never recommended Lewis's work on the electron pair for the prize; in retrospect, this may have been Lewis's most significant achievement, particularly if his work on acid–base theory and photochemistry is viewed as part of the same effort as his discovery of the covalent bond. Svedberg's (1932) and Ramberg's (1940) dismissals of Lewis's work on the chemical bond can be attributed to their evaluation of Lewis's 1916 paper using hindsight based on the quantum mechanics of the 1920s and 1930s, and to a continental European disdain for mechanical models as explanations. But much of the importance of Lewis's work, particularly his work on acid–base theory and on photochemistry, became understood only after his death, so it is historically unfair to judge Svedberg and Ramberg too harshly. Lewis was seen as a physical chemist, and if he were to have been awarded a prize, he would have needed the backing of the physical chemists on the committee. The prevailing attitude among Swedish physical chemists after the quantum revolution was clearly not in favor of simplified nonmathematical models based on the electron pair, no matter how useful they might have been to other areas of chemistry. But physical organic chemists like Louis Hammett and Christopher Ingold, both of whom nominated Lewis for the prize, would not have agreed with Svedberg and Ramberg. Ironically, Lewis's greatest contributions may have been to organic chemistry, a subject in which he had received a "D" in his undergraduate days at Harvard. Lewis's explanation of the nonpolar bond was important to the investigation of reaction mechanisms; his theory of acids and bases in nonprotonic solvents is used to explain catalysis of organic reactions; his thermodynamics is the common language of organic reaction equilibria and kinetics; his isolation and characterization of short-lived intermediates provided one of the first ways to look inside organic reactions; and deuterium labeling of organic compounds is now a standard investigatory technique in organic chemistry.

This sort of prejudice against Lewis's ideas on the electron pair has persisted. William Lipscomb, who won the Nobel Prize in Chemistry in 1976, has been quoted as saying that although Lewis discovered the covalent bond, he did not understand it[14] (presumably because he did not understand quantum mechanics). This is of course true, but only in the sense that Mendeleev did not understand the periodic law (being unaware of the concept of atomic number, the electron, or the existence of the inert gases) and that Newton did not understand Newtonian mechanics (being unaware of the principle of the conservation of energy). A joint prize for Lewis with Linus Pauling, who extended Lewis's theory using a quantum treatment that included directional bonds, would have been appropriate. Pauling, like Lewis, made electronic theory accessible to a large number of chemists, and Pauling did receive a Nobel Prize in Chemistry in 1954, nine years after Lewis's death, "for his research into the nature of the chemical bond and its application to the elucidation of the structure of complex substances."[15] Pauling dedicated his 1939 book, *The Nature of the Chemical Bond*, to Lewis.

The Nobel Prizes in the early years tended to favor experimental work over theoretical work: Einstein never received a prize for his theory of relativity, and Max Planck had to wait until 1918 to receive the Nobel Prize in Physics for his 1900 quantum theory. This preference for experiment over theory was even truer for chemistry than for physics. There was a general prejudice against electronic theory within the Nobel chemistry committee, with the idea that theory was really something more appropriate for a physics prize. Other pioneers in chemical electronic theory, including Walter Heitler, Fritz London, John Slater, and Erich Hückel, were also passed over for the prize.

But even had Lewis's work on the electron pair—the covalent bond, acids and bases, and photochemistry—all been considered as a body of work, it is still unlikely that Lewis would have received a Nobel Prize, at least in 1944. In 1929, Hans von Euler-Chelpin had joined the committee. Euler-Chelpin had a strong bias toward biochemistry,[16] and after 1936, the chemistry awards took a pronounced biochemical tilt. The awards for 1937–39 were all biochemical, and the awards for 1940–42 were reserved because of the war. In 1943 and 1944 the awards were for isotope research, but by 1945 and 1946 the attention had returned to biochemistry, with an award for 1945 for an improved method of fodder preservation! After his last serious evaluation in 1934, Lewis's physical chemistry was out of step with the interests of the committee.

Owen Richardson, who had been awarded the 1926 Nobel Prize in Physics for his work on thermionic emission, nominated Lewis for the chemistry prize repeatedly beginning in 1940. His last few nominating letters all concluded with the same plea: "All I wish to add to this is to point out to the committee that Lewis is a man well on in years and if they wish to honour him in this way (and I think the work he has done thoroughly deserves it) the time is now right to do so."[17] Lewis died in his laboratory at the age of seventy in 1946, while performing a photochemical experiment, two years after Fredga's report on photochemistry. The committee's 1944 option to "to wait for further development in this area" had expired.

Many of Lewis's problems in achieving the Nobel Prize were undoubtedly of his own making. He had isolated himself in Berkeley, where he had built himself a scientific support system in which he could work confidently, but in the broader arena he was often defensive and overly sensitive to slights. Promoting one's own work may be a necessary evil if one is to achieve scientific recognition, and Lewis was not always willing to do this. Where Walther Nernst had beaten the drum of "my heat theorem" for years, Lewis was always ready to move on to the next challenge, and he started his work in acids and bases and in photochemistry while in his sixties. But when he started work in a new field, he stopped publishing in the old field, and part of self-promotion may be keeping a nominal presence in a field even after one has tired of it. Lewis did not attach his name to papers on thermodynamics or electronic theory that were written by others in his department after he had left those fields. By the time the Nobel committee began to evaluate his work in thermodynamics, he had long left it behind—Svedberg's 1926 recommendation that the committee wait to evaluate new results in thermodynamics was not going to find anything new from Lewis, who had left the field in 1923. Similarly, he left his theory of the chemical bond behind as well in 1923, picking it up again only in 1938 when he returned to his theory of acids and bases. In the meantime, Pauling and others had integrated the new quantum theory into chemistry.

The effects of isolation upon Lewis are most evident in his proclivity to make enemies. He had gratuitously attacked Nernst, and Nernst's friend Palmaer seems to have made him pay for it. But there were doubtless others who had no great affection for Lewis. Bigeleisen, Lewis's student, has described Lewis's approach when entering a new field of research: he would make sure to cite all prior work and would then stake claims to as much territory as possible. If others then failed to recognize his claims, or began to work in the field without citing his contributions, he would see them as enemies.[18] Some of those he viewed in this fashion were undoubtedly completely unaware of his attitude. Bigeleisen reports that Lewis would disparage the physical organic chemist Louis Hammett in private conversations in 1941–43 because he believed that

Hammett had ignored his acid–base theory. When Bigeleisen met Hammett in 1943, he found that Hammett in fact had a high opinion of Lewis. Hammett nominated Lewis for the Nobel Prize in 1944.[19]

But Lewis's isolation went beyond his enemies. He had not spent time in Europe or at international meetings to build support in Europe or Scandinavia as had Irving Langmuir, who was a friend of Niels Bohr and Max Born, for example. And even Lewis's American supporters were shaky. While in theory the committee assessed the work of a nominated scientist without regard to the number, names, or nationalities of the nominators, and while the nationality of the scientist was not supposed to be a consideration, a solid block of home-country nominators could only help. When Arrhenius blocked Nernst's Nobel Prize for many years, seven German scientists nominated Nernst in 1921, and Arrhenius finally gave up when the Germans were joined by fourteen Swedish and Norwegian scientists who cosigned two Swedish nominations (with Palmaer one of the chief organizers of this support).[20] Lewis's friends never managed to achieve this solidarity, possibly in part because of his prickly nature and his isolation. In 1934, Lewis resigned from the National Academy of Sciences, which may have been prompted by Harold Urey winning the 1934 Nobel Prize in Chemistry (chapter 8). This petulant resignation must have discouraged many of his supporters, and the American Nobel nominations for Lewis fell off after 1934.

Nationalism does not seem to have been a factor in passing Lewis over, but national outlook does. Any discipline segments itself into different subdisciplines, based on national identity or area of specialization, for example, and chemistry is no exception. Although the Nobel Prizes are international awards, the chemistry prizes are awarded by the Royal Swedish Academy of Sciences, and all the committee members during Lewis's candidature were Swedes,[21] with the exception of Euler-Chelpin, a German national who had been Nernst's student and Arrhenius's assistant and had then taken a professorship at the University of Stockholm. Lewis's Swedish evaluators were almost certainly more influenced by the attitudes of German scientists (with whom many of them had trained) than they were by British or American scientists. The Germans generally preferred Nernst's formulation of thermodynamics and were never supporters of Lewis's and Langmuir's ideas on electronic theory.

Simple longevity matters in Nobel recognition; there are no posthumous Nobel Prizes. Had Lewis lived a few years longer and been able to continue his promising photochemical research, it is possible that he would have received the prize for his work in this area, as the committee's interests swung back to physical chemistry after 1946. An award for thermodynamics might also have been possible: William Giauque, a Berkeley graduate and faculty member, received the 1949 chemistry prize "for his contributions in the field of chemical thermodynamics, particularly concerning the behavior of substances at extremely low temperatures."[22] With Palmaer gone from the committee, Svedberg might have pressed for the 1949 prize to be shared with Lewis and thus finally implemented his recommendation from 1926, awarding a prize for thermodynamics to the two Berkeley faculty members.

Lewis and His Last Students, Bigeleisen and Kasha

In 1941, Lewis was sixty-six years old. He had been both dean of the College of Chemistry and chemistry department chairman since 1912 and had been a continuing irritant to the university administration all those years.[23] While the administration was not prepared to attempt to force a scientist of his stature to retire as a professor, it was made clear that he could not continue in his leadership roles past the standard retirement age. Wendell Latimer was made dean, and Joel Hildebrand department chairman.[24] While both were Lewis's friends, they began correcting some of what they had viewed as Lewis's idiosyncratic behavior—his neglect of chemical engineering, his extreme emphasis on physical chemistry, and his preference for hiring only Berkeley graduates for the faculty.

When Jacob Bigeleisen enrolled at Berkeley in 1941, Lewis had just given up his faculty posts. Bigeleisen had come from Washington State, where he had completed a master's degree with Otto Redlich, who had translated Lewis and Randall's *Thermodynamics* into German. Bigeleisen had enrolled at Berkeley intending to study nuclear chemistry with Sam Ruben, the codiscoverer of carbon-14. When Bigeleisen met with Wendell Latimer, Latimer tried to dissuade him from working with Ruben, saying that there was no future in nuclear chemistry (this was in 1941!) and that he would be better off working with William Bray, a physical inorganic chemist who had come to the Berkeley faculty from MIT with Lewis in 1912. Bigeleisen listened politely, but none of Latimer's advice made much sense to him, and he was determined to work for Ruben. But as Redlich's student, he felt that he first needed to make a courtesy call on Lewis. He arranged the meeting with Mabel Kittredge, Lewis's longtime secretary, to whom Lewis had long delegated much of the day-to-day running of the chemistry department. Lewis received Bigeleisen in his office, talked about his own current research efforts, gave him a few reprints of recent publications, and asked him to come back the next week. When Bigeleisen returned, Lewis said, "I'm accepting you as a graduate student. I'll expect you to work 14 hours a day, six days a week. Is that agreeable? See Dr. Lipkin. He'll get you a spot and get you started on a problem." Bigeleisen felt trapped. He could see no way to turn Lewis down and stay at Berkeley, so he accepted despite his desire to work for Ruben. In December 1941, Lipkin resigned as Lewis's research assistant to go into war work, and Bigeleisen was the only one left in Lewis's lab. He was supporting himself through a position as a teaching assistant. Lewis came into the lab and said, "Bigeleisen, are you prepared to give up your [position as a teaching assistant] and be my research assistant?" When Bigeleisen agreed, Lewis doubled his salary.[25]

Bigeleisen began working with Lewis on his photochemical experiments, attempting to understand the relationship of fluorescence to molecular orientation. Bigeleisen told me of his work:

> I had worked out the physics of the relationship the last night, and had been explaining it to [Samuel] Weissman and Lipkin when Lewis came in. Of course Lewis wanted to hear it all. I started by drawing my coordinate system on the board, with the y and z axes in the plane of the board and the x axis perpendicular to the board, and with light traveling along the x axis. Lewis stopped me immediately: "In this laboratory, light travels along the z axis." Of course this meant I had to change all my notation, which was no simple task. I went through it as best I could, and we broke for lunch. Lewis came back after lunch and had me go through the whole derivation again very slowly. At the completion he said, "I want you to write this up, just as you have presented it to me. Put it in your thesis. It doesn't matter what else you have in your thesis; it will be accepted." My Ph.D. thesis was only 28 pages.
>
> Lewis later told me that he had trouble thinking in three dimensions, which is why it took him a while to get my derivation. When Einstein had published his relativity theory, he and [Richard] Tolman had spent evenings trying to understand and refute it, only to finally accept it after all efforts at refutation failed. He had then presented Einstein's work to a group of physicists, only to be booed off the stage as they refused to accept it. I asked him why, if he could think in four dimensions [as relativity required], three dimensions were difficult. He answered that three or four dimensions were equally difficult for him—it was the transition from two to three dimensions that was difficult.[26]

At that time Lewis was not working very hard, at least not on physical chemistry—he would come in about 10:30 each morning, meet briefly with Bigeleisen to review the last day's work, and then go to lunch at the faculty club, where he would play bridge. He would return at about 2:00, work in the laboratory until 4:30, and go home. Bigeleisen, of course, had been there since 9:00 A.M. and would stay until 11:00 P.M. But this may be unfair. Lewis was actively engaged in work in two areas that were only tangentially related to his university work—an explanation of the ice ages, and his ideas on the anthropology of the Americas. Bigeleisen coauthored eight papers with Lewis and described what it was like to write a paper with Lewis—Lewis simply dictated them without any notes:

> Lewis didn't like to cut into research time for papers. We would meet in the evenings in his office, where there was a large unabridged dictionary on a

pedestal. We would usually work until about 10:00 or 11:00. Lewis would simply dictate, and I would write. I would then read it back to him, and he would check words in the dictionary and make some changes. At this point he would say "Get it typed up." I would take it to Mabel Kittredge, his secretary, the next morning, and she would have it typed by the time Lewis came in. He would edit the document from her typed copy. On one of my eight papers Lewis allowed me to write the experimental section, which I considered a great honor.[27]

Bigeleisen told me that shortly before he finished his work with Lewis, his rooming house caught fire. When he told Lewis that he would need to take a couple of days off to find somewhere to live, Lewis returned after lunch with a key to a room at the faculty club, which he had rented out of his own pocket. When I said that this seemed very generous, Bigeleisen said that he thought that Lewis just didn't want him missing time in the laboratory.[28]

Just before Bigeleisen left in June 1943, Lewis's last student, Michael Kasha, arrived. Unlike Bigeleisen, Kasha had come with the definite intention of working for Lewis:

My friend and I graduated from college with the highest honors, and the department chairman called us in to congratulate us. He told us, "Boys, I'm very proud of you. Arthur, I'm sending you to Squeedunk University, and Michael, you're going to Podunk University"—two totally unheard-of places. We went out in the hall and looked at each other. He said, "I'm going to work for Peter Debye," and I said, "OK, I'm going to work for G. N. Lewis." And we did. When I got there, the whole prewar crew was leaving. Bigeleisen was there another month, Weissman was only there a couple of days. They were all gone, and I was to be his only graduate student. People told me he was old, what can you get out of that? But we had an enormously good time.

I had read all of Lewis's papers. It was natural that I would do what he suggested. He had a whole laboratory full of dyes. I wasn't so excited by that—I thought that dyes had been studied to the hilt. But he had found this new type of dye. It was a gigantic molecule, a polycyclic hydrocarbon. There was one electronic instrument in our department. It was a Beckman DU, operated by batteries.* He wanted me to get complete spectra. I knew what complete meant—from 20,000 angstroms to 2,000 angstroms. No one had even thought to go to 20,000. What is there out there? I found a

* The Beckman DU is a spectrophotometer that can measure the absorption of visible and ultraviolet (UV) light—both the color and the intensity. Spectrophotometers are now common in paint stores for matching the color of paint chips. The Beckman DU could measure the colors not only of visible light, but also of ultraviolet light (wavelengths down to 2,000 angstroms) and

band that no one had ever seen before. Lewis was astounded. He had found the lowest band coming down into the infrared. How could there be a band in the near-infrared? Within two days we had a theory.... Perhaps for a very large polycyclic molecule, the electrons could move in a circle, which would give a much lower energy band. But then I found one in every molecule I looked in. Lewis said, "Let's test this." We tried a long, sticklike molecule. It had the same band. There had to be a problem with the monochromator. We tested it at each frequency, observing the colors. When we got to the infrared, Lewis said, "I'm the first man to see the infrared, and it's green!" We had wasted months finding this out.[29]

Lewis's Depression

When Bigeleisen left Berkeley in 1943, he thought that Lewis was very depressed. Lewis's chemistry department, which he had built into one of the world's centers of physical chemistry, had been pulled apart by the war and the Manhattan Project. As Lewis and Bigeleisen finished the last of their papers together one Sunday morning, Bigeleisen thanked Lewis for the opportunity to work with him. Lewis replied, "It's not the way it was," choked up, and had to leave the room.[30]

The extent of Lewis's depression at the time of his death is not clear. Bigeleisen left Berkeley in 1943, just after Michael Kasha arrived. Kasha has a completely different recollection of Lewis than does Bigeleisen, remembering him as upbeat and enjoying both his work and his life in general. But Lewis's colleagues Kenneth Pitzer and Joel Hildebrand commented on his depression after his death, and as we shall see, Lewis had several possible reasons for depression. For one, he was cut out of defense work in World War II.

In World War I, Lewis had been able to make significant contributions, although he was already forty-two years old when the United States entered that war (see chapter 4). But during World War II he was left alone to shuffle around the chemistry department—he was too old and too much of a loner to find a place in the war effort. He said that he did not want to be a "flying briefcase" going to meetings. Glenn Seaborg—Lewis's research associate only a few years earlier—did his initial work on plutonium only a few doors down from Lewis's Gilman Hall laboratory. By 1942, Seaborg was running the plutonium purification effort for the Manhattan Project. Lewis was not even supposed to know what was happening, but he did. Seaborg tried to get him hired as a consultant for the Manhattan Project, but nothing came of it (see

---

near-infrared light (wavelengths up to 20,000 angstroms), which the human eye cannot see.

chapter 8). Robert Connick, then a young faculty member at Berkeley, was managing the plutonium purification effort at Berkeley and had this to say:

> The plutonium project was headquartered in Chicago. Seaborg went there to head the chemistry part, but a group was left here. I don't know how that was all arranged, but I believe Latimer managed to persuade Seaborg to leave a group here to work on plutonium. Latimer and [Gerhard] Rollefson were co-chairs, and Rollefson died as it was still going, and then Latimer ran it. Leo Brewer was on it, and Jack Gofman was on it. Art Wahl was slated to go to Los Alamos—he was the chemist who really did the work on the discovery of plutonium. This was the latter part of the summer of 1942, and Seaborg had already gone to Chicago. Art was a good friend of mine—we had been in the same doctoral class—and told me everything he could about the separation of plutonium before he left for Los Alamos. Gofman and I then headed a small group continuing that work. And I'll tell you a story about Lewis.
>
> I don't remember who was doing it—maybe it was Kennedy—had a couple of guys—maybe undergraduates—on the top floor of Gilman Hall where Lewis's office was who were doing ether extractions of uranyl nitrate. Lewis was working with Bigeleisen then, I think, and they had a darkroom up there for the fluorescence experiments. It had a two-step entrance to keep it dark. The place where the extractions were going on was a long room under the eaves. As was characteristic of Lewis, he would go into the labs to see what was going on. I think the only way into the darkroom was through the long room where the ether extractions were going on. And the two young fellows doing the uranyl nitrate extractions were becoming very alarmed because Lewis would go through this room with his cigar amid all the [highly explosive] ether fumes. Finally one of them worked up his courage to approach Lewis and say that they were concerned about his cigar possibly causing an explosion. Lewis reportedly said, "Oh. Yes. I must be more careful." And from then on he would walk through the room holding the cigar behind his back.[31]

David Lipkin and Samuel Weissman, two young Berkeley scientists who were working in the Manhattan Project, went to Robert Oppenheimer to suggest that he enlist Lewis for the project. Oppenheimer had spent years on the Berkeley physics faculty and knew Lewis well. He just looked at them and replied, "And with whom would *he* collaborate?"[32]

When Harold Johnston became dean of the College of Chemistry in 1966, he moved into Lewis's old office. While moving his papers in, he found the manuscript of a war novel, written by Lewis, involving pitched battles between American and Japanese troops on Pacific islands. Johnston says that Lewis had tried to have it published—Johnston also found correspondence with an agent or publisher—but that in his judgment, Lewis was a better physical chemist than pulp novelist. The manuscript has not been preserved.

Figure 11-1. Lewis on his seventieth birthday, shortly before his death. Photo by Michael Kasha.

Johnston viewed the novel as Lewis's attempt to make a contribution to the war effort.[33]

And Lewis was still nursing grudges. Bigeleisen tells a story that makes it clear that even in 1942, more than twenty years after the spat over the "Lewis-Langmuir theory," Lewis had still not forgiven Langmuir:

> Carlos, the chemistry department janitor, neither read nor wrote English. Carlos was however a noted philatelist, and was in communication with other stamp collectors from all over the world. Each year he would select a new graduate student to act as his secretary to handle his correspondence, and I was the most recent one. Carlos was devoted to Lewis, and the two of them had become friends. Late at night, Carlos would call me into Lewis's office and show me Lewis's medals from World War I, which Lewis kept in his desk drawer. In fact, Lewis's office was the only room in the chemistry department Carlos actually cleaned—everywhere else, he would simply push the dirt around, leaving the room neither cleaner nor dirtier than when he had entered. He also mined Lewis's wastebasket for cigar butts, which he would grind up for his pipe. Carlos knew all about all the

department legends involving Lewis, including those involving Langmuir. One Monday evening Carlos came into my lab and said, "Big trouble tomorrow. This afternoon a man came in. I said, 'Can I help you?' He says his name is 'Langmuir,' I send him to Latimer. The old man go to France, win war, Langmuir steal his ideas and get the Nobel Prize. Big trouble tomorrow." Carlos was right. Langmuir had been invited to speak at the departmental seminar the following day. Anyway, four o'clock rolled around, Langmuir was ready to speak, and no Lewis. Hildebrand, who had likely invited Langmuir, was pacing nervously. He had a problem: Should he start now and offend Lewis, or wait and offend Langmuir? Lewis finally arrived. Lewis, cigar in hand, introduced Langmuir as follows: "Our speaker today is a man about whom we have heard so much, and from whom we have seen so little."[34]

Lewis's Death

On Saturday afternoon, 23 March 1946, Lewis died in a laboratory full of hydrogen cyanide. The cause of Lewis's death has never been clear. No autopsy was done, and it was in the interest of both the university and the Lewis family to have a coroner's verdict of death by natural causes. The coroner's verdict was just that—attributed to arteriosclerosis,[35] which is strange as a cause of death—but some of Lewis's colleagues in the chemistry department thought that Lewis had been depressed and that he had killed himself. In his history of the Berkeley chemistry department, William Jolly wrote:

> It is possible that the melancholy associated with the end of the war, disappointments about the Nobel prize, and deterioration in health may have led Lewis to despair in his last year. One of Lewis's bridge-playing cronies, Gerald Marsh, said that on the afternoon of March 23, 1946, Lewis appeared to be morose while playing cards at the Faculty Club. He then went to his laboratory in Gilman Hall, where he was later found dead near a broken ampoule of hydrogen cyanide. Now the fact that Lewis had an ampoule of the deadly poison, hydrogen cyanide, in his laboratory is not as sinister as it might at first appear. He had been studying the variation of the absorption spectra of dyes with solvent dielectric constant, and hydrogen cyanide, which has an extremely high dielectric constant, was one of the solvents he was using. In a retrospective symposium honoring G. N. Lewis, Michael Kasha attempted to quash the suggestion that Lewis committed suicide, but his arguments were not compelling. Hildebrand believed he took his own life. Pitzer has pointed out that somebody as smart as Lewis might well commit suicide in a way as to make it appear accidental.[36]

Kasha says that Lewis almost never came into the laboratory on a Saturday, but that he had a lunch meeting scheduled that day at the faculty club. Kasha remembers that Lewis was cheery while discussing their research plans that morning, but that he came back somber from his meeting with his lunch guest and went immediately into his laboratory. Shortly thereafter, Kasha found him lying dead on the laboratory floor:

> I was working with G. N. Lewis on the Saturday afternoon, March 23, 1946 when he died, and I find it worth recording the events of the last day of his life, particularly because there has been misinterpretation of the circumstances of his death. The Saturday morning was a particularly sunny one scientifically speaking. We had an unusually fruitful discussion, and I especially remember being filled with so many ideas on research that they seemed enough to sustain a year of work. I had some new ideas on triplet-triplet absorption, and Lewis described more of his ideas on photomagnetism which I was then to undertake.
>
> Lewis had a particular experiment he planned to do by himself at the vacuum bench that afternoon. A few days before, he had read in the latest issue of *Transactions of the Faraday Society* a paper which he showed me containing a graph indicating that the dielectric constant of liquid hydrogen cyanide changes by a factor of over 100 in a certain accessible temperature range. Lewis said, "That would be a very interesting medium in which to test the effect of dielectric constant on the color of dyes!" He planned that experiment for late Saturday afternoon.
>
> Lewis went to lunch with a distinguished guest and returned at 2 P.M. It was unusual for Lewis to go to the Faculty Club on a Saturday. When he returned, he went to his vacuum bench in his lab, which was at the opposite end of the hall from his office, my laboratory being in between.
>
> I was working on the spectrophotometer in my laboratory. About every 20 minutes or so I walked by to see if everything was all right. Around 4 P.M. when I passed the vacuum bench room on my way downstairs, I glanced in and noticed Lewis missing. I began to step into the laboratory and got a noticeable whiff of HCN [hydrogen cyanide, which has an almond odor]. Stepping back into the hall, I saw Lewis' feet just visible behind the bench. I gave out a yell to the lab at the end where I knew Daniel Cubiciotti was probably working, and I ran toward the hood with my nose clamped shut with my left hand. I threw a brick, which we kept as a weight in the fume hood, through the window. Returning to the hall I noticed the bottle of sodium bicarbonate in the hood and rushed into the lab again, and covered the liquid on the vacuum bench table with bicarbonate, the active bubbling of which suggested that liquid HCN had just spilled out. Shortly afterwards, Cubiciotti and I dragged Lewis into the hall and called for medical help. He had a serious welt in the middle of his forehead, indicating that he had fallen forward and hit his head on a vacuum bench clamp. Lewis was dead on arrival in the University Hospital, and a medical autopsy indicated clearly that he had died of a heart attack. [In fact, no autopsy was done.] We concluded that many minutes after he had died, the pressure had built up in the container of liquid HCN, from which the Dewar [the

enclosing flask that had been chilling it, probably filled with dry ice and acetone] had been removed, and the vessel dropped to the vacuum bench, spilling the contents.[37]

Ted Geballe and Frances Connick also independently discovered Lewis's body that day. Geballe says that he passed Lewis's lab and saw his feet sticking out near the hood. He yelled, and Frances Connick answered his call. While she ran the few hundred yards to Cowell Hospital for help, Geballe ran to the telephone to call William Giauque, his research adviser. It is likely that Kasha discovered Lewis while Geballe was on the telephone. When Connick got to the hospital, she was told that the hospital's policy was not to send anyone, and that she would need to call an ambulance from the neighboring city, Oakland. By the time the ambulance arrived, Lewis was dead.[38]

When I interviewed Michael Kasha in 2004, he told me of Lewis's lunch meeting that day, and he then told me something that astounded me. Almost sixty years later, Kasha had remembered who Lewis's lunch guest was on the day of his death: Irving Langmuir.

It was understandable that with all the confusion and grief of that day, the last thing on Kasha's mind would have been Lewis's lunch guest. But I admit that I was skeptical about a memory suddenly recovered after all that time. Did Langmuir really come from Schenectady, New York, to Berkeley, California, for lunch with Lewis on the day of his death? I went to the Library of Congress, where Langmuir's papers are collected, and found the evidence. On 23 March 1946—the day of Lewis's death—Langmuir received an honorary degree from Berkeley.[39] Langmuir was also invited to give the Hitchcock Lectures over the next several weeks, which are an annual series of talks at Berkeley given by an invited scientist.

Langmuir had arrived the night before, 22 March. It is not clear whether Lewis and Langmuir lunched alone, or whether Wendell Latimer and Joel Hildebrand, then dean and department chairman, respectively, joined them, but the latter seems likely. Latimer had presented Langmuir his award before lunch, and Lewis seems to have decided not to attend the award ceremony, so Latimer and Hildebrand may have twisted Lewis's arm and arranged the lunch in a private room at the faculty club, as no one reported seeing them in the common room. Something seems to have happened at the lunch that upset Lewis. He went from his lunch with Langmuir to drop in on his usual bridge game before returning to the laboratory, where Gerald Marsh observed that Lewis appeared morose while playing cards. Lewis then went back to Gilman Hall, where he seemed somber according to Kasha, and entered the lab, where he died about an hour later. Kasha said of that meeting between Lewis and Langmuir:

> When Lewis came back from that lunch, he was very quiet. Whatever they had talked about, whatever he had thought about, the trauma of meeting

> the man who in some sense outdid him, got to him. I think great personalities sometimes wither from the things that they have hidden in themselves. Lewis would never talk about the Nobel Prize, and no one ever brought it up in front of him. But he cared about it—anybody would have cared about it. He must have felt that he was fully of the caliber to deserve the prize.[40]

As he was saying this, Kasha had trouble controlling his emotions, and tears were in his eyes, almost sixty years after that day.

Did Lewis kill himself? The question is one that is unlikely to be answered definitively. Lewis might have had reason to do so. He was seventy years old, Kasha says that his short-term memory was failing, his close friend William Bray had died the month before, and he was down to a single part-time research assistant (Kasha) who would likely be leaving soon. His work had been everything to him, and that appeared to be nearing an end. Lewis was a proud man, and he might have chosen to control the time and place of his death; he would not have wanted to be seen as a suicide. So Lewis could have carefully planned the experiment with hydrogen cyanide with the intention of staging an accident, then come in to the research meeting with Kasha that morning, cheerful and enthusiastic.

But I believe that this new evidence, that Lewis met Langmuir that day, makes a planned suicide unlikely. Lewis was almost certainly not looking forward to his lunch with Langmuir, a man whom he did not like. If Lewis had carefully planned all of this, wouldn't he have chosen a different day? It is more likely that if it were a suicide, it was an impulsive suicide, due to frustration and depression caused by his meeting with Langmuir. To put it another way, if Lewis had carefully planned his suicide to make it appear to be a laboratory accident, and if for some reason he had decided it must be done that day, wouldn't he have killed himself *before* lunch?

It is possible that Lewis's death was accidental. Fatal laboratory accidents were not uncommon. Sam Ruben, one of the discoverers of carbon-14, had died three years earlier at Berkeley in a laboratory accident involving phosgene. Chemists were not nearly as safety-conscious as they are today, and laboratory ventilation and vacuum-line techniques were primitive by modern standards. An accident might have been made more likely by the stress of Lewis's lunch with Langmuir.

But given Kasha's and the other students' observations that Lewis's body was not cyanotic—they say that they did not see his lips and fingernails as blue—I think that it is more likely that his death was due to natural causes. The hydrogen cyanide might have been released after his death when the pressure built up in the flask and it dropped to the bench, as Kasha has suggested, or Lewis might have knocked the flask askew as he fell after a heart attack or stroke. Lewis was a seventy-year-old man who did not watch his diet, refused to see doctors, had chain-smoked cigars for over forty years, and took no exercise. Why shouldn't

Figure 11-2. Lewis in the laboratory in 1944. Courtesy of the College of Chemistry, University of California, Berkeley.

he have a heart attack or stroke, particularly after the stress of the meeting with Langmuir? Perhaps he was depressed and was upset by his meeting with Langmuir, but people who are depressed and upset have heart attacks, too.

Three questions remain: what was Lewis doing in the laboratory with hydrogen cyanide that day, who had arranged for Langmuir to visit, and why was the luncheon with Langmuir not disclosed?

Kasha described Lewis as working from an article in *Transactions of the Faraday Society*. I was unable to find such an article, but I did find a two-year-old paper in another British journal that exactly matched Kasha's description.[41] The experiment that Kasha describes is very Lewis-like and gives a good explanation of why he was working with hydrogen cyanide that day.

Who had been behind the honorary degree for Langmuir, and who had organized the lunch? Given Langmuir's status as a Nobel Prize–winning chemist, and the long-term enmity of Lewis toward Langmuir, it must have been someone senior from the chemistry department—no one else would have had the nerve. The evidence points to Joel Hildebrand, Lewis's old friend and colleague, who had served with Lewis in the army during the first war and had taken over as department chairman at the time of Lewis's retirement. The Faculty Senate's Committee on Honorary Degrees had nominated Langmuir, and Hildebrand was on that committee. Hildebrand later said that Lewis and Langmuir were the scientists who had most stimulated his mind,[42] and he might have thought that it was time for Lewis to put his twenty-five-year-old resentment behind him. By inviting Langmuir to Berkeley for a few weeks, he would have given Langmuir and Lewis time to reconcile. When Lewis died that afternoon, Hildebrand must have felt horrible. As Lewis's friend, he had almost certainly been concerned about Lewis's depression, and he later said that he thought Lewis's death was a suicide. Hildebrand had likely seen Lewis become upset at lunch, and he must have felt that his invitation had led to Lewis's death. This would explain the concealment of the meeting.

In the weeks that followed, Hildebrand and Langmuir became friends. During his stay in Berkeley for the Hitchcock Lectures, Hildebrand took Langmuir up to a ski lodge in the Sierras and, after Langmuir's death, Hildebrand wrote a memoir on Langmuir. Here are the first two paragraphs from that memoir: "My acquaintance with Irving Langmuir began about 1908, soon after each of us had returned from study in Germany, he from Göttingen, 1906, I from Berlin, 1907. Our subsequent meetings were occasional and casual until 1945, when he came to Berkeley to deliver the Hitchcock Lectures."[43] Hildebrand got the date of Langmuir's visit wrong—it was not 1945, but 1946. It is possible that this is only an error, but it seems more likely that it was meant to obscure any connection between Langmuir's visit and Lewis's death, or that the change of date was at least a psychologically significant slip. Hildebrand was a careful man, and it would have been hard in any case to confuse 1945 and 1946, especially in the Berkeley chemistry department. In March 1945, the country was at war and the chemistry department was off in the military or the Manhattan Project—undergraduates were purifying plutonium in the chemistry laboratories at Berkeley in 1945. In 1946, the department was coming back together.

I have been through Latimer's papers, through Hildebrand's papers and oral history interviews, and through Langmuir's papers and have found no mention of that day. Like Hildebrand, Langmuir also fudged the date of his visit to Berkeley. In his paper "Pathological Science," delivered in 1953 when he was seventy-two years old, Langmuir said, "[I]n 1945 or 46, I was at the University of California." He then proceeded to tell a story from an earlier date, about Lewis's bet with Latimer over the Allison effect (see chapter 10), without mentioning the circumstances of Lewis's death.

I believe that all the participants in the lunch meeting concealed Langmuir's presence for many years. But I also believe that Hildebrand and Langmuir were wrong—that Lewis did not commit suicide and that Hildebrand bore no responsibility for his death.

The View from the Cathedral

In 1883, when Arrhenius completed his doctoral work shortly after Lewis's 1875 birth, chemistry and physics were disjoint disciplines. Jacobus van't Hoff was just beginning to apply thermodynamics to the study of chemical affinity, but working chemists did not use thermodynamics. For them, predicting the course of chemical reactions was a matter of trial and error rather than calculation. Similarly, physics and chemistry had no common theory of the chemical bond. Even in 1916, Lewis's valence theory and Bohr's quantum theory attempted to describe the same physical objects, but they were not reconcilable—Lewis could explain the chemical bond, and Bohr could explain the spectrum of the hydrogen atom. But by the time of Lewis's death, the original mission of physical chemistry—integrating the theories of chemistry and physics—had been completed. Assigning a precise date to that integration is difficult. For thermodynamics, we might take it as 1905, when Nernst proclaimed what would become the third law of thermodynamics, opening the field to thermal measurements of free energy; or 1923, when Lewis and Randall published *Thermodynamics and the Free Energy of Chemical Substances*, which made thermodynamics accessible to ordinary chemists. For electronic theory, we might take the date as 1926, when Wolfgang Pauli announced his exclusion principle, which brought Lewis's electron pair into the physicists' new quantum theory; or 1927, when Heitler and London made the first quantum calculation showing a stable chemical bond in the hydrogen molecule; or 1939, when Pauling published *The Nature of the Chemical Bond*, which showed ordinary chemists how to use the concepts of quantum mechanics.

The problem of determining chemical affinity had been solved in two ways: first, by macroscopic measurements of mass, temperature, pressure, voltage, and so forth, using thermodynamics; and second, through an understanding of what was going on *inside* the molecule, employing Lewis's valence theory and eventually the physicists' quantum theory. In composing his *Elements of Chemistry* in 1789, Antoine Lavoisier wrote:

> The rigorous law from which I have never deviated, of forming no conclusions which are not fully warranted by experiment, and of never supplying the absence of facts, has prevented me from comprehending in this work the branch of chemistry which treats of affinities... [because] the principal data are still wanting, or, at least, those we have are not sufficiently defined, or not sufficiently proved, to become the foundation upon which to build so very important a branch of chemistry.[44]

By the time of Lewis's death more than 150 years later, that foundation had been completed.

Epilogue

George Eliot ends *Middlemarch* with a chapter that begins, "Every limit is a beginning as well as an ending. Who can quit young lives after being long in company of them, and not desire to know what befell them in their after-years? For the fragment of a life, however typical, is not the sample of an even web: promises may not be kept, and an ardent outset may be followed by declension; latent powers may find their long-waited opportunity; a past error may urge a grand retrieval." In her spirit, I take up the later lives of a few of the actors in this book. Not all of the stories are happy.

The Ionists: Wilhelm Ostwald had an incredible personal vitality, publishing more than 20,000 pages in his career on such diverse subjects as the history of electrochemistry, the philosophy of science, color perception, a synthetic world language, and a formula for happiness. He was a passionate amateur painter who made his own paints. While his personal contributions to the theory of physical chemistry are less than those of Svante Arrhenius, Jacobus van't Hoff, or Walther Nernst, his work as a leader and spokesperson drove the physical chemistry revolution. By the time he retired, thirty-four of his students had become professors.[1] He died on the eve of the Nazi era in 1932 at the age of seventy-nine. Van't Hoff, the first person awarded a Nobel Prize in Chemistry, died of tuberculosis in 1911 at the age of fifty-eight.

The Nazi proponents of "German science" did not fare well after the war. Bernhard Rust killed himself in May 1945. Johannes Stark was classified as a "Major Offender" by the Allied war crimes tribunal, was tried, and served four years. Philipp Lenard was eighty-one years old at the end of the war. He was stripped of his emeritus status on the Heidelberg faculty by the Allies and died two years later.

Given what was to follow for Jews in Germany, the Jewish scientists who were dismissed in 1933 were lucky to have been forced to emigrate. Many of them became prominent in America and England; James Franck and Franz (later Sir Francis) Simon were active in the American and English bomb projects, respectively. Max Planck, who had tried to protect Jewish scientists, suffered horribly in the war. His son was involved in the July 1944 assassination plot against Hitler and was executed despite Planck's desperate attempts to save him. At the end of the war, Planck was found with his wife by the Americans. The eighty-seven-year-old Planck had been sleeping in open fields, and was in great pain from fused vertebrae in his back. He died in 1946. The former Kaiser Wilhelm Institutes were soon renamed the "Max Planck Institutes" within the American, British, and French occupied zones of Germany.

Fritz Haber's horrible family story echoed through at least the next two generations. Haber's son Hermann, who had found his mother's body in the garden of their house after her suicide, moved to France after his father's death in 1934. He was interned by the French as an enemy alien when World War II started but was allowed to join the French Foreign Legion. Once the French surrendered to the Germans, he and his family were Jews in occupied France. Einstein organized exit visas for the family, and they made it to America. Hermann's wife, Magda, died of leukemia shortly after the end of the war. Hermann took his own life in 1946, and Hermann's oldest daughter did the same a short time later.[2]

During World War I, Haber developed a crystalline cyanide-based insecticide with the brand name "Zyklon." It was odorless, and there was risk that those using it might walk into a gassed room unwittingly. So in 1919, his institute modified Zyklon, adding a foul-smelling gas to warn people off, and renamed it "Zyklon B." After Haber's death, the Nazis used Zyklon B to gas millions in the death camps. The victims included Haber's relatives.[3]

William Harkins died in 1951. Possibly because of Gilbert Lewis's intervention on his behalf, he had been allowed to stay at the University of Chicago instead of being forced out as a "scientific bounder." His magnum opus, *The Physical Chemistry of Surface Films*, appeared shortly after his death. While his reputation in the field never equaled that of Irving Langmuir, and while few seemed to have liked him much, he had had a productive career.

Harold Urey retired from the University of Chicago in 1958 at age sixty-five. He spent the next twenty-four years, until his death in 1981, at the University of California, San Diego, where he published an additional 105 papers, many of them on the origin of the planets, or what he called "cosmochemistry." When he was asked why he continued to work so hard, he would reply, "Well, I'm no longer tenured."[4]

When we left Dorothy Wrinch, she had found a position on the Smith College faculty in Northampton, Massachusetts, with the help of Otto Glaser, a biologist at nearby Amherst College, whom she later married. Glaser died in 1950. Wrinch continued to work on her cyclol theory well into the 1960s, long after the structure of proteins had been established to everyone else's satisfaction. In the 1950s, chemical researchers discovered a structure in an alkaloid that contained a bond similar to her cyclol bond. She took this as proof that she was right and that Linus Pauling (who had argued that the bond was thermodynamically unstable and had not been found in nature) was wrong; she was convinced that both she and her cyclol theory had been vindicated, telling Marjorie Senechal, "First they said my structure couldn't exist. Then when it was found in nature they said it couldn't be synthesized in a laboratory. Then when it was synthesized, they said it wasn't important anyway."[5] (The cyclol structure has never been observed in proteins but was found in some small cyclic polypeptides five years before Wrinch's death.)[6] Wrinch became obsessive, drafting bitter unsent letters to Pauling and Pauling's colleague Robert Corey[7] and putting a physicist friend, William Scott, up to corresponding with Pauling in the hope of trapping him into admitting that she had been right.[8] Her tendentiousness wore down her friends and supporters: David Harker, Pauling's student who had supported Wrinch against Pauling, finally wrote an angry letter to her: "For almost twenty years I have been defending your picture of protein structure as one worthy of consideration, in the face of arguments that no one as biased as you are could produce a sound theory. At present the experimental evidence weighs so heavily against...your cyclols that I feel such structures can make only a minor contribution, if any, to the structure of any substance whatsoever."[9] Wrinch's daughter Pamela earned a Ph.D. from Yale in international relations in 1954 and became a lecturer at Milton College. Pamela died in a fire on Cape Cod in December 1975,[10] and Marjorie Senechal says that "Dorothy essentially never spoke thereafter."[11] She died in 1976.

Linus Pauling became much more active politically after he won the Nobel Prize in Chemistry for 1954. He had been on the political left before the war but had moderated his activities due to pressure from the CalTech administration. After the test of a hydrogen bomb in 1954, he realized that the bomb had been intentionally designed to cause massive radioactive fallout, and he became very active in opposing nuclear testing. A Senate committee issued a pamphlet calling him a leading sponsor of Communist fronts, and the State Department eventually pulled his passport. CalTech's president and trustees were unhappy with the political heat from its donors and alumni, and with what it saw as Pauling's inattention to research and teaching, and pushed him out as chairman of the chemistry department. He was subpoenaed by the Senate Internal Security Subcommittee, where he refused to disclose the names of those who had prepared and circulated anti-testing petitions. In 1963, the United States and the Soviet Union signed an atmospheric test-ban treaty, and Pauling received the Nobel Peace Prize for that year.

Pauling seemed to grow personally odder and more estranged from the scientific community. Because of his peace activities, his scientific efforts had slowed, and many of his colleagues at CalTech resented his refusal to give up his valuable lab space even though he was not doing much research. And what research he was doing was outside the scientific mainstream—he became convinced that vitamin C could prevent colds, and that vitamin deficiencies were the cause of much of the world's mental illness. He came up with a theory, which he called "orthomolecular psychiatry," that explained a great deal of physiology in terms of simple chemical processes—reaction rates, catalysts, the amounts of starting materials and products, and so forth. Pauling reduced human physiology to what he knew well—chemistry. While the idea that chemistry drove human physiology was true on a fundamental level, it was not at all clear in the early 1960s that an exclusively chemical focus would be a productive research approach. Arrhenius had done this sort of thing in the early 1900s with his attempt to reduce immunochemistry to the law of mass action, but it had not led to the sort of practical results that Paul Ehrlich had achieved in the field. Just as Wrinch had asserted the primacy of mathematics over chemistry, Pauling asserted the primacy of chemistry over biology, physiology, and medicine. And, as Wrinch had done, he seemed unwilling to accept experimental evidence that his hypotheses were flawed; like Wrinch, he seemed to believe that he had a deeper understanding of the subject than did his critics, and that their experiments were inconclusive irritants. The medical establishment rejected his papers and refused to fund his research. He left CalTech and wandered through appointments at the University of California, Santa Barbara, the University of California at San Diego, and Stanford. He then founded his own institutes, the Institute of Orthomolecular Medicine and the Linus Pauling Institute of Science and Medicine, and eventually became involved in a messy and expensive lawsuit with a former student whom he had selected to run the latter. Pauling died in 1994 at age ninety-three, promoting his theories of physiology and medicine until the end.[12]

Irving Langmuir remained active scientifically well into his retirement. He never lost his interest in children's science education. I found a letter written by Langmuir from 1945, when he was a sixty-four-year-old Nobel laureate and one of America's most famous scientists. The letter was to Elihu Lubkin, a ten-year-old from Brooklyn, who had sent Langmuir a booklet he had prepared on his "Sub-Atomic Theories." Most eminent scientists would have had a secretary send an encouraging note to a precocious child, telling him to keep up the good work. Not Langmuir, who replied with enormous respect, treating Elihu as he would a junior colleague. Langmuir sent young Elihu a three-page letter with page references to his booklet, telling him where he thought he was on track and where he needed more work:

> I have received your very interesting booklet, "Sub-Atomic Theories." I think this shows remarkably good understanding of many features of modern physics and shows many original ideas.
>
> I am particularly interested in the statement in the forward that you are only ten years of age and did not acquire your knowledge through schooling. There is no date given on that page so I am not clear as to how old you

are now. I would very much like to hear from you again to learn more about how you did acquire your knowledge, what books you have read, etc.

You asked for a frank criticism. I shall try to make the kind of criticism you will make yourself a few years from now when you look back over what you have written. When you go deeply into the structure of matter as you are trying to do, you need very badly a good mathematical training. You must start to learn algebra, and I think you would not have much difficulty in getting the foundations of the knowledge of differential and integral calculus. I started both of these things myself in my second year in high school before receiving formal schooling in them. I found later that this is the very best way to learn things of this kind. The things you learn by yourself will always be more valuable to you than the things that somebody else teaches you.

In your booklet you use practically no mathematics. The modern quantum theory, which is hard to express except in terms of mathematics, is very necessary for any profound understanding of the structure of matter. You do not appear as yet to make much out of the quantum theory.

In general, your ideas on fundamental particles are sound although in many details they are undoubtedly wrong. I shall take up a few of these details.

1. On page 2 you say that size and health are decreasing and you think this may lead to the disappearance of life on earth. I don't believe there is any good basis for your thoughts that health is decreasing; in fact, I think there is every indication that the opposite is true.
2. Again on page 2 you say that the sun is gradually cooling off. I think that the opinion of the best astronomers is that the temperature of the sun will gradually rise, and toward the end of its life will become much hotter than at present. It is rather an obsolete idea that the sun is cooling off.
3. On page 3—if you have a copy of the article "Atoms and Molecules," I would like to receive it.
4. Later, on page 6, you talk about negative and positive energies, but I am not clear what you mean. It is true that Dirac has talked about negative energy levels and this is tied in with his theory of the positron. Do you know of his work?
5. The idea that gravity is a push instead of a pull is a theory that was suggested a great many years ago, but I think the more modern view is that it is neither a push nor a pull but it is an effect produced by the "curvature of space." At least, this is Einstein's theory, which seems to have been well confirmed.
6. On page 18 you think that the fact that tides on the opposite sides of the earth have the same height proves the push theory as against the pull theory. I do not think that this is so. Remember that the force acting between two bodies varies inversely as the square of the distance between them. You seem to have overlooked the fact that the earth is pulled towards the moon more than the tide on the far side so that the

> earth does move away from the latter tide leaving it as a lump. This theory would become clearer to you if you analyze it mathematically.
>
> I would be glad to hear from you in regard to these matters and also any of your further ideas. I am sending you herewith some reprints of some articles which I think would be of general interest to you.[13]

I found Elihu Lubkin through an Internet search, and he told me that after he received this letter, his father took him on the train to Schenectady to visit Langmuir. Lubkin went on to become a professor of physics at the University of Wisconsin in Milwaukee.

Langmuir died at age seventy-seven in 1957, at home with his family after a series of heart attacks. The American Chemical Society has named its journal of surface chemistry *Langmuir.*

Although Gilbert Lewis did not win a Nobel Prize, he left a remarkable Nobel legacy: many who served on the Berkeley faculty or received their doctorates from Berkeley during his tenure did receive prizes: Harold Urey (1934), William Giauque (1949), Glenn Seaborg (1951), Willard Libby (1960), Melvin Calvin (1961), and Henry Taube (1983).

Joel Hildebrand, Lewis's colleague in the Berkeley chemistry department who had invited Langmuir to Berkeley and possibly arranged the lunch on the day of Lewis's death, retired from teaching at age seventy-one in 1952. He was active in research almost until his death at age 101 in 1983, publishing more papers *after* his retirement than he had before. He was one of the few living people to have a building named after him on the Berkeley campus—he said that the regents got tired of waiting for him to die.

Jacob Bigeleisen and Michael Kasha, Lewis's last two students, have had productive research careers, Bigeleisen in isotope chemistry and Kasha in photochemistry. Both are members of the National Academy of Sciences.

Mabel Kittredge, who was Lewis's and successive deans' personal secretary, wielded great authority within the Berkeley chemistry department for years. Young professors were convinced that they would never be promoted if they got on her bad side.[14] In the mid-1950s, William Jolly, a young Berkeley professor and a former student of Wendell Latimer's, tried to change the textbook for the inorganic chemistry course he was teaching. The old textbook was by Latimer, who had died, and Jolly felt it was time for a newer text. Kittredge viewed this as sacrilege and got Kenneth Pitzer, then dean, to overrule Jolly.[15] Kittredge had revered Lewis. When Bigeleisen returned to the Berkeley campus in the 1980s and was working on a paper on the history of isotope chemistry at Berkeley, he remembered that she had typed all Lewis's papers on heavy water. He called her at the retirement home at which she was then living, attempting to get some information. When he told her that he writing an article about deuterium, she said, "Professor Lewis discovered deuterium." Bigeleisen made the mistake of saying, "No, Harold Urey discovered deuterium." She then said, "I have nothing more to say to you," and hung up the phone.[16]

Lewis's personal papers—his laboratory notebooks and any personal diaries or letters—have not been preserved. I made many attempts to find these, without success. Kasha said that after Lewis's death he packed up carton after carton of Lewis's papers to be sent to his family.[17] When Roger Hahn, a professor of history at Berkeley, went to ask Mrs. Lewis to donate the papers to Berkeley's Bancroft Library in the 1960s, he came back with only a shoebox of odds and ends.[18] She said that there was nothing else, and Lewis's son Edward told me that he had never heard of the family receiving any such papers.[19] Since Kittredge would have been the intermediary between Kasha's packing and the delivery to the family, I suspect that she either confiscated or destroyed them.

Carlos, the Berkeley chemistry department janitor and stamp collector, continued in his position for a few years after Lewis's death. But his heart was no longer in his work. He had never gotten along with Mabel Kittredge, and he missed his friend Lewis and the cigar butts that he had scavenged for his pipe. As he was nearing retirement, he was carrying his dirty push broom and preparing to enter Latimer's office when Kittredge said, "Stop! You can't go in there with that!" Carlos gave her a look, went in, shoved the broom into Latimer's hands, and said, "Here Latimer! Now you clean the floors!"[20]

Sources and Acknowledgments

Sources

The bulk of my own research has been on Gilbert Lewis and Irving Langmuir, and most of the accounts in this book regarding other scientists are based on the work of others. When quoting from letters or manuscripts noted in secondary sources, I have not cited the archive where the original document may be found but have referred the reader to the secondary source. These sources are usually more accessible than the primary sources and often embed the document in a larger context than I have done. For the reader interested in going more deeply into the subject matter of this book, I describe some of the secondary sources that I have found especially useful. Full references are given in the section "Suggested Reading."

Daniel Charles's biography of Fritz Haber and Thomas Hager's biography of Linus Pauling are two of the best biographies I have read of any scientist. Charles's biography of Haber relies on Margit Szöllösi-Janze's longer biography, which unfortunately has not yet been translated from German. Diana Barkan's biography of Walther Nernst contains important material, particularly on Nernst's lamp and on the difficulties that he had with Svante Arrhenius and with the Nobel committee. Elisabeth Crawford's biography of Arrhenius is an excellent source on Wilhelm Ostwald and Jacobus van't Hoff, as well. Ostwald's *History of Electrochemistry* is a contemporaneous account by one of the participants in the physical chemical revolution.

Kurt Mendelssohn's biography of Nernst and Albert Rosenfeld's biography of Langmuir are valuable but have some problems as historical sources. Mendelssohn was a scientist in Nernst's institute, and his book is often based on his own recollections or on stories he has heard from others. His anecdotes about Nernst are entertaining and seem true in their essence but have possibly been improved, as Mendelssohn enjoyed telling a good story. I have included several of them, but I cannot vouch for their historical accuracy.

Rosenfeld was science editor of *Life* magazine, and his biography of Langmuir was included in a collection of Langmuir's works that was published with the financial support of General Electric in 1962, five years after Langmuir's death. I have included many of Rosenfeld's quotations from Langmuir's journals and from his interviews with Langmuir's colleagues and family. The biography is hagiographic, omitting almost all material about Langmuir that might be considered negative—his predecessors and competitors in surface chemistry are largely ignored, as is his collaboration with Dorothy Wrinch on proteins. Rosenfeld's book is not footnoted, and I have spent many hours finding the sources for some of his quotes, which I have included in the endnotes to this book. Where I have not been able to find the source, I have referred the reader back to Rosenfeld. What Rosenfeld had that was especially valuable was the cooperation of Langmuir's family and colleagues, and partial access to his private journals as redacted by Langmuir's son Kenneth. Langmuir's journals seem not to have been preserved—they are not part of his collected papers at the Library of Congress, and his grandson Roger Summerhayes does not know their whereabouts. And if they were to be found, Summerhayes would likely be the one to find them—he has written and produced a documentary film, *Langmuir's World*, containing home movies made by Langmuir and of interviews with Langmuir's relatives and associates, including Bernard and Kurt Vonnegut.

Pnina Abir-Am has written the most complete work to date on Dorothy Wrinch's personal history, and Pierre Laszlo has covered the scientific controversy over her cyclol theory. The Wikipedia article on cyclols, at least at the time of this writing in 2008, is more complete than any other summary of the subject that I have found. Marjorie Senechal, who was a colleague of Wrinch's at Smith College, is working on a book on Wrinch that should be very useful.

Edward Lewis's biography of his father Gilbert is valuable, especially for family information. The definitive historical studies on Lewis's electronic theory and on the Lewis-Langmuir controversy are by Robert Kohler, Jr.; Roger Stuewer admirably described Lewis's theory of radiation; and William Jolly has written a history of the Berkeley chemistry department during Lewis's time. Lewis's personal notebooks and papers have unfortunately not been preserved, but his professional correspondence can be found at the Bancroft Library at the University of California, Berkeley. Some of the best secondary source material on Lewis came out of a 1983 symposium in Las Vegas, published in several issues in 1984 of the *Journal of Chemical Education*. The paper by Jacob Bigeleisen on isotope research at Berkeley from that symposium is valuable in describing Lewis's work with heavy water.

I have relied on Ludwig Fritz Haber's and Charles Heller's accounts of chemical warfare in World War I, Ute Deichmann's history of the expulsion of Jewish scientists by the Nazis, John Heilbron's biography of Max Planck, John Heilbron and Robert Seidel's history of Ernest Lawrence's laboratory (available in part at www.lbl.gov/Science-Articles/Research-Review/Magazine/1981/index.html), and John Servos's history of physical chemistry in America. The last book is especially important for anyone who wants a broader perspective on the field.

Acknowledgments

Above all, I thank my wife, Ellen Pulleyblank Coffey, who had provided both editorial and emotional support. Jacob Bigeleisen has been an invaluable help throughout the process, and John Heilbron has both critiqued the work and helped me understand what it means to write history; I thank the three of them especially. Many have helped by reviewing the iterations of manuscripts, papers, and talks that led to this book, including my editor Jeremy Lewis, my copy editor Trish Watson, my production editor Heather Hartman, John Prausnitz, Michael Whitt, Roald Hoffmann, Dudley Herschbach, Sason Shaik, John Servos, Daniel Coffey, Jonathan Omer-Man, Roger Hahn, Edward Lewis, Gil Lewis, Marjorie Senechal, Joel Leventhal, and Susan Rabiner. Those who agreed to share their memories in interviews include Ted Geballe, Frances Connick, Robert Connick, Harold Johnston, Michael Kasha, Leo Brewer, Charles Auerbach, David Altman, Ed Zebroski, and Sam Weissman. In addition, I thank the historians Cathryn Carson, Erwin Hiebert, Guido Bacciagaluppi, Pnina Abir-Am, Göran Bergson, Alan Beyerchen, Andreas Kleinert, and Diana Buchwald; the science writers David Lindley, George Johnson, Susan Quinn, and Nancy Greenspan; my Swedish translator, Britt-Marie Gizerik, and Nobel Archives researcher Adrian Thomasson; Diana Wear at the Berkeley Office for History of Science and Technology; Karl Grandin at the Nobel Archives; Tammy Westgren at the Swedish Royal Institute of Technology; Janice Goldblum at the National Academy of Sciences; Len Bruno at the Library of Congress; Mary Ann Moore and David Farrell at the UC Berkeley libraries; and Phil Geisler, Bill Jolly, Clayton Heathcock, Jane Scheiber, Don Tilley, Michael Barnes, and Charles Harris at the UC Berkeley chemistry department. I thank Theodore Conant and the Pusey Library, the Bancroft Library, Smith College, and the Nobel Archives for permission to publish selections from their collections. I thank those who noted errors and omissions in the first printing of this book, including Yehuda Haas, Karl Jung, Rollie Myers, and Lee Sobotka.

Suggested Reading

What follows is a select bibliography that I believe will be of most interest to readers looking for a broader understanding of the ideas, people, and events described in this book.

Abir-Am, Pnina. "Disciplinary and Marital Strategies in the Career of Mathematical Biologist Dorothy Wrinch." In *Uneasy Careers and Intimate Lives, Women in Science 1789–1979*, edited by Pnina Abir-Am and D. Outram. New Brunswick, N.J.: Rutgers University Press, 1987; 239–280.

Arnold, James, Jacob Bigeleisen, and Clyde Hutchinson. "Harold Clayton Urey: April 24, 1893–January 5, 1981." In *National Academy of Sciences Biographical Memoirs*. Washington, D.C.: National Academy of Sciences, 1995; 363–412.

Barkan, Diana Kormos. *Walther Nernst and the Transition to Modern Physical Science*. Cambridge: Cambridge University Press, 1999.

Beyerchen, Alan D. *Scientists Under Hitler: Politics and the Physics Community in the Third Reich*. New Haven, Conn.: Yale University Press, 1977.

Bigeleisen, Jacob. "Gilbert N. Lewis and the Beginnings of Isotope Chemistry." *Journal of Chemical Education* 61 (1984): 108–116.

Charles, Daniel. *Master Mind: The Rise and Fall of Fritz Haber, the Nobel Laureate Who Launched the Age of Chemical Warfare*. New York: Ecco, 2005.

Coffey, Patrick. "Chemical Free Energies and the Third Law of Thermodynamics." *Historical Studies in Physical and Biological Sciences* 36 (2006): 365–396.

Crawford, Elisabeth. *Arrhenius: From Ionic Theory to the Greenhouse Effect*. Canton, Ohio: Science History Publications/USA, 1996.

Deichmann, Ute. "The Expulsion of German-Jewish Chemists and Biochemists and Their Correspondence with Colleagues in Germany after 1945: The Impossibility of Normalization?" In *Science in the Third Reich*, edited by Margit Szöllösi-Janze. New York: Oxford University Press, 2000; 243–280.

Greenspan, Nancy. *The End of the Uncertain World: The Life and Science of Max Born*. New York: Basic Books, 2005.

Haber, Ludwig Fritz. *The Poisonous Cloud: Chemical Warfare in the First World War*. Oxford: Clarendon Press, 1986.

Hager, Thomas. *Force of Nature: The Life of Linus Pauling*. New York: Simon and Schuster, 1995.

Heilbron, J.L. *The Dilemmas of an Upright Man: Max Planck as Spokesman for German Science*. Berkeley: University of California Press, 1986.

——. "The Nobel Science Prizes of World War I." In *Historical Studies in the Nobel Archives*, edited by Elisabeth Crawford. Tokyo: Universal Academy Press, 2002; 19–38.

Heilbron, J.L., and Robert W. Seidel. *Lawrence and His Laboratory*. Berkeley: University of California Press, 1989. Available in part at www.lbl.gov/Science-Articles/Research-Review/Magazine/1981/index.html.

Heller, Charles. *Chemical Warfare in World War I: The American Experience, 1917–1918, Leavenworth Papers*. Fort Leavenworth, Tex.: Combat Studies Institute, U.S. Army Command and General Staff College, 1984. Available at www-cgsc.army.mil/carl/resources/csi/Heller/HELLER.asp.

Hewlett, Richard G., and Oscar E. Anderson, Jr. *A History of the United States Atomic Energy Commission*: Vol. 1, *The New World, 1939–1945*. Philadelphia: University of Pennsylvania Press, 1962.

Jolly, William. *From Retorts to Lasers*. Berkeley: William Jolly, distributed by the College of Chemistry, University of California, Berkeley, 1987.

Kohler, Robert E., Jr. "G.N. Lewis's Views on Bond Theory 1900–1916." *British Journal of the History of Science* 8 (1975): 233–239.

——. "Irving Langmuir and the 'Octet' Theory of Valence." *Historical Studies in Physical Sciences* 4 (1972): 39–97.

——. "The Lewis-Langmuir Theory of Valence and the Chemical Community, 1920–1928." *Historical Studies in Physical Sciences* 6 (1975): 431–468.

——. "The Origin of G.N. Lewis's Theory of the Shared Pair Bond." *Historical Studies in Physical Sciences* 3 (1971): 343–376.

Kuhn, Thomas. *The Structure of Scientific Revolutions*. Chicago: University of Chicago Press, 1962.

Langmuir, Irving. *Phenomena, Atoms, and Molecules*. New York: Philosophical Library, 1950.

Laszlo, Pierre. *A History of Biochemistry: Molecular Correlates of Biological Concepts*. Vol. 34A in *Comprehensive Biochemistry*, edited by A. Neuberger and L.L.M. Van Deemen. Amsterdam: Elsevier, 1986.

Lewis, Edward. *A Biography of Distinguished Scientist Gilbert Newton Lewis*. New York: Edwin Mellen Press, 1998.

Lewis, Gilbert N. *The Anatomy of Science*. New Haven, Conn.: Yale University Press, 1926.

——. "The Atom and the Molecule." *Journal of the American Chemical Society* 32 (1916): 752–785.

Lewis, Gilbert N., and Merle Randall. *Thermodynamics and the Free Energy of Chemical Substances*. New York: McGraw-Hill, 1923.

Lindley, David. *Boltzmann's Atom: The Great Debate That Launched a Revolution in Physics*. New York: Free Press, 2001.

Luttenberger, Franz. "Arrhenius vs. Ehrlich on Immunochemistry: Decisions about Scientific Progress in the Context of the Nobel Prize." *Theoretical Medicine and Bioethics* 13 (2005): 137–173.

Medawar, Jean, and David Pyke. *Hitler's Gift: The True Story of the Scientists Expelled by the Nazi Regime*. New York: Arcade Publishing, 2001.

Mendelssohn, Kurt M. *The World of Walther Nernst: The Rise and Fall of German Science*. London: Macmillan, 1973.

Nernst, Walther. *The New Heat Theorem*. London: Methuen and Co., 1926.

Ostwald, Wilhelm. *Elektrochemie: Ihre Geschichte und Lehre*. 2 vols. Leipzig: Verlag von Veit and Comp., 1896. English translation, Smithsonian Institution and National Science Foundation. New Delhi: Amerind Pub. Co., 1980.

Pauling, Linus. *The Nature of the Chemical Bond*. Ithaca, N.Y.: Cornell University Press, 1939.

Rosenfeld, Albert. "The Quintessence of Irving Langmuir." In *The Collected Works of Irving Langmuir*, Vol. 12, edited by C. Guy Suits and Harold Way. New York: Pergamon Press, 1962; 5–229.

Servos, John W. *Physical Chemistry from Ostwald to Pauling: The Making of a Science in America*. Princeton, N.J.: Princeton University Press, 1990.

Smil, Vaclav. *Enriching the Earth: Fritz Haber, Carl Bosch, and the Transformation of World Food Production*. Cambridge, Mass.: MIT Press, 2001.

Smyth, Henry DeWolf. *Atomic Energy for Military Purposes*. Princeton, N.J.: Princeton University Press, 1945.

Stoltzenberg, Dietrich. *Fritz Haber: Chemist, Nobel Laureate, German, Jew*. Translated by Charles Passage. Philadelphia: Chemical Heritage Press, 2004.

Stuewer, Roger. "G. N. Lewis on Detailed Balancing, the Symmetry of Time, and the Nature of Light." *Historical Studies in Physical Sciences* 6 (1975): 469–511.

Summerhayes, Roger. *Langmuir's World*. Film. CustomFlix, 1996.

Szöllösi-Janze, Margit. *Fritz Haber 1868–1934: Eine Biographie*. Munich: Verlag C.H. Beck, 1998.

Taylor, Hugh. "Irving Langmuir, 1881–1957." *Biographical Memoirs of Fellows of the Royal Society* 4 (1957): 167–184.

Widmalm, Sven. "Science and Neutrality: The Nobel Prizes of 1919 and Scientific Internationalism in Sweden." *Minerva* 33 (1995): 339–360.

Endnotes

Abbreviations

CCP	Records of the College of Chemistry, 1874–1955, Gilbert Newton Lewis collection, Bancroft Library, University of California, Berkeley
DWP	Dorothy Wrinch Papers 1919–1975, Sophie Smith Collection, Smith College Archives, Northampton, Mass.
ILP	Irving Langmuir Papers, Library of Congress, Washington, D.C.
KVA	Nobel Archive of the Royal Swedish Academy of Sciences, Stockholm
NBL	Niels Bohr Library, American Institute of Physics, College Park, Md.
TRP	Theodore Richards Papers, Pusey Library, Harvard University, Cambridge, Mass.

Prologue

1. Author's interview with Ted Geballe, 2005.

2. Arthur Koestler, *The Sleepwalkers: A History of Man's Changing Vision of the Universe* (London: Hutchinson, 1959), 10.

3. Roald Hoffmann, "What Might Philosophy of Science Look Like If Chemists Built It?" *Synthese* 155, no. 3 (2007): 321–336.

4. Thomas Kuhn, *The Structure of Scientific Revolutions* (Chicago: University of Chicago Press, 1962).

5. Nicholas Wade, "Scientists Transplant Genome of Bacteria." *New York Times*, 29 June 2007.

Chapter 1

1. Arrhenius to Ostwald, 14 Dec. 1925. Quoted in Elisabeth Crawford, *Arrhenius: From Ionic Theory to the Greenhouse Effect* (Canton, Mass.: Science History Publications/USA, 1996), 43.

2. Olof Arrhenius, "Svante Arrhenius: Det Första Kvartseklet (Svante Arrhenius: The First Quarter Century)," in *Svante Arrhenius Till 100-Årsminnet av hans Födelse* (Svante Arrhenius, on the 100th Anniversary of his Birth) *Kongliga Ventenskap-Akademiens Årsbok 1959 Bilaga* (Stockholm: Almquist and Wicksell, 1959), 43–64, on 63. Quoted in Crawford, *Arrhenius*, 43.

3. Royal Swedish Academy of Sciences, *Le Prix Nobel en 1903* (Stockholm, 1906), 67. Quoted in Crawford, *Arrhenius*, 205.

4. Hermann Kolbe, "Zeichen der Zeit: II," *Journal für Praktische Chemie* 123 (1877): 473–477, on 474. Quoted in Robert Root-Bernstein, "The Ionists: Founding Physical Chemistry, 1872–1890" (Ph.D. thesis, Princeton University, 1980), 195–196.

5. Root-Bernstein, "The Ionists," 31.

6. Crawford, *Arrhenius*, 19.

7. Ibid., 14.

8. Ibid., 20.

9. Ibid.

10. Svante Arrhenius, "Recherches sur la Conductibilité Galvanique des Electrolytes," *Bihang till Kongliga Svenska Ventenskaps-Akademiens Handlingar* 8, no. 13 (part 1), 14 (part 2) (1884). Quoted in Root-Bernstein, "The Ionists," 103.

11. Root-Bernstein, "The Ionists," 29.

12. Ibid., 44.

13. Henri le Chatelier, *Recherches Expérimentales et Théoriques sur les Équilibres Chimiques* (Paris: Ve Ch. Dunod, 1888), 11–12. Also published in *Annales des Mines* 13, 157–382. Quoted in Gilbert N. Lewis and Merle Randall, *Thermodynamics and the Free Energy of Chemical Substances* (New York: McGraw-Hill, 1923), 3.

14. Arrhenius, "Recherches," quoted in Root-Bernstein, "The Ionists," 82.

15. Crawford, *Arrhenius*, 59–60.

16. Ostwald to his wife, 17 and 22 Aug. 1884. Quoted in Crawford, *Arrhenius*, 53.

17. Crawford, *Arrhenius*, 60–61.

18. Robert Millikan, "Walther Nernst, a Great Physicist, Passes," *Scientific Monthly* 51 (1942): 84–86.

19. Kurt M. Mendelssohn, *The World of Walther Nernst: The Rise and Fall of German Science* (London: Macmillan, 1973), 23.

20. Arrhenius to Eric Edlund, 19 Dec. 1886. Quoted in Crawford, *Arrhenius*, 67.

21. Diana Kormos Barkan, *Walther Nernst and the Transition to Modern Physical Science* (Cambridge: Cambridge University Press, 1999), 41.

22. Ibid., 34.

23. Ibid., 51.

24. Arrhenius, "Recherches," part 1, 6. Quoted in Crawford, *Arrhenius*, 30.

25. John W. Servos, *Physical Chemistry from Ostwald to Pauling: The Making of a Science in America* (Princeton, N.J.: Princeton University Press, 1990), 53.

26. Arrhenius to Ostwald, 13 Aug. 1889. Quoted in Crawford, *Arrhenius*, 96.

27. Edvard Hjelt, "Den Nya Electrokemiska Teorin," *Finska Ventenskaps-Societetens Översikt* 34 (1892): 10–11. Quoted in Crawford, *Arrhenius*, 97.

28. Nernst to Ostwald, 17 May 1890. Quoted in Crawford, *Arrhenius*, 101.

29. Crawford, *Arrhenius*, 81–108.

30. Hermann von Helmholtz, *Sitzungsberichte Berlin Akademie Ges. Abhandl.*, Vol. 2 (1882). Quoted in Wilhelm Ostwald, *Elektrochemie: Ihre Geschichte und Lehre*, 2 vols. (Leipzig: Verlag von Veit and Comp., 1896). English trans., Smithsonian Institution and National Science Foundation (New Delhi: Amerind Pub. Co., 1980), 1007–1008.

31. Jacobus van't Hoff, *Physical Chemistry in the Service of the Sciences* (Chicago: University of Chicago Press, 1903), 33.

32. Jacobus van't Hoff, "Lois de l'Equilibre Chimique dans l'Etat Dilué, Gazeux ou Dissous," *Kongliga Svenska Vetenskaps-Akademiens Handlingar* 21, no. 17 (1885).

33. Barkan, *Walther Nernst*, 50.

34. Thomas Kuhn, *The Structure of Scientific Revolutions* (Chicago: University of Chicago Press, 1962).

35. Mendelssohn, *The World of Walther Nernst*, 40–41; Crawford, *Arrhenius*, 84–85.

36. Elisabeth Crawford, *The Beginnings of the Nobel Institution: The Science Prizes, 1901–1915* (Cambridge: Cambridge University Press, 1984), 52.

37. Crawford, *Arrhenius*, 125–126.

38. Svante Arrhenius, "Naturens Värmehushållning," *Nordisk Tidskrift* 14 (1896): 121–130, on 130. Quoted in Crawford, *Arrhenius*, 154.

39. James Franck, oral history interview by Thomas Kuhn and Maria Goeppert Mayer, 9–14 July 1962, *NBL*.

40. Mendelssohn, *The World of Walther Nernst*, 49.

41. Ibid., 78.

42. Crawford, *Arrhenius*, 43–44.

43. Barkan, *Walther Nernst*, 43.

44. Van't Hoff to Ostwald, around 8 May 1896. Quoted in Barkan, *Walther Nernst*, 63.

45. Mendelssohn, *The World of Walther Nernst*, 45–47, 43. Also recounted in Albert Rosenfeld, "The Quintessence of Irving Langmuir," in *The Collected Works of Irving Langmuir*, Vol. 12, ed. C. Guy Suits and Harold Way (New York: Pergamon Press, 1962), 60.

46. Mendelssohn, *The World of Walther Nernst*, 49–50.

47. Crawford, *The Beginnings of the Nobel Institution*, 63.

48. Ibid., 81–82.

49. Ibid., 69–76.

50. Crawford, *Arrhenius*, 144.

51. Nernst to Ostwald, 24 Feb. 1901. Quoted in Barkan, *Walther Nernst*, 222–223.

52. Crawford, *The Beginnings of the Nobel Institution*, 116–122.

53. Royal Swedish Academy of Sciences, *Le Prix Nobel en 1903* (Stockholm: 1906), 67. Quoted in Crawford, *Arrhenius*, 205.

54. Ibid.

55. Crawford, *Arrhenius*, 205.

56. Crawford, *The Beginnings of the Nobel Institution*, 125–126.

57. Ibid., 126.

58. Arthur M. Silverstein, *Paul Ehrlich's Receptor Immunology: The Magnificent Obsession* (New York: Academic Press, 2002), 3–4.

59. Paul Ehrlich, in *Proceedings VII International Congress of Hygiene and Demography*, Vol. 2 (London: 1891), 211. Quoted in Silverstein, *Paul Ehrlich's Receptor Immunology*, 12.

60. Silverstein, *Paul Ehrlich's Receptor Immunology*, 76–84.

61. Arrhenius to Thorvald Madsen, 13 June 1907. Quoted in Franz Luttenberger, "Arrhenius vs. Ehrlich on Immunochemistry: Decisions about Scientific Progress in the Context of the Nobel Prize," *Theoretical Medicine and Bioethics* 13 (2005): 137–173, on 155.

62. Silverstein, *Paul Ehrlich's Receptor Immunology*.

63. Luttenberger, "Arrhenius vs. Ehrlich."

64. Silverstein, *Paul Ehrlich's Receptor Immunology*, 71–72.

65. Luttenberger, "Arrhenius vs. Ehrlich," 143.

66. Crawford, *Arrhenius*, 231.

67. Paul Ehrlich, "Partial Cell Functions," in *Nobel Lectures: Physiology or Medicine* (Stockholm: Nobel Foundation, 1967), 305. Quoted in Crawford, *Arrhenius*, 232.

Chapter 2

1. Langmuir card catalog, *ILP*.

2. *Harvard University Catalog 1902–1903*, 408, Harvard University Archives.

3. John W. Servos, *Physical Chemistry from Ostwald to Pauling: The Making of a Science in America* (Princeton, N.J.: Princeton University Press, 1990), 54–55.

4. Ibid., 54.

5. Edward Lewis, *A Biography of Distinguished Scientist Gilbert Newton Lewis* (New York: Edwin Mellen Press, 1998), 7.

6. Lewis student file, Harvard University Archives.

7. Richards to Julius Stieglitz, 17 Nov. 1916. Quoted in Servos, *Physical Chemistry*, 118.

8. Servos, *Physical Chemistry*, 81.

9. Thomas Kuhn, interview with J. Robert Oppenheimer, 18 Nov. 1963. Quoted in Alice Kimball Smith and Charles Wiener, eds., *Robert Oppenheimer: Letters and Recollections* (Cambridge: Cambridge University Press, 1980), 6.

10. Manuscripts may be found in the Gilbert Newton Lewis Papers, 1904–1946, Bancroft Library, University of California, Berkeley.

11. For an extensive history of Le Sage's and related theories (although not including Lewis's), see Matthew Edwards, ed., *Pushing Gravity: New Perspectives on Le Sage's Theory of Gravitation* (Montreal: Apeiron, 2002).

12. Lewis to R.A. Millikan, 28 Oct. 1919, *CCP*. Quoted in Servos, *Physical Chemistry*, 119.

13. Michael Kasha, "The Triplet State: An Example of G. N. Lewis' Research Style," *Journal of Chemical Education* 61 (1984): 204–215. Kasha believed that "The Electron and the Molecule" was Lewis's master's thesis, but the Harvard catalog from that period specifies that the master's degree did not require a thesis.

14. Gilbert N. Lewis, "A General Equation for Free Energy and Physico-chemical Equilibrium, and Its Application" (Ph.D. thesis, Harvard University, 1899).

15. Theodore Richards and Gilbert N. Lewis, "Some Electrochemical and Thermochemical Relations of Zinc and Cadmium Amalgams," *Proceedings of the American Academy of Arts and Sciences* 34 (1898): 87–99.

16. Gilbert N. Lewis, "The Development and Application of a General Equation for Free Energy and Physico-chemical Equilibrium," *Proceedings of the American Academy of Arts and Sciences* 35 (1900): 3–38.

17. J.J. Waterston, *The Collected Scientific Papers of John James Waterston*, ed. J.S. Haldane, *Philosophical Transactions of the Royal Society of London*, Series A (Edinburgh: Oliver and Boyd, 1928), 209–210.

18. Wilder Bancroft, "Review of the Development and Application of a General Equation for Free Energy and Physico-chemical Equilibrium, by Gilbert Lewis," *Journal of Physical Chemistry* 5 (1901): 405.

19. Theodore Richards, "The Significance of Changing Atomic Volume," *Proceedings of the American Academy of Arts and Sciences* 38 (1902): 291–341. In this paper, Richards concludes that free energy is *not* the measure of affinity.

20. Gilbert N. Lewis and Merle Randall, *Thermodynamics and the Free Energy of Chemical Substances* (New York: McGraw-Hill, 1923), 386.

21. Henri le Chatelier, *Recherches Expérimentales et Théoriques sur les Équilibres Chimiques* (Paris: Ve Ch. Dunod, 1888), 184. Also published in *Annales des Mines* 13, 157–382. Quoted in Lewis and Randall, *Thermodynamics*, 435–436.

22. For a more mathematical treatment of the subject, see Patrick Coffey, "Chemical Free Energies and the Third Law of Thermodynamics," *Historical Studies in Physical and Biological Sciences* 36 (2006): 365–396, on 382–390.

23. Lewis to Richards, 5 Apr. 1899, 3 July 1899, 30 July 1899, 8 Aug. 1899, *TRP.*

24. Gilbert N. Lewis, "A New Conception of Thermal Pressure and a Theory of Solutions," *Proceedings of the American Academy of Arts and Sciences* 36 (1900): 145–168.

25. In the kinetic theory of an ideal gas, the temperature is proportional to the average kinetic energy of the molecules. Lewis proposed that for a nonideal gas, the temperature should be considered proportional to the product of the average momentum of the molecules multiplied by the number of molecules passing in one second through unit area. Lewis, "A New Conception," 154.

26. Lewis to Richards, 13 Jan. 1901, *TRP.* Quoted in Coffey, "Chemical Free Energies," 393.

27. Richards to Lewis, 16 Mar. 1901, *TRP.* Quoted in Coffey, "Chemical Free Energies," 385.

28. Richards, undated scrap, apparently sent to Lewis with the letter of 16 Mar. 1901, *TRP.* Quoted in Coffey, "Chemical Free Energies," 386.

29. Theodore Richards, "The Driving Tendency of Physico-chemical Reaction, and Its Temperature Coefficient," *Proceedings of the American Academy of Arts and Sciences* 35 (1900): 471–481.

30. Lewis to Richards, 5 Apr. 1901, *TRP.* Quoted in Coffey, "Chemical Free Energies," 387.

31. Ibid.

32. Gilbert N. Lewis, "The Law of Physico-chemical Change," *Proceedings of the American Academy of Arts and Sciences* 37 (1901): 49–69, on 66.

33. Ostwald to Richards, 30 Mar. 1901, *TRP.*

34. Lewis to Richards, 13 Jan. 1901, *TRP.* Quoted in Coffey, "Chemical Free Energies," 391–392.

35. Lewis to Richards, 5 Apr. 1901, *TRP.* Quoted in Coffey, "Chemical Free Energies," 392.

36. Servos, *Physical Chemistry*, 79.

37. Gilbert N. Lewis, "The Atom and the Molecule," *Journal of the American Chemical Society* 32 (1916).

38. Theodore Richards, Chemistry 8 notebook, 1902, Harvard University Archives. Quoted in Servos, *Physical Chemistry*, 119.

39. Author's interview with Lewis's son Edward Lewis, 2005.

40. Louis Hammett, oral history interview by Leon Gortler, 1 May 1978, *NBL.*

41. "Report of the National Academy of Sciences, 1906" (Washington, D.C.: National Academy of Sciences, 1906), 19.

42. Elisabeth Crawford, J.L. Heilbron, and Rebecca Ullrich, *The Nobel Population, 1901–1937* (Berkeley, Calif.: Office for History of Science and Technology, 1987), 228.

43. Lewis to Richards, 6 Dec. 1915, *CCP.*

44. Author's interview with Jacob Bigeleisen, 2004.

45. Lewis to Hartley, 8 Oct. 1928, *CCP.*

46. J.M. Cottrell to Lewis, 10 May 1928, *CCP.*

47. George Baxter, "Theodore William Richards," *Science* 68 (1928): 338–339.

48. Lewis personnel file, Harvard University Archives.

49. Gilbert N. Lewis, "Outlines of a New System of Thermodynamic Chemistry," *Proceedings of the American Academy of Arts and Sciences* 43 (1907): 259–293.

50. Ibid., 259.

51. Gilbert N. Lewis, *The Anatomy of Science* (New Haven, Conn.: Yale University Press, 1926), 181. Lewis quoted Poincaré but did not provide a citation. The quote is from Henri Poincaré, *Science and Hypothesis* (New York: Science Press, 1913), 31.

52. For a review of Lewis's work on strong electrolytes, see Kenneth Pitzer, "Gilbert Lewis and the Thermodynamics of Strong Electrolytes," *Journal of Chemical Education* 61 (1984): 104–107.

53. Albert Rosenfeld, "The Quintessence of Irving Langmuir," in *The Collected Works of Irving Langmuir,* Vol. 12, ed. C. Guy Suits and Harold Way (New York: Pergamon Press, 1962), 42.

54. Langmuir to his mother, 25 Jan. 1903, *ILP.*

55. Langmuir to his mother, 21 July 1903, *ILP.*

56. Langmuir to his mother, 26 July 1903, *ILP.*

57. Edward Lewis, *Biography*, 7.

58. Rosenfeld, "The Quintessence of Irving Langmuir," 51–55.

59. Ibid., 9.

60. Probably Walther Nernst, "Über die Bildung von Stickoxyd bei Hohen Temperaturen," *Nachrichten von der Koniglichen Gesellschaft der Wissenschaften zu Göttingen, Mathematisch-Physikalische Klasse*, no. 4 (1904): 261–276.

61. Langmuir to his mother, 16 May 1904, *ILP.*

62. Langmuir to his mother, 7 June 1904, *ILP.*

63. Langmuir to his mother, 13 June 1904, *ILP.*

64. Percy Bridgman, "Some of the Physical Aspects of the Work of Irving Langmuir," in Suits and Way, *The Collected Works of Irving Langmuir,* Vol. 12,

433–457, on 434; Hugh Taylor, "Irving Langmuir, 1881–1957," *Biographical Memoirs of Fellows of the Royal Society* 4 (1957): 167–184, on 168.

65. Langmuir to his mother, 13 Dec. 1904, *ILP*.

66. Irving Langmuir to Arthur Langmuir, 7 Apr. 1905, *ILP*.

67. Langmuir to his mother, 7 July 1905, *ILP*.

68. Langmuir to his mother, 7 Nov. 1905, *ILP*.

69. Irving Langmuir, "The Dissociation of Water Vapor and Carbon Dioxide at High Temperatures," *Journal of the American Chemical Society* 28 (1906): 1357–1379.

70. Irving Langmuir, "The Velocity of Reactions in Gases Moving through Heated Vessels and the Effect of Convection and Diffusion," *Journal of the American Chemical Society* 30 (1908): 1742–1754.

71. Herbert Langmuir to Irving Langmuir, 13 Dec. 1904, *ILP*

72. Langmuir to F.J. Pond, 2 July 1909, *ILP*.

73. Cable from F.J. Pond to Langmuir, 14 July 1909, *ILP*.

74. Irving Langmuir, "Atomic Hydrogen as an Aid to Industrial Research," *Science* 67 (1928): 201–208.

Chapter 3

1. Daniel Charles, *Master Mind: The Rise and Fall of Fritz Haber, the Nobel Laureate Who Launched the Age of Chemical Warfare* (New York: Ecco, 2005), 16.

2. Kurt M. Mendelssohn, *The World of Walther Nernst: The Rise and Fall of German Science* (London: Macmillan, 1973), 84.

3. Quoted in Charles, *Master Mind*, 30.

4. Haber to Max Hamburger, 11 Apr. 1889. Quoted in Charles, *Master Mind*, 19.

5. Haber to Max Warburg, 29 July 1891. Quoted in Dietrich Stoltzenberg, *Fritz Haber: Chemist, Nobel Laureate, German, Jew*, trans. Charles Passage (Philadelphia: Chemical Heritage Press, 2004), 28–29.

6. Hannah Arendt, *The Origins of Totalitarianism* (New York: Harvest Books, 1968), 56; following discussion, 56–68.

7. Ibid., 67.

8. Charles, *Master Mind*, 29.

9. Ibid., 30–32.

10. Ibid., 32.

11. Wilhelm Ostwald, *Lebenslinien. Eine Selbstbiographie*, Vol. 2 (Berlin: Klasing, 1926–1927), 252. Quoted in Margit Szöllösi-Janze, *Fritz Haber 1868–1934: Eine Biographie* (Munich: Verlag C.H. Beck, 1998), 111. Also quoted in Charles, *Master Mind*, 37–38.

12. Szöllösi-Janze, *Fritz Haber*, 151.

13. Theodore Richards, "The Significance of Changing Atomic Volume," *Proceedings of the American Academy of Arts and Sciences* 38 (1902): 291–341.

14. Jacobus van't Hoff, "Einfluss der Änderung der Spezifischen Wärme auf die Umwandlungsarbeit," in *Festschrift der Ludwig Boltzmann* (Leipzig: Verlag von Johann Ambrosium Barth, 1904), 233–247.

15. Fritz Haber, *Thermodynamik Technischer Gasreaktionen* (Munich: Oldenbourg Verlag, 1905). English trans., Fritz Haber, *Technical Gas Reactions* (London: Longmans, Green, 1908), 48.

16. Walther Nernst, "Über die Berchnung Chemischer Gleichgewichte aus Thermischen Messungen," *Nachrichten von der Koniglichen Gesellschaft der Wissenschaften zu Göttingen, Mathematisch-Physikalische Klasse*, no. 1 (1906): 1–40.

17. Walther Nernst, *Experimental and Theoretical Applications of Thermodynamics to Chemistry* (New Haven, Conn.: Yale University Press, 1913), 56–57.

18. Elisabeth Crawford, J.L. Heilbron, and Rebecca Ullrich, *The Nobel Population, 1901–1937* (Berkeley, Calif.: Office for History of Science and Technology, 1987), 180, 184.

19. Gilbert N. Lewis and Merle Randall, *Thermodynamics and the Free Energy of Chemical Substances* (New York: McGraw-Hill, 1923), 436–438.

20. Lewis to Sir Harold Hartley, 23 Apr. 1929, *CCP*.

21. Charles, *Master Mind*, 65.

22. William Crookes, *The Wheat Problem* (London: Longmans, Green, 1917), 1–41. Quoted in Charles, *Master Mind*, 80.

23. Vaclav Smil, *Enriching the Earth: Fritz Haber, Carl Bosch, and the Transformation of World Food Production* (Cambridge, Mass.: MIT Press, 2001), 64.

24. Ibid., 24.

25. Fritz Haber, "Über die Nutzbarmachung des Stickstoffs," *Verhandlungen des Naturwissenschaftlichen Vereins in Karlsruhe* 23 (1909/1910). Quoted in Stoltzenberg, *Fritz Haber*, 90.

26. Koppel to the Kaiser, 10 Mar. 1910. Quoted in Szöllösi-Janze, *Fritz Haber*, 218.

27. Jeffrey Johnson, *The Kaiser's Chemists: Science and Modernization in Imperial Germany* (Chapel Hill: University of North Carolina Press, 1990), 126.

28. Richard Willstätter, *Aus Meinem Leben* (Weinheim: Verlag Chemie, 1943), 263. Quoted in Charles, *Master Mind*, 123.

29. Smil, *Enriching the Earth*, 85–86.

30. Szöllösi-Janze, *Fritz Haber*, 165.

31. Richard Weidlich, "Erinnerungen" [Memoirs]. Quoted in Stoltzenberg, *Fritz Haber*, 95.

32. Stoltzenberg, *Fritz Haber*, 95.

33. Smil, *Enriching the Earth*, 102.

Chapter 4

1. Dietrich Stoltzenberg, *Fritz Haber: Chemist, Nobel Laureate, German, Jew*, trans. Charles Passage (Philadelphia: Chemical Heritage Press, 2004), 127.

2. Kurt M. Mendelssohn, *The World of Walther Nernst: The Rise and Fall of German Science* (London: Macmillan, 1973), 78.

3. Lance Sergeant Elmer Cotton, 1915. Quoted in "Chlorine Gas," www.spartacus.schoolnet.co.uk/FWWchlorine.htm, accessed 23 Jan. 2008.

4. Mendelssohn, *The World of Walther Nernst*, 80–81.

5. Daniel Charles, *Master Mind: The Rise and Fall of Fritz Haber, the Nobel Laureate Who Launched the Age of Chemical Warfare* (New York: Ecco, 2005), 177.

6. Stoltzenberg, *Fritz Haber*, 128–132; Charles, *Master Mind*, 175–177.

7. John Horne and Alan Kramer, *German Atrocities, 1914: A History of Denial* (New Haven, Conn.: Yale University Press, 2006), 8–86.

8. "Manifesto of the Ninety-Three German Intellectuals to the Civilized World," 1914. Reprinted in *The World War I Document Archive*, net.lib.byu.edu/~rdh7/wwi/1914/93intell.html, accessed 21 June 2007.

9. Sven Widmalm, "Science and Neutrality: The Nobel Prizes of 1919 and Scientific Internationalism in Sweden," *Minerva* 33 (1995): 339–360, on 342.

10. Ludwig Fritz Haber, *The Poisonous Cloud: Chemical Warfare in the First World War* (Oxford: Clarendon Press, 1986), 18–19.

11. Ibid., 24–25.

12. Ibid., 25.

13. Charles, *Master Mind*, 155.

14. Haber, *The Poisonous Cloud*, 27.

15. Margit Szöllösi-Janze, *Fritz Haber 1868–1934: Eine Biographie* (Munich: Verlag C.H. Beck, 1998), 270.

16. Haber, *The Poisonous Cloud*, 27.

17. Ibid., 34.

18. Ibid., 25.

19. Mendelssohn, *The World of Walther Nernst*, 90–91.

20. Haber, *The Poisonous Cloud*, 25–26.

21. Ibid., 38.

22. Ibid., 38 n. 69.

23. Ibid., 34 n. 55.

24. Ibid., 89 n. 17.

25. J. E. Coates, quoted in Charles, *Master Mind*, 153.

26. Haber, *The Poisonous Cloud*, 107.

27. Ibid., 128.

28. Ibid., 128–130.

29. Clara Haber to Abegg, 25 Apr. 1909. Quoted in Stoltzenberg, *Fritz Haber*, 174–175.

30. Mrs. Noack, daughter of a friend of Clara Haber. Quoted in Stoltzenberg, *Fritz Haber*, 175.

31. Charles Heller, *Chemical Warfare in World War I: The American Experience, 1917–1918*, Leavenworth Papers No. 10 (Fort Leavenworth, Kans.: Combat Studies Institute, U.S. Army Command and General Staff College, 1984). Available online at www-cgsc.army.mil/carl/resources/csi/Heller/HELLER.asp, accessed 12 Feb. 2008.

32. Stoltzenberg, *Fritz Haber*, 175.

33. Charles, *Master Mind*, 178.

34. Stoltzenberg, *Fritz Haber*, 184.

35. Haber, *The Poisonous Cloud*, 267.

36. Ibid., 57.

37. Ibid., 53.

38. Stoltzenberg, *Fritz Haber*, 143.

39. Wilfred Owen, "The Complete Poems and Fragments," ed. Jon Stallworthy (New York: W.W. Norton, 1986), 117.

40. Fritz Haber, *Fünf Vorträge* (Berlin: Verlag von Julius Springer, 1924), 36. Quoted in Charles, *Master Mind*, 173.

41. Charles, *Master Mind*, 173–174.

42. Haber, *The Poisonous Cloud*, 192.

43. Stoltzenberg, *Fritz Haber*, 147.

44. Ibid., 148–150.

45. Mendelssohn, *The World of Walther Nernst*, 91.

46. Walther Nernst, *The New Heat Theorem* (London: Methuen and Co. Ltd., 1926), v.

47. Mendelssohn, *The World of Walther Nernst*, 90.

48. Ibid.

49. Lloyd Scott, *Naval Consulting Board of the United States* (Washington, D.C.: Government Printing Office, 1920).

50. Albert Rosenfeld, "The Quintessence of Irving Langmuir," in *The Collected Works of Irving Langmuir*, Vol. 12, ed. C. Guy Suits and Harold Way (New York: Pergamon Press, 1962), 103.

51. Rosenfeld, "The Quintessence of Irving Langmuir," 103–104.

52. G.H. Fonda to Willis Whitney 20 Nov. 1918, copied to Langmuir, *ILP*.

53. Heller, *Chemical Warfare in World War I*, chapter 3.

54. Haber, *The Poisonous Cloud*, 107.

55. Joel Hildebrand, "Gilbert Newton Lewis 1865–1946," *Obituary Notices of Fellows of the Royal Society* 5 (1947): 491–506, on 494.

56. Ibid., 494.

57. Arthur Lachman, *Borderland of the Unknown: The Life Story of Gilbert Newton Lewis, One of the World's Great Scientists* (New York: Pageant Press, 1955), 69; Edward Lewis, *A Biography of Distinguished Scientist Gilbert Newton Lewis* (New York: Edwin Mellen Press, 1998), 25.

58. Haber, *The Poisonous Cloud*, 291.

59. Ibid., 278.

60. Nancy Greenspan, *The End of the Uncertain World: The Life and Science of Max Born* (New York: Basic Books, 2005), 93–95.

61. J.L. Heilbron, "The Nobel Science Prizes of World War I," in *Historical Studies in the Nobel Archives*, ed. Elisabeth Crawford (Tokyo: Universal Academy Press, 2002), 19–38.

62. Widmalm, "Science and Neutrality," 343.

63. Elisabeth Crawford, J.L. Heilbron, and Rebecca Ullrich, *The Nobel Population, 1901–1937* (Berkeley, Calif.: Office for History of Science and Technology, 1987).

64. Recommendation of the Nobel Committee for Chemistry, 1916. Quoted in Widmalm, "Science and Neutrality," 349.

65. Widmalm, "Science and Neutrality," 350.

66. Ibid., 354.

67. Ibid., 360.

68. Haber to the Swedish Royal Academy of Sciences, 24 Nov. 1919. Quoted in Widmalm, "Science and Neutrality," 351.

Chapter 5

1. Lewis to Langmuir, 17 Mar. 1917, *CCP*.

2. William Jolly, *From Retorts to Lasers* (Berkeley: William Jolly, distributed by the College of Chemistry, University of California, Berkeley, 1987), 54; Arthur Lachman, *Borderland of the Unknown: The Life Story of Gilbert Newton Lewis, One of the World's Great Scientists* (New York: Pageant Press, 1955), 68.

3. John Gofman, "Human Radiation Studies: Remembering the Early Years. Oral History of Dr. John W. Gofman, M.D., Ph.D., Conducted December 20, 1994," U.S. Department of Energy, www.hss.energy.gov/healthsafety/ohre/roadmap/histories/0457/0457toc.html, accessed 12 Feb. 2008.

4. Joel Hildebrand, oral history interview by Thomas Kuhn and John Heilbron, 6 Aug. 1962, *NBL*.

5. Jolly, *From Retorts to Lasers*, 62.

6. Glenn Seaborg, "The Research Style of Gilbert N. Lewis," *Journal of Chemical Education* 61 (1984): 93–100, on 93.

7. Joel Hildebrand, "Gilbert Newton Lewis 1865–1946," *Obituary Notices of Fellows of the Royal Society* 5 (1947): 491–506, on 493.

8. Albert Rosenfeld, "The Quintessence of Irving Langmuir," in *The Collected Works of Irving Langmuir*, Vol. 12, ed. C. Guy Suits and Harold Way (New York: Pergamon Press, 1962), 5–229, on 99–100.

9. Ibid., 117–123.

10. Ibid., 121–122.

11. Harold Meade Mott-Smith, oral history interview by George Wise, 1 and 2 Mar. 1977, *NBL*.

12. Rosenfeld, "The Quintessence of Irving Langmuir," 93.

13. Mott-Smith, oral history interview.

14. Irving Langmuir, *Phenomena, Atoms, and Molecules* (New York: Philosophical Library, 1950), 144.

15. Percy Bridgman, "Some of the Physical Aspects of the Work of Irving Langmuir," in Suits and Way, *The Collected Works of Irving Langmuir*, Vol. 12, 433–457, on 437–441.

16. Mott-Smith, oral history interview.

17. Humboldt Leverenz, oral history interview by Michael Wolff, 10 July 1979, *NBL*.

18. Benjamin Franklin, *Philosophical Transactions of the Royal Society of London* 64 (1774); Franklin, *The Works of Benjamin Franklin*, ed. John Bigelow (New York: G.P. Putnam and Sons, 1904), 235–250.

19. Agnes Pockels, "Surface Tension," *Nature* 43 (1891): 437–439.

20. Irving Langmuir, "The Constitution and Fundamental Properties of Solids and Liquids. II," *Journal of the American Chemical Society* 39 (1917): 1848–1906.

21. Langmuir, *Phenomena, Atoms, and Molecules*, 43–44; Irving Langmuir, "Surface Chemistry. Nobel Lecture 14 Dec. 1932," in *Nobel Lectures, Chemistry 1922–1941* (Amsterdam: Elsevier, 1966).

22. Irving Langmuir, "The Constitution of Liquids with Especial Reference to Surface Tension Phenomena," *Metallurgical and Chemical Engineering* 15 (1916): 468.

23. Langmuir, "Constitution and Fundamental Properties. II."

24. William Harkins, F.E. Brown, and E.C.H. Davies, "The Structure of the Surfaces of Liquids, and Solubility as Related to the Work Done by the Attractor of Two Liquid Surfaces as They Approach Each Other. [Surface Tension. V.]," *Journal of the American Chemical Society* 39 (1917): 354–361.

25. William Harkins, Earl Davies, and George Clark, "The Orientation of Molecules in the Surfaces of Liquids, the Energy Relations at the Surfaces, Solubility, Adsorption Emulsification, Molecular Association, and the Effects of Acids and Bases on Interfacial Tension," *Journal of the American Chemical Society* 39 (1917): 541–596, footnote on 541; emphasis in original.

26. George Kauffman, "William Draper Harkins (1873–1951)," *Journal of Chemical Education* 62 (1985): 758–761.

27. Frederick M. Fowkes, "William Harkins," in *Dictionary of Scientific Biography*, Vol. 6 (New York: Charles Scribner and Sons, 1981), 117–119.

28. Author's interview with Jacob Bigeleisen, 2006.

29. Langmuir to Harkins, 13 Jan. 1917; Harkins to Langmuir, 15 Jan. 1917; Langmuir to Harkins, 18 Jan. 1917; Langmuir to Harkins, 9 Feb. 1917; Harkins to Langmuir, 25 Feb. 1917; Langmuir to Harkins, 9 Mar. 1917, *ILP*.

30. Harkins to Langmuir, 25 Feb. 1917, *ILP*.

31. Langmuir to Harkins, 9 Mar. 1917, *ILP*.

32. Langmuir, "Constitution and Fundamental Properties," footnote 1, 1848–1850.

33. Gilbert N. Lewis, "The Activity of the Ions and the Degree of Dissociation of Strong Electrolytes," *Journal of the American Chemical Society* 34 (1912): 1631–1644. Here Lewis is trying to determine whether the anomalies are due to increased mobility of the ions or to a change in the dissociating power of the solvent.

34. Gilbert N. Lewis, "The Atom and the Molecule," *Journal of the American Chemical Society* 38 (1916): 762–785.

35. Robert E. Kohler, Jr., "G.N. Lewis's Views on Bond Theory 1900–1916," *British Journal of the History of Science* 8 (1975): 233–239.

36. Gilbert N. Lewis, "Valence and Tautomerism," *Journal of the American Chemical Society* 35 (1913): 1448–1455.

37. Gilbert N. Lewis, *Valence and the Structure of Atoms and Molecules* (New York: Chemical Catalog Company, 1923), 147.

38. See J.L. Heilbron, "J. J. Thomson and the Bohr Atom," *Physics Today* 30 (Apr. 1977): 23–30.

39. Niels Bohr, "On the Constitution of Atoms and Molecules. Part III—Systems Containing Several Nuclei," *Philosophical Magazine* 26 (1913): 857–875, on 862.

40. Robert Millikan, "Atomism in Modern Physics," *Journal of the Chemical Society* 125 (1924): 1405–1417.

41. Langmuir, "Constitution and Fundamental Properties," 1853–1854.

42. Robert E. Kohler, Jr., "Irving Langmuir and the 'Octet' Theory of Valence," *Historical Studies in Physical Sciences* 4 (1972): 39–87, on 40.

43. Langmuir to Lewis, 22 Apr. 1919, *ILP*.

44. Kohler, "'Octet' Theory of Valence," 54.

45. Ibid., 63.

46. Lewis to Langmuir, 9 July 1919, *CCP*. Letter may be found in Kohler, "'Octet' Theory of Valence," 73–74.

47. Lewis to W.A. Noyes, 13 July 1926, *CCP*. Quoted in Kohler, "'Octet' Theory of Valence," 63.

48. Lewis to Lamb, 13 Jan. 1920, *CCP*. Letter may be found in Kohler, "'Octet' Theory of Valence," 74–76.

49. Langmuir to Lewis, 3 Apr. 1920, *ILP*. Letter may be found in Kohler, "'Octet' Theory of Valence," 76–87.

50. Langmuir to Lamb, 4 Apr. 1920, *ILP*.

51. Irving Langmuir, "Theories of Atomic Structure," *Science* 105 (1920): 104–105.

52. Born to Langmuir, 11 May 1920, *ILP*.

53. Sommerfeld to Langmuir, 3 Sept. 1920, *ILP*.

54. Langmuir to Bohr, 25 Oct. 1920; Bohr to Langmuir, 3 Dec. 1920, *ILP*.

55. Max Born, "On Cubic Atomic Models," *Verhandlungen der Deutschen Physikalischen Gesellschaft* 20 (1918): 230–239.

56. Alfred Landé, "Cubed Atoms, Periodic Table and Molecule Formation," *Zeitschrift für Physik* 2 (1920): 380–404.

57. Irving Langmuir, "The Structure of the Helium Atom," *Science* 51 (1920): 605–607.

58. Langmuir to his mother, 9 Mar. 1921, *ILP*. Quoted in Kohler, "'Octet' Theory of Valence," 65.

59. Lewis, *Valence*, 55–56.

60. Langmuir, "Theories of Atomic Structure."

61. Harold Urey, oral history interview by John Heilbron, 24 Mar. 1964, *NBL*.

62. Linus Pauling, *The Nature of the Chemical Bond* (Ithaca, N.Y.: Cornell University Press, 1939).

63. Lewis, *Valence*, 87.

64. Ibid., 101.

65. Robert E Kohler, Jr., "The Lewis-Langmuir Theory of Valence and the Chemical Community, 1920–1928," *Historical Studies in Physical Sciences* 6 (1975): 431–468.

66. Kohler, "'Octet' Theory of Valence."

67. Walther Heitler and Fritz London, *Zeitschrift für Physik* 44 (1927): 455.

68. E.B. Wilson to Lewis, 9 Jan. 1922, *CCP*.

69. Lewis to E.B. Wilson, 27 Jan. 1922, *CCP*.

70. Harkins to Lamb, undated copy sent to Langmuir, 1923 folder, *ILP*.

71. Hildebrand to Lamb, undated copy sent to Langmuir, 1923 folder, *ILP*.

72. Langmuir to Lamb, 8 Sept. 1923, *ILP*.

73. Harkins to Lewis, 31 Mar. 1924, *CCP*.

74. Stieglitz to Lewis, 12 Aug. 1926, *CCP*.

75. Lewis to Stieglitz, 3 Sept. 1926, *CCP*.

Chapter 6

1. Kurt M. Mendelssohn, *The World of Walther Nernst: The Rise and Fall of German Science* (London: Macmillan, 1973), 146–147. Mendelssohn has Nernst

arriving at Haber's laboratory looking for a job, which is unlikely given Nernst's pride, age, and previous relations with Haber. The explanation that he wanted to confer with Haber over how best to help their Jewish colleagues seems more likely.

2. Alan D. Beyerchen, *Scientists under Hitler: Politics and the Physics Community in the Third Reich* (New Haven, Conn.: Yale University Press, 1977), 9.

3. Jean Medawar and David Pyke, *Hitler's Gift: The True Story of the Scientists Expelled by the Nazi Regime* (New York: Arcade Publishing, 2001), 12.

4. Wilhelm Jost, "The First Forty-Five Years of Physical Chemistry in Germany," *Annual Reviews of Physical Chemistry* 17 (1966): 1–14.

5. James Franck, oral history interview by Thomas Kuhn and Maria Goeppert Mayer, 9–14 July 1962, *NBL*.

6. Ibid.

7. Daniel Charles, *Master Mind: The Rise and Fall of Fritz Haber, the Nobel Laureate Who Launched the Age of Chemical Warfare* (New York: Ecco, 2005), 137.

8. Max Planck, *Vorlesungen über Thermodynamik*, trans. Alexander Ogg, 7th German, 3rd English ed. (Leipzig: Walter de Gruyter and Co, 1922), xi–xii.

9. Mendelssohn, *World of Walther Nernst*, 137.

10. Max Planck, "Recent Thermodynamic Theories (Nernst Heat-Theorem and Energy Unit Hypothesis)," *Berichte der Deutschen Chemischen Gesellschaft* 45 (1912): 5–23.

11. Gilbert N. Lewis and George Gibson, "The Entropy of the Elements and the Third Law of Thermodynamics," *Journal of the American Chemical Society* 39 (1917): 2554–2581; Gilbert N. Lewis and George Gibson, "The Third Law of Thermodynamics and the Entropy of Solutions and of Liquids," *Journal of the American Chemical Society* 42 (1920): 1529–1533.

12. Walther Nernst, *The New Heat Theorem* (London: Methuen, 1926), 227.

13. Letter from Richards to Arrhenius, 23 Mar. 1916, *TRP*. Cited in Sheldon Jerome Kopperl, "The Scientific Work of Theodore William Richards" (Ph.D. thesis, University of Wisconsin, 1970), 229.

14. Theodore Richards, "Die Bedeutung der Änderung des Atomvolums III," *Zeitschrift für Physikalische Chemie* 42 (1902): 129–154.

15. Theodore Richards, 1912 autobiographical statement, *TRP*. Quoted in Kopperl, "Theodore William Richards," 229.

16. Letter from Richards to Arrhenius, 14 Dec. 1915, *TRP*. Cited in Kopperl, "Theodore William Richards," 228.

17. Theodore Richards, "Note concerning the Third Law of Thermodynamics," *TRP*. Quoted in Kopperl, "Theodore William Richards," 230.

18. Theodore Richards, "The Present Aspect of the Hypothesis of Compressible Atoms," *Journal of the American Chemical Society* 36 (1914), 2417–2439, on 2433.

19. Nernst, *The New Heat Theorem*, 227–231. Cited in Kopperl, "Theodore William Richards," 230–231.

20. Diana Kormos Barkan, *Walther Nernst and the Transition to Modern Physical Science* (Cambridge: Cambridge University Press, 1999), 217.

21. Letter from Arrhenius to Tamman, 22 Dec. 1910. Quoted in Barkan, *Walther Nernst*, 227.

22. Robert Marc Friedman, *The Politics of Excellence: Behind the Nobel Prizes in Science* (New York: Times Books, Henry Holt, 2001), 91–92.

23. Barkan, *Walther Nernst*, 232.

24. Ibid., 234–236.

25. Letter from Arrhenius to Theodore Richards, 21 May 1921, *TRP*. Quoted in Kopperl, "Theodore William Richards," 229.

26. Letter from Arrhenius to Richards, 29 Oct. 1921, *TRP*. Quoted in Kopperl, "Theodore William Richards," 229.

27. Barkan, *Walther Nernst*, 208–240.

28. Mendelssohn, *World of Walther Nernst*, 139.

29. Walther Nernst memorial Web site, www.nernst.de/, accessed 12 June 2007.

30. Mendelssohn, *World of Walther Nernst*, 113.

31. Franck, oral history interview.

32. Author's interview with Jacob Bigeleisen, 2004.

33. Author's interview with Samuel Weissman, 2004.

34. Mendelssohn, *World of Walther Nernst*, 138.

35. Ibid.

36. Dietrich Stoltzenberg, *Fritz Haber: Chemist, Nobel Laureate, German, Jew*, trans. Charles Passage (Philadelphia: Chemical Heritage Press, 2004), 159–170.

37. Ibid., 241–249.

38. Charles, *Master Mind*, 206–207.

39. Haber to Richard Wilstätter, 24 Feb. 1933, quoted by Charles, *Master Mind*, 216.

40. Ute Deichmann, "The Expulsion of German-Jewish Chemists and Biochemists and Their Correspondence with Colleagues in Germany after 1945: The Impossibility of Normalization?" in *Science in the Third Reich*, ed. Margit Szöllösi-Janze (New York: Oxford University Press, 2000), 243–280, on 245.

41. Ruth Lewin Sime, *Lise Meitner, a Life in Physics* (Berkeley: University of California Press, 1996), 141.

42. Primo Levi, *The Drowned and the Saved*, trans. Raymond Rosenthal (New York: Summit Books, 1988), 128.

43. Medawar and Pyke, *Hitler's Gift*, 37.

44. Beyerchen, *Scientists under Hitler*, 4.

45. Mendelssohn, *World of Walther Nernst*, 149.

46. Franck, oral history interview.

47. Medawar and Pyke, *Hitler's Gift*, 35–36.

48. Dieter Hoffmann,"Lenard Fülöp—Philipp Lenard: Von Pressburg nach Heidelberg" in Marc Schalenberg and Peter Walther, *"Immer im Forschen Bleiben". Rüdiger vom Bruch zum 60. Geburtstag* (Stuttgart: Franz Steiner Verlag 2004), 337–350,

49. Joseph Needham, *The Nazi Attack on International Science* (London: Watts, 1941), 21.

50. Mendelssohn, *World of Walther Nernst*, 103.

51. Franck, oral history interview. Quoted in Medawar and Pyke, *Hitler's Gift*, 36.

52. Johannes Stark, "International Status and Obligations of Science," *Nature* 133 (1934): 290. This was in reply to A.V. Hill, "International Status and Obligations of Science," *Nature* 132 (1933): 952–954. See also Hill, "International Status and Obligations of Science," *Nature* 133 (1934): 290; and Johannes Stark, "The Attitude of the German Government towards Science," *Nature* 133 (1934): 614.

53. Needham, *Nazi Attack on International Science*, 12.

54. Ibid., 18.

55. Mendelssohn, *World of Walther Nernst*, 146.

56. Ibid., 154.

57. Ibid., 106.

58. John L Heilbron, *The Dilemmas of an Upright Man: Max Planck as Spokesman for German Science* (Berkeley: University of California Press, 1986), 150–151.

59. Planck to Paul Langevin, 24 Sept. 1924. Quoted in Heilbron, *Dilemmas*, 152.

60. *New York Times*, 12 Jan. 1936. See Planck to R.B. Goldschmidt, 12 July 1936. Quoted in Heilbron, *Dilemmas*, 164.

61. Heisenberg to Born, 2 June 1933. Quoted in Heilbron, *Dilemmas*, 154.

62. Deichmann, "Expulsion," 26–29.

63. Medawar and Pyke, *Hitler's Gift*, 29.

64. Charles, *Master Mind*, 222–224.

65. Mendelssohn, *World of Walther Nernst*, 146.

66. Stoltzenberg, *Fritz Haber*, 282–286.

67. Einstein to Haber, 19 May 1933. Quoted in Charles, *Master Mind*, 225–226.

68. Einstein to Haber, 9 Aug. 1933. Quoted in Charles, *Master Mind*, 229–230.

69. Haber to Sir William Pope, 4 Aug. 1933. Quoted in Charles, *Master Mind*, 228.

70. Charles, *Master Mind*, 241.

71. Ibid., 243–245.

72. Heilbron, *Dilemmas*, 162.

73. Einstein to Hermann and Magda Haber, undated, presumably 1934. Quoted in Charles, *Master Mind*, 239–241.

74. Mendelssohn, *World of Walther Nernst*, 157.

75. Ibid., 152, 8.

76. Ibid., 160–162.

77. Barkan, *Walther Nernst*, 246.

78. Mendelssohn, *World of Walther Nernst*, 160–162.

79. *New York Times*, 20 Nov. 1941.

80. Franck, oral history interview.

81. Barkan, *Walther Nernst*, 185.

82. Robert Millikan, "Walther Nernst, a Great Physicist, Passes," *Scientific Monthly* 51 (1942): 84–86; italics original.

83. Author's interview with Bigeleisen, 2004.

84. Albert Einstein, "Walther Nernst," *Scientific Monthly* 54 (1942): 195–196.

85. Medawar and Pyke, *Hitler's Gift*, 29; Deichmann, "Expulsion," 245.

86. Levi, *The Drowned and the Saved*, 37.

87. Deichmann, "Expulsion," 253.

88. Levi, *The Drowned and the Saved*, 73.

89. Charles, *Master Mind*, 234–235.

90. Sime, *Lise Meitner*, 144.

91. Freudenberg to Barger, 17 July 1933. Quoted in Deichmann, "Expulsion," 240.

92. Franz (Sir Francis) Simon to Karl Bonhöffer, 22 Mar. 1951. Quoted in Deichmann, "Expulsion," 271.

Chapter 7

1. Einstein to Lewis, 26 Aug. 1926, *CCP*. This letter has been translated into English and published by Roger Stuewer, "G. N. Lewis on Detailed Balancing, the Symmetry of Time, and the Nature of Light," *Historical Studies in Physical Sciences* 6 (1975): 469–511, on 494–496.

2. Author's interview with Michael Kasha, 2004.

3. William Jolly, *From Retorts to Lasers* (Berkeley: William Jolly, distributed by the College of Chemistry, University of California, Berkeley, 1987), 99.

4. Gilbert N. Lewis and Merle Randall, *Thermodynamics and the Free Energy of Chemical Substances* (New York: McGraw-Hill, 1923), ix.

5. Gilbert N. Lewis and Merle Randall, *Thermodynamics*, 2nd ed., rev. Kenneth S. Pitzer and Leo Brewer (New York: McGraw-Hill, 1961); Kenneth Pitzer, *Thermodynamics*, 3rd ed. (New York: McGraw-Hill, 1995).

6. Gilbert N. Lewis and Merle Randall, "The Activity Coefficient of Strong Electrolytes," *Journal of the American Chemical Society* 43 (1921): 1112–1154.

7. Peter Debye and Erich Hückel, "The Theory of Electrolytes. I. Lowering of Freezing Point and Related Phenomena," *Physikalische Zeitschrift* 24 (1923): 185–206. Debye and Hückel made no reference to Lewis's ionic strength in their original paper, but Debye did note that Lewis's work was in agreement with their theory in a later paper: Debye, "Osmotic Equation of State and the Activity of Strong Electrolytes in Dilute Solutions," *Physikalische Zeitschrift* 25 (1924): 97. For a more complete explanation of the relation of the Debye-Hückel theory to Lewis's work, see Kenneth Pitzer, "Gilbert Lewis and the Thermodynamics of Strong Electrolytes," *Journal of Chemical Education* 61 (1984): 104–107.

8. Lewis and Randall, *Thermodynamics and the Free Energy of Chemical Substances*, 448; emphasis original.

9. Gilbert N. Lewis, *Valence and the Structure of Atoms and Molecules* (New York: Chemical Catalog Company, 1923), 51. The ellipsis in the quote removes a discussion of whether it is possible that the wire would have exhibited zero resistance, which Lewis rejected.

10. I thank Cathryn Carson and John Heilbron for their comments on this point.

11. Lewis, *Valence*, preface.

12. Linus Pauling, "The Nature of the Chemical Bond. Application of Results Obtained from the Quantum Mechanics and from a Theory of Paramagnetic

Susceptibility to the Structure of Molecules," *Journal of the American Chemical Society* 53 (1931): 1367–1400, from the abstract on 1367.

13. Hendrik Lorentz, "The Radiation of Light," *Nature* 113 (1924): 608–611, on 611. As quoted in Stuewer, "G. N. Lewis," 469–511, on 470.

14. Stuewer, "G. N. Lewis," 471.

15. Gilbert N. Lewis, "A New Principle of Equilibrium," *Proceedings of the National Academy of Sciences* 11 (1925): 179–183, on 181.

16. Stuewer, "G. N. Lewis," 472.

17. Gilbert N. Lewis, "The Distribution of Energy in Thermal Radiation and the Law of Entire Equilibrium," *Proceedings of the National Academy of Sciences* 12 (1925): 422–428.

18. R.H. Fowler to Lewis, 6 June 1925, 30 Aug. 1925, *CCP*. Quoted in Stuewer, "G. N. Lewis," 472.

19. Stuewer, "G. N. Lewis," 503.

20. Gilbert N. Lewis, "The Nature of Light," *Proceedings of the National Academy of Sciences* 12 (1926): 22–29, on 25.

21. Gilbert N. Lewis, *The Anatomy of Science* (New Haven, Conn.: Yale University Press, 1926), 85.

22. Gilbert N. Lewis, "A Revision of the Fundamental Laws of Matter and Energy," *Technology Quarterly* 21 (1908): 212–225; Lewis and Richard Tolman, "The Principles of Relativity, and Non-Newtonian Mechanics," *Proceedings of the American Academy of Arts and Sciences* 44 (1909): 711–728; Lewis, "The Fundamental Laws of Matter and Energy," *Science* 30 (1909): 84–86.

23. Lewis to Einstein, 23 Feb. 1921, *CCP*. The meeting was probably in 1911 or earlier, because the letter mentions that Georg Bredig was also present, and Bredig left Zurich to take a position in Karlsruhe that year.

24. Gilbert N. Lewis, "Light Waves and Light Corpuscles," *Nature* 117 (1926): 236–238, on 237.

25. Lewis, *The Anatomy of Science*, 113.

26. Arthur Lachman, *Borderland of the Unknown: The Life Story of Gilbert Newton Lewis, One of the World's Great Scientists* (New York: Pageant Press, 1955), 117.

27. Stuewer, "G. N. Lewis," 510.

28. Langmuir to Lewis, 2 July 1930, *CCP*; Lewis to Langmuir, 5 Aug. 1930, *ILP*.

29. Lewis, *The Anatomy of Science*.

30. Ibid., 1–7.

31. Ibid., 180–189.

32. Ibid., 158, 191–217

33. Ibid., 181–182.

34. I thank Scott Denmark for pointing this out.

35. See, e.g., Yukio Saito and Hiroyuki Hyuga, "Complete Homochirality Induced by Nonlinear Autocatalysis and Recycling," *Journal of the Physical Society of Japan* 73 (2004): 33–35.

36. Gilbert N. Lewis, "A Plan for Stabilizing Prices," *Economic Journal* 35 (1925): 40–46; Lewis, "Europas Skulder och Myntfoten," *Finsk Tidskrfit* Dec. (1924): 373–390.

37. See, e.g., Gilbert N. Lewis, "Ultimate Rational Units and Dimensional Theory," *Philosophical Magazine* 49 (1925): 739–750.

38. Jolly, *From Retorts to Lasers*, 125.

39. Albert Rosenfeld, "The Quintessence of Irving Langmuir," in *The Collected Works of Irving Langmuir*, Vol. 12, ed. C. Guy Suits and Harold Way (New York: Pergamon Press, 1962), 5–229, on 6.

40. *A General Electric Scrapbook History* (Schenectady, N.Y.: General Electric Company, 1953).

41. Rosenfeld, "The Quintessence of Irving Langmuir," 153.

42. Ibid., 126.

43. Max Born, "Erinnerungen an Fritz Haber," written for Johannes Jänicke's uncompleted biography of Haber. Quoted in Nancy Greenspan, *The End of the Uncertain World: The Life and Science of Max Born* (New York: Basic Books, 2005), 93.

44. Rosenfeld, "The Quintessence of Irving Langmuir," 127–128.

45. See, for example, ibid., 126–131.

46. Ibid., 125–126.

47. Langmuir to his mother, 23 Oct. 1927, *ILP*.

48. Brillouin to Langmuir, 6 Sept. 1927, *ILP*.

49. Guido Bacciagaluppi and Antony Valentini, *Quantum Theory at the Crossroads: Reconsidering the 1927 Solvay Conference* (Cambridge: Cambridge University Press, forthcoming).

50. Langmuir to Whitney, 6 Nov. 1927, *ILP*.

51. Available at www.maxborn.net/index.php?page=filmnews, accessed 12 Feb. 2008. This link has been provided by Nancy Greenspan, the author of *The End of the Certain World*; Roger Summerhayes, Langmuir's grandson, provided the footage for Greenspan. Additional films made by Langmuir may be found in a documentary DVD by Roger Summerhayes, *Langmuir's World* (CustomFlix, 1996).

52. Rosenfeld, "The Quintessence of Irving Langmuir," 129–131.

53. Ibid., 146–150.

54. Langmuir to his mother, 2 Nov. 1930, *ILP*.

55. Rice to Langmuir, 31 Dec. 1928, *ILP*.

56. Herman Skolnik and Kenneth Reese, *A Century of Chemistry: The Role of Chemists and the American Chemical Society* (Washington, D.C.: American Chemical Society, 1976), 456.

57. Rosenfeld, "The Quintessence of Irving Langmuir," 154.

58. Svante Arrhenius, "1924 Special Report on Gilbert Lewis to the Nobel Chemistry Committee (Report 1)," *KVA*.

59. Ibid.

60. Ibid.

61. Robert Marc Friedman, *The Politics of Excellence: Behind the Nobel Prizes in Science* (New York: Times Books, Henry Holt, 2001), 189, 334.

62. Gilbert N. Lewis, "Outlines of a New System of Thermodynamic Chemistry," *Proceedings of the American Academy of Arts and Sciences* 43 (1907): 259–293, on 259.

63. For example, "In his papers on this subject, brief as they have been, Arrhenius has with great fairness and extraordinary acumen states, as far as our present knowledge permits, the truth, the whole truth, and nothing but the truth about ionic dissociation" and "I venture to predict that the later and better theories will not be substitutes for, but rather development of the simple hypothesis of Arrhenius." Gilbert N. Lewis, "The Use and Abuse of the Ionic Theory," *Science* 30 (1909): 1–6. Also, "The dissociation of acetic acid is of a type first explained by Arrhenius's brilliant theory of electrolytic dissociation, which, in the half century that has elapsed since its inception, has been the subject of bitter contention, but is now universally accepted." Lewis and Randall, *Thermodynamics and the Free Energy of Chemical Substances*, 308.

64. Theodor Svedberg, "1926 Special Report on Gilbert Lewis to the Nobel Chemistry Committee (Report 8)," *KVA*.

65. Ibid.

66. Nobel Chemistry Committee, "1926 Report of the Nobel Chemistry Committee to the Royal Swedish Academy of Sciences," *KVA*.

67. Patrick Coffey, "Chemical Free Energies and the Third Law of Thermodynamics," *Historical Studies in Physical and Biological Sciences* 36 (2006): 365–396.

68. Lewis and Randall, *Thermodynamics and the Free Energy of Chemical Substances*, 5, 454. A full list of Lewis's attacking references to Nernst within *Thermodynamics* can be found in Coffey, "Chemical Free Energies," 365–396, on 394.

69. Friedman, *The Politics of Excellence*, 193.

70. Elisabeth Crawford, J.L. Heilbron, and Rebecca Ullrich, *The Nobel Population, 1901–1937* (Berkeley, Calif.: Office for History of Science and Technology, 1987).

71. Diana Kormos Barkan, *Walther Nernst and the Transition to Modern Physical Science* (Cambridge: Cambridge University Press, 1999), 215, 216, 232, 234.

72. Walther Nernst, "Studies in Chemical Thermodynamics," in *Nobel Lectures* (*KVA*, 1921).

73. Nernst to Palmaer, 2 Sept. 1939, Palmaer papers, Royal Swedish Academy of Sciences. Cited in Barkan, *Walther Nernst*, 245.

74. Crawford, Heilbron, and Ullrich, *The Nobel Population, 1901–1937*, 260.

75. Karl Grandin, Royal Swedish Academy of Sciences, Stockholm, private communication, 2004.

76. Wilhelm Palmaer nominating letter to the Nobel committee, 30 Jan. 1932. Nobel Archive of the Royal Swedish Academy of Sciences, Stockholm.

77. Wilhelm Palmaer, "1932 Special Report on Gilbert Lewis to the Nobel Chemistry Committee (Report 12)," *KVA*.

78. Ibid.

79. Ibid.

80. Ibid.

81. Ibid.

82. Ibid.

83. Theodor Svedberg, "1932 Special Report on Gilbert Lewis to the Nobel Chemistry Committee (Report 11)," *KVA*.

84. Ibid.

85. Ibid.

86. Svedberg to Arne Westgren, 10 May 1932, *KVA*.

87. Westgren to Svedberg, 14 May 1932, *KVA*.

88. Nobel Chemistry Committee, "1932 Report of the Nobel Chemistry Committee to the Royal Swedish Academy of Sciences," *KVA*.

89. Ibid.

90. Rosenfeld, "The Quintessence of Irving Langmuir," 154.

91. Langmuir to his mother, 18 Dec. 1932, *ILP*.

92. Nobel Chemistry Committee, "1933 Report of the Nobel Chemistry Committee to the Royal Swedish Academy of Sciences," *KVA*.

93. Karl Grandin, Royal Swedish Academy of Sciences, Stockholm, private communication, 2005.

94. Nobel Chemistry Committee, "1933 Report."

95. Wilhelm Palmaer, "1934 Supplementary Report on Gilbert Lewis to the Nobel Chemistry Committee (Report 6)," *KVA*.

96. Ibid.

97. Ibid.

98. Ibid.

99. Nobel Chemistry Committee, "1934 Report of the Nobel Chemistry Committee to the Royal Swedish Academy of Sciences," *KVA*.

100. I thank John Prausnitz for this analogy.

101. Friedman, *The Politics of Excellence*, 209.

102. Ibid., 208.

Chapter 8

1. The Oakland *Post-Intelligencer* has long been out of business, but archival copies may be found on microfilm at the Oakland, California, public library. The microfilm image did not copy well, and this is a reproduction of the article.

2. Edward Lewis, *A Biography of Distinguished Scientist Gilbert Newton Lewis* (New York: Edwin Mellen Press, 1998), 13.

3. Author's interview with Jacob Bigeleisen, 2006.

4. Ibid.

5. Arthur Lachman, *Borderland of the Unknown: The Life Story of Gilbert Newton Lewis, One of the World's Great Scientists* (New York: Pageant Press, 1955), 142.

6. Author's interview with Michael Kasha, 2004. Kasha was told this story by Simon Freed.

7. Ferdinand Brickwedde, "Harold Urey and the Discovery of Deuterium," *Physics Today* 35, Sept. (1982): 34–39, on 35.

8. Harold Urey, "Heat Capacities and Entropies of Diatomic and Polyatomic Gases," *Journal of the American Chemical Society* 45 (1923): 1445–1455.

9. Urey to Lewis, 9 Sept. 1923, 1 Jan. 1924, *CCP*.

10. Author's interview with Bigeleisen, 2004.

11. Lewis to Urey, 2 Feb. 1924; Urey to Lewis, 8 Apr. 1924, *CCP*.

12. Raymond Birge and D.H. Menzel, "The Relative Abundance of the Oxygen Isotopes, and the Basis of the Atomic Weight System," *Physical Review* 37 (1931): 1669–1671.

13. Brickwedde, "Harold Urey," 35.

14. K.P. Cohen et al., "Harold Clayton Urey, 29 April 1893–5 January 1981," *Biographical Memoirs of Fellows of the Royal Society* 29 (1983): 622–659, on 629.

15. Harold Urey, Ferdinand Brickwedde, and George Murphy, "A Hydrogen Isotope of Mass 2," *Physical Review* 39 (1932): 164.

16. Harold Urey, "Some Thermodynamic Properties of Hydrogen and Deuterium, 14 February 1935," in *Nobel Lectures, Chemistry 1922–1941* (Amsterdam: Elsevier, 1966).

17. Gilbert N. Lewis, *Valence and the Structure of Atoms and Molecules* (New York: Chemical Catalog Company, 1923), 17.

18. Gilbert N. Lewis and Ronald T. Macdonald, "Concentration of H_2 Isotope," *Journal of Chemical Physics* 1 (1933): 341–344, on 342.

19. Author's interview with Bigeleisen, 2004.

20. Edward Washburn and Harold Urey, "Concentration of the H-2 Isotope of Hydrogen by the Fractional Electrolysis of Water," *Proceedings of the National Academy of Sciences* 18 (1932): 496–498.

21. Cohen et al., "Harold Clayton Urey," 630.

22. Gilbert N. Lewis and Robert Cornish, "Separation of the Isotopic Forms of Water by Fractional Distillation," *Journal of the American Chemical Society* 55 (1933): 2616–2617.

23. Edward Washburn, Edgar Smith, and Mikkel Frandsen, "The Isotopic Fractionation of Water," *Journal of Chemical Physics* 1 (1933): 288.

24. Merle Randall and Wells Allen Webb, "Separation of Isotopes by Fractional Distillation of Water," *Journal of Industrial and Engineering Chemistry* 31 (1939): 227–230.

25. Gilbert N. Lewis, "Different Kinds of Water" (paper presented at the IX Congreso Internacional de Quimica Pura y Aplicada, Madrid, 1934), on 14.

26. For a detailed review of isotope research by Lewis, see Jacob Bigeleisen, "Gilbert N. Lewis and the Beginnings of Isotope Chemistry," *Journal of Chemical Education* 61 (1984): 108–116.

27. Gilbert N. Lewis, "The Biology of Heavy Water," *Science* 79 (1934): 151–153, on 152–153.

28. Nuel Pharr Davis, *Lawrence and Oppenheimer* (New York: Simon and Schuster, 1968), 29–30.

29. Davis, *Lawrence and Oppenheimer*, 53–56.

30. Lewis to Langmuir, 9 July 1919, *CCP*. Quoted in Robert E. Kohler, Jr., "Irving Langmuir and the 'Octet' Theory of Valence," *Historical Studies in Physical Sciences* 4 (1972): 73–74.

31. Lewis to Urey, 13 May 1933; Urey to Lewis, 16 May 1933, 18 May 1933, 29 May 1933, *CCP*.

32. Urey to Lewis, 1 June 1933, *CCP*.

33. Lord Rutherford, "Heavy Hydrogen," *Nature* 132 (1933): 955–956.

34. Rutherford to Urey, 6 Jan. 1934, *CCP*; undated copy to Lewis of communication by Urey to *Nature*, *CCP*.

35. "Deuterium vs. Diplogen," *Time*, 19 Feb. 1934. For a more extensive account of the naming dispute, see Roger Stuewer, "The Naming of the Deuteron," *American Journal of Physics* 54 (1986): 206–218.

36. Nobel Chemistry Committee, "1934 Report of the Nobel Chemistry Committee to the Royal Swedish Academy of Sciences," *KVA*.

37. Theodor Svedberg, "1934 Special Report on Gilbert Lewis to the Nobel Chemistry Committee (Report 10)," *KVA*.

38. Svedberg to Westgren, 18 May 1934, *KVA*.

39. Theodor Svedberg, "1934 Supplementary Report on Gilbert Lewis to the Nobel Chemistry Committee (Report 11)," *KVA*.

40. Brickwedde, "Harold Urey," 34.

41. James Arnold, Jacob Bigeleisen, and Clyde Hutchinson, "Harold Clayton Urey: April 24, 1893–January 5, 1981," in *National Academy of Sciences Biographical Memoirs* (Washington, D.C.: National Academy of Sciences, 1995), 363–412, on 375.

42. Mildred Cohn, "Harold Urey: A Personal Remembrance, Part I," *Chemical Heritage* 23, no. 4 (2005): 8–11, 48, on 48.

43. Mildred Cohn, "Harold Urey: A Personal Remembrance. Part II," *Chemical Heritage* 23, no. 5 (2005): 9–13, on 11.

44. Author's interview with David Altman, 2006.

45. Jacob Bigeleisen, private communication, 2005.

46. Bigeleisen, "Gilbert N. Lewis," 108–116.

47. John W. Servos, *Physical Chemistry from Ostwald to Pauling: The Making of a Science in America* (Princeton, N.J.: Princeton University Press, 1990), 316–320.

48. Gilbert N. Lewis, "The Chemical Bond," *Journal of Chemical Physics* 1 (1933): 17–28.

49. Lewis and Macdonald, "Concentration of H_2 Isotope."

50. Gilbert N. Lewis, Melvin Calvin, and Michael Kasha, "Photomagnetism. Determination of the Paramagnetic Susceptibility of a Dye in Its Phosphorescent State," *Journal of Chemical Physics* 17 (1947): 804–812.

51. Urey to Lewis, 30 Oct. 1935, 3 Dec. 1935, 14 Feb. 1936, *CCP*.

52. Glenn Seaborg, *Journals of Glenn Seaborg, 1946–1958*, Vol. 1 (Berkeley: Lawrence Berkeley Labs, University of California, 1990), 72.

53. Lewis to Home Secretary, National Academy of Sciences, 29 Dec. 1934, *CCP*.

54. Wilson to Lewis, 18 Feb. 1935, *CCP*.

55. Edward Lewis, *A Biography*, 64.

56. Gilbert N. Lewis, "A Theory of Orbital Neutrons," *Physical Review* 50 (1936): 857–860.

57. Gilbert N. Lewis and P. Schutz, "Refraction of Neutrons," *Physical Review* 50 (1937): 369.

58. Bethe to R. Buchta, 6 May 1936. Quoted in J.L. Heilbron and Robert W. Seidel, *Lawrence and His Laboratory , Volume I* (Berkeley: University of California Press, 1989), 182.

59. Maurice Nahamias to Frederic Joliot, 28 Apr. 1937. Quoted in Heilbron and Seidel, *Lawrence and His Laboratory*, 182.

60. Glenn Seaborg, *Journal of Glenn Seaborg, August 11, 1934–June 30, 1939*, entry for 20 Apr. 1937, (Berkeley: Lawrence Berkeley Labs, University of California, 1982), 228.

61. Glenn Seaborg, *Journal of Glenn Seaborg, August 11, 1934–June 30, 1939*, entry for 24 May 1937, 242.

62. Glenn Seaborg, "The Research Style of Gilbert N. Lewis," *Journal of Chemical Education* 61 (1984): 93–100, on 93–94.

63. "Meet Glenn Seaborg," isswprod.lbl.gov/Seaborg/bio.htm, accessed 12 Feb. 2008.

64. Lewis, *Valence*, 142; italics in original.

65. Gilbert N. Lewis, "Acids and Bases," *Journal of the Franklin Institute* 226 (1938): 293–313.

66. W.F. Luder and Saverio Zuffanti, *The Electronic Structure of Acids and Bases* (New York: John Wiley and Sons, 1947).

67. Ibid., vii.

68. Seaborg, "Research Style of Gilbert N. Lewis," 94.

69. Author's interview with Bigeleisen, 2004.

70. Hisashi Yamamoto, *Lewis Acids in Organic Synthesis* (New York: Wiley VCH, 2000).

71. Heilbron and Seidel, *Lawrence and His Laboratory*, 461.

72. Glenn Seaborg, Ronald Kathren, and Jerry Gough, *The Plutonium Story: The Journals of Professor Glenn T. Seaborg 1939–1946* (Columbus, Ohio: Battelle Press, 1994), 20.

73. Seaborg to McMillan, 8 Mar. 1941. Quoted in Heilbron and Seidel, *Lawrence and His Laboratory*, 463. The parenthetical exclamation marks are theirs.

74. Seaborg, Kathren, and Gough, *The Plutonium Story*, 45.

75. Ibid., 30.

76. Ibid., 41.

77. Ibid., 45.

78. Ibid.

79. See Richard G. Hewlett and Oscar E. Anderson, Jr., *The New World: A History of the United States Atomic Energy Commission*: Vol. 1, *1939–1946*. (Philadelphia: University of Pennsylvania Press, 1962), 33–34; Heilbron and Seidel, *Lawrence and His Laboratory*, 456–464.

80. Seaborg, Kathren, and Gough, *The Plutonium Story*, 109–110.

81. Glenn Seaborg, *A Chemist in the White House* (Washington, D.C.: American Chemical Society, 1998), 7–8.

82. Author's interview with Kasha, 2004.

83. Seaborg, Kathren, and Gough, *The Plutonium Story*, 141.

84. Ibid., 246.

85. Seaborg, *A Chemist in the White House*, 10.

86. Washington State Department of Health, Hanford Health Information Network, "An Overview of Hanford and Health Radiation Effects" (2004), www.doh.wa.gov/Hanford/publications/overview/overview.html, accessed 12 June 2007.

87. Henry DeWolf Smyth, *Atomic Energy for Military Purposes* (Princeton, N.J.: Princeton University Press, 1945), 121.

88. Simone Turchetti, "For Slow Neutrons, Slow Pay," *Isis* 97 (2006): 1–27.

89. Seaborg, Kathren, and Gough, *The Plutonium Story*, 698–699.

90. "Chemists Reminisce on 50th Anniversary of the Atomic Bomb," *Chemical and Engineering News*, 17 July 1995, 62.

91. Seaborg, Kathren, and Gough, *The Plutonium Story*, 725–728.

92. Richard B. Frank, *Downfall: The End of the Imperial Japanese Empire* (New York: Random House, 1999).

93. Seaborg, Kathren, and Gough, *The Plutonium Story*, 745.

94. Arnold, Bigeleisen, and Hutchinson, "Harold Clayton Urey," 382.

95. Harold Urey, oral history interview by John Heilbron, 24 Mar. 1964, *NBL*.

96. Arnold, Bigeleisen, and Hutchinson, "Harold Clayton Urey," 390.

97. Cohn, "Harold Urey, Part II," 12.

98. Jessica Wang, "Science, Security, and the Cold War: The Case of E. U. Condon," *Isis* 83 (1992): 238–269.

99. Arnold, Bigeleisen, and Hutchinson, "Harold Clayton Urey," 390.

Chapter 9

1. Linus Pauling, "Notes on talks between Dr. D. M. Wrinch and Linus Pauling, January 26 and 27, 1938, Cornell," quoted in Thomas Hager, *Force of Nature: The Life of Linus Pauling* (New York: Simon and Schuster, 1995), 227.

2. Linus Pauling and Carl Niemann, "The Structure of Proteins," *Journal of the American Chemical Society* 61 (1939): 1860–1867.

3. Alex Todd to Pauling, 28 July 1939, quoted in Hager, *Force of Nature*, 230.

4. Hager, *Force of Nature*, 56.

5. William Jolly, *From Retorts to Lasers* (Berkeley: William Jolly, distributed by the College of Chemistry, University of California, Berkeley, 1987), 60.

6. Related by Pauling in a 1964 oral history interview with Harriet Zuckerman. Quoted in Hager, *Force of Nature*, 89.

7. Linus Pauling and Richard Tolman, "The Entropy of Supercooled Liquids at the Absolute Zero," *Journal of the American Chemical Society* 47 (1925): 2148–2156.

8. Hager, *Force of Nature*, 200–211.

9. Ibid., 100–101.

10. Linus Pauling, "Arthur Amos Noyes," in *Proceedings of the Robert A. Welch Foundation Conferences on Chemical Research*: Vol. 20, *American Chemistry—Bicentennial*, 8–10 Nov. 1976, Houston, Tex., on 100–101. Quoted with permission of Oregon State University.

11. Linus Pauling, "The Nature of the Chemical Bond. Application of Results Obtained from the Quantum Mechanics and from a Theory of Paramagnetic Susceptibility to the Structure of Molecules," *Journal of the American Chemical Society* 53 (1931): 1367–1400.

12. Wendell Latimer and Worth Rodebush, "Polarity and Ionization from the Standpoint of the Lewis Theory of Valence," *Journal of the American Chemical Society* 52 (1920): 1419–1423.

13. W.A. Noyes to Lewis, 30 Apr. 1932, *CCP*.

14. Lewis to W.A. Noyes, 20 May 1932, *CCP*.

15. Lewis to Pauling, 20 May 1932, *CCP*.

16. Copy of Lewis's letter to Pauling, 20 May 1932, *CCP*.

17. Gilbert N. Lewis, "The Chemical Bond," *Journal of Chemical Physics* 1 (1933), 17–28.

18. Pierre Laszlo, *A History of Biochemistry: Molecular Correlates of Biological Concepts*, Vol. 34A in *Comprehensive Biochemistry*, ed. A. Neuberger and L.L.M. Van Deemen (Amsterdam: Elsevier, 1986), 263; emphasis in original.

19. Hager, *Force of Nature*, 192.

20. A.E. Mirsky and Linus Pauling, "On the Structure of Native, Denatured, and Coagulated Proteins," *Proceedings of the National Academy of Sciences* 22 (1936): 439–447, on 442.

21. John D. Bernal and H.D. Megaw, "The Function of Hydrogen in Intermolecular Forces," *Proceedings of the Royal Society* A151 (1935): 384–420.

22. Dorothy Wrinch, "Chromosome Micelle and the Banded Structure of Chromosomes in the Salivary Gland," *Nature* 136 (1935): 68–69.

23. Hardy to Wrinch, Dec. 1929, *DWP*. Quoted in Marjorie Senechal, "Hardy as Mentor," *Mathematical Intelligencer* 29 (2007): 16–21.

24. Pnina Abir-Am, "Disciplinary and Marital Strategies in the Career of Mathematical Biologist Dorothy Wrinch," in *Uneasy Careers and Intimate Lives, Women in Science 1789–1979*, ed. Pnina Abir-Am and D. Outram (New Brunswick, N.J.: Rutgers University Press, 1987), 239–280, on 257.

25. D'Arcy Thompson, *On Growth and Form* (Cambridge: Cambridge University Press, 1917).

26. Dorothy Wrinch and F. Jordan Lloyd, "The Hydrogen Bond and the Structure of Proteins," *Nature* 138 (1936): 758–759.

27. Dorothy Wrinch, "On the Pattern of Proteins," *Proceedings of the Royal Society of London*, Series A, *Mathematical and Physical Sciences* 160, no. 900 (1937): 59–86.

28. Dorothy Wrinch, "The Cyclol Hypothesis and the 'Globular' Proteins," *Proceedings of the Royal Society of London*, Series A, *Mathematical and Physical Sciences* 161, no. 907 (1937): 505–524.

29. Wrinch, "On the Pattern of Proteins," 76.

30. Irving Langmuir and Vincent Schaeffer, "Properties and Structures of Protein Monolayers," *Chemical Reviews* 24 (1939): 181–202.

31. Pauling to Weaver, 6 Mar. 1937. Quoted in Sibilla Kennedy, "Dorothy Wrinch and the Rockefeller Foundation Grants," in *National History of Science Meeting* (Norwalk, Conn.: 1983), *DWP*.

32. Tisdale to Weaver, 24 Jan. 1936. Quoted in Kennedy, "Dorothy Wrinch."

33. Weaver diary, Oxford, 15 June 1936. Quoted in Kennedy, "Dorothy Wrinch."

34. Personal communication by Jones to Marjorie Senechal, May 2006.

35. Weaver diary, 2 Oct. 1937. Quoted in Kennedy, "Dorothy Wrinch."

36. *Time*, 20 Sept. 1937.

37. Weaver diary, 1 Nov. 1937. Quoted in Kennedy, "Dorothy Wrinch."

38. Hager, *Force of Nature*, 226.

39. Pauling to Weaver, 11 Apr. 1938. Quoted in Kennedy, "Dorothy Wrinch."

40. Weaver to Pauling, 11 Apr. 1938. Quoted in Kennedy, "Dorothy Wrinch."

41. *New York Times*, 19 Apr. 1940, 14.

42. *New York Post*, 3 June 1941. Quoted in Abir-Am, "Disciplinary and Marital Strategies," 265.

43. Max Bergmann and Carl Niemann, "On the Structure of Proteins. Cattle Hemoglobin, Egg Albumin, Cattle Fibrin, and Gelatin," *Journal of Biological Chemistry* 118 (1937): 301–314.

44. Dorothy Wrinch, "Structure of Insulin," *Transactions of the Faraday Society* 33 (1937): 1368–1380.

45. Wrinch to Bergmann, 1 June 1937, *DWP*.

46. Bergmann to Wrinch, 16 June 1937, *DWP*.

47. Bergmann to Wrinch, 19 Nov. 1937, *DWP*.

48. Wrinch to Bergmann, 14 Dec. 1937; Bergmann to Wrinch, 23 Dec. 1937, *DWP*.

49. Dorothy Wrinch, "Structure of the Insulin Molecule," *Science* 88 (1938): 148–149.

50. Dorothy Wrinch and Irving Langmuir, "The Structure of the Insulin Molecule," *Journal of the American Chemical Society* 60 (1938): 2247–2255, on 2253.

51. H. Lipson and W. Cochran, *The Determination of Crystal Structures* (London: G. Bell and Sons, 1953), 163. I thank Marjorie Senechal for pointing this out.

52. John D. Bernal, "The Structure of Proteins," *Nature* 142 (1939): 631–636.

53. Pauling and Niemann, "The Structure of Proteins," 1860.

54. Langmuir to Pauling, 12 May 1939, 17 May 1939, *DWP*.

55. Elisabeth Crawford, *The Nobel Population, 1901–1950* (Tokyo: Universal Academic Press, 2002), 306.

56. Langmuir to Lamb, 22 May 1940, *DWP*.

57. Dorothy Wrinch, "Geometrical Attack on Protein Structure," *Journal of the American Chemical Society* 63 (1941): 330–333.

58. Irving Langmuir, "Molecular Films in Chemistry and Biology," in *Molecular Films, the Cyclotron, and the New Biology: Essays*, ed. H.S. Taylor, Irving Langmuir, and E.O. Lawrence (New Brunswick, N.J.: Rutgers University Press, 1942), 61.

59. Wrinch to Neville, Oct. 1940, *DWP*. Quoted in Abir-Am, "Disciplinary and Marital Strategies," 268.

60. Tisdale's log, Copenhagen, 29 Oct. to 3 Nov. 1938. Quoted in Kennedy, "Dorothy Wrinch."

61. Dorothy Wrinch and David Harker, "Lengths and Strengths of Atomic Bonds," *Journal of Chemical Physics* 8 (1940): 502–503.

62. Pauling to Harker, 8 July 1940, *DWP*.

63. Harker to Pauling, 16 July 1940, *DWP*.

64. John D. Bernal, "Structure of Proteins," *Nature* 143 (1939): 663–667.

65. Laszlo, *A History of Biochemistry*, 277.

66. Marjorie Senechal, "A Prophet without Honor: Dorothy Wrinch, Scientist, 1894–1976," *Smith Alumnae Quarterly* 48 (1977): 18–23, on 21.

67. Dorothy Crowfoot Hodgkin to Pierre Laszlo, 1984, as quoted in Laszlo, *A History of Biochemistry*, 226.

68. Marjorie Senechal, personal communication, 2007.

69. Wrinch, "The Cyclol Hypothesis," 506–524, on 506. Quoted in Laszlo, *A History of Biochemistry*, 217.

70. Dorothy Wrinch, "The Structure of Globular Proteins," *Nature* 143 (1939): 482–483, on 483. Quoted in Hager, *Force of Nature*, 225.

71. Laszlo, *A History of Biochemistry*, 236.

72. Paul Dirac, "The Evolution of the Physicist's Picture of Nature," *Scientific American* 208 no. 5 (1963): 45–53, on 47.

73. In an obituary by Freeman Dyson, "Prof. Hermann Weyl, For. Mem. R.S.," *Nature* 177 (1956): 457–458, on 458.

74. See Laszlo, *A History of Biochemistry*, 261–287.

75. Pauling and Niemann, "The Structure of Proteins." Quoted in Laszlo, *A History of Biochemistry*, 269.

76. Hager, *Force of Nature*, 203.

77. John D. Bernal, "The Pattern of Linus Pauling's Work in Relation to Molecular Biology," in *Structural Chemistry in Molecular Biology*, ed. Alexander Rich and Norman Davidson (New York: W.H. Freeman, 1968), 370–379, on 372. Quoted in Laszlo, *A History of Biochemistry*, 279.

78. Linus Pauling, "Modern Structural Chemistry. Nobel Lecture 11 December 1954," in *From Nobel Lectures, Chemistry 1942–1962* (Amsterdam: Elsevier, 1964).

79. Linus Pauling, L.O. Bockway, and J.Y. Beach, "The Dependence of Interatomic Distance on Single Bond-Double Bond Resonance," *Journal of the American Chemical Society* 57 (1935): 2705–2709.

80. Linus Pauling, "A Theory of the Structure and Process of Formation of Antibodies," *Journal of the American Chemical Society* 62, no. 10 (1940): 2643–2657.

81. Laszlo, *A History of Biochemistry*, 269.

82. Hager, *Force of Nature*, 372.

83. Lawrence Bragg, J.C. Kendrew, and M.F. Perutz, "Polypeptide Chain Configurations in Crystalline Proteins," *Proceedings of the Royal Society of London, Series A, Mathematical and Physical Sciences* 203, no. 1074 (1950): 321–357, on 331. Quoted in Laszlo, *A History of Biochemistry*, 281.

84. Hager, *Force of Nature*, 374.

85. M.F. Perutz, "New X-Ray Evidence on the Configuration of Polypeptide Chains," *Nature* 167 (1951): 1053–1054.

86. Hager, *Force of Nature*, 378.

87. Linus Pauling and Robert B. Corey, *Proceedings of the National Academy of Sciences of the United States of America* 37, no. 5 (1951): "Atomic Coordinates and Structure Factors for Two Helical Configurations of Polypeptide Chains," 235–240; "The Structure of Synthetic Polypeptides," 241–250; "The Pleated Sheet,

a New Layer Configuration of Polypeptide Chains," 251–256; "The Structure of Feather Rachis Keratin," 256–261; "The Structure of Hair, Muscle, and Related Proteins," 261–271; "The Structure of Fibrous Proteins of the Collagen-Gelatin Group," 272–281; and "The Polypeptide-Chain Configuration in Hemoglobin and Other Globular Proteins," 282–285.

Chapter 10

1. Langmuir's diary, 25 Jan. 1950, quoted in Albert Rosenfeld, "The Quintessence of Irving Langmuir," in *The Collected Works of Irving Langmuir*, Vol. 12, ed. C. Guy Suits and Harold Way (New York: Pergamon Press, 1962), 5–229, on 209–210.
2. Transcribed by Rupert N. Hall from an audio recording, 25 Jan. 1950, *ILP*. Published as Irving Langmuir, "Pathological Science," Report No. 68-C-035 (Schenectady, N.Y.: General Electric Company, 1968), and as Langmuir, "Pathological Science," *Physics Today* (Oct. 1989): 36–48. Available online at www.cs.princeton.edu/~ken/Langmuir/langmuir.htm, accessed 12 Feb. 2008.
3. Dennis Rousseau and S.P.S. Porto, "Polywater: Polymer or Artifact?" *Science* 167 (1970): 1715–1719.
4. See, e.g., *New Energy Times* at www.newenergytimes.com, accessed 20 Feb. 2007.
5. David Goodstein, "Whatever Happened to Cold Fusion?" *American Scholar* 63 (1994): 527–541. Available online at www.its.caltech.edu/~dg/fusion_art.html, accessed 14 Feb. 2008.
6. Langmuir, "Pathological Science," GE Report No. 68-C-035.
7. Langmuir, "Pathological Science," *Physics Today*.
8. Fred Allison and Edgar Murphey, "A Magneto-optic Method of Chemical Analysis," *Journal of the American Chemical Society* 52 (1930): 3796–3806, on 3802.
9. Rosenfeld, "The Quintessence of Irving Langmuir," 127.
10. Pierre Laszlo, *A History of Biochemistry: Molecular Correlates of Biological Concepts*, Vol. 34A in *Comprehensive Biochemistry*, ed. A. Neuberger and L.L.M. Van Deemen (Amsterdam: Elsevier, 1986), 233.
11. Rosenfeld, "The Quintessence of Irving Langmuir," 177.
12. Ibid., 180.
13. Ibid., 180–185.
14. Ibid., 186–191.
15. *Time*, "Weather or Not," 28 Aug. 1950. Quoted in Rosenfeld, "The Quintessence of Irving Langmuir," 203.
16. Quoted in Rosenfeld, "The Quintessence of Irving Langmuir," 204.
17. Irving Langmuir, "The Production of Rain by a Chain Reaction in Cumulus Clouds at Temperatures above Freezing," *Journal of Meteorology* 5 (1948): 175–192, on 191.
18. Bernard Vonnegut, "The Nucleation of Ice Formation by Silver Iodide Smokes," *Journal of Applied Physics* 44 (1949): 277–289.

19. James McDonald, "Physics of Cloud Formation," *Advances in Geophysics* 5 (1958): 225–365, on 236.

20. *Time*, "Weather or Not."

21. Irving Langmuir, "Control of Precipitation from Cumulus Clouds by Various Seeding Techniques," *Science* 112 (1950): 35–41.

22. Ferguson Hall, Gardner Emmons, Bernhard Haurwitz, George P. Wadsworth, and Hurd C. Willett, "Dr. Langmuir's Article on Precipitation Control," *Science* 113 (1951): 189–192.

23. Langmuir's journal, 25 Jan. 1950. Quoted in Rosenfeld, "The Quintessence of Irving Langmuir," 209–210.

24. McDonald, "Physics of Cloud Formation," 296.

25. American Meteorological Society: Policy Statement, "Planned and Inadvertent Weather Modification," *Bulletin of the American Meteorological Society* 73 (1992): 331–137, available at www.ametsoc.org/AMS/policy/wxmod.html; Policy Statement, "Planned and Inadvertent Weather Modification," *Bulletin of the American Meteorological Society* 79 (1998): 2771–2772, available at www.ametsoc.org/AMS/policy/wxmod98.html; "Scientific Background for the AMS Policy Statement on Planned and Inadvertent Weather Modification," *Bulletin of the American Meteorological Society* 79 (1998): 2773–2778. As quoted in Ronald B. Standler, "History and Problems in Weather Modification" (21 Jan. 2003), www.rbs2.com/w2.htm, accessed 28 Jan. 2008.

26. Glenn Seaborg, Ronald Kathren, and Jerry Gough, *The Plutonium Story: The Journals of Professor Glenn T. Seaborg 1939–1946* (Columbus, Ohio: Battelle Press, 1994), 765.

27. Irving Langmuir, "Testimony on Atomic Energy Control," Senate Hearings on Atomic Energy, 30 Sept. 1945, reprinted in C. Guy Suits and Harold Way, eds., *The Collected Works of Irving Langmuir*, Vol. 12 (New York: Pergamon Press, 1962), 365–370.

28. Irving Langmuir, Joint Hearings on Science Bills before Senate Military Affairs and Commerce Committees, 8 Oct. 1945, reprinted in Suits and Way, *The Collected Works of Irving Langmuir*, 353–364, on 359.

Chapter 11

1. Gilbert N. Lewis, "Über Komplexbildung, Hydration und Farbe," *Zeitschrift für Physikalische Chemie* 56 (1906).

2. Gilbert N. Lewis and Melvin Calvin, "The Color of Organic Substances," *Chemical Reviews* 25 (1939): 273–328.

3. Author's interview with Robert Connick, 2004.

4. Gilbert N. Lewis, Theodore Magel, and David Lipkin, "The Absorption and Re-emission of Light by Cis- and Trans-Stilbenes and the Efficiency of Their Photochemical Isomerization," *Journal of the American Chemical Society* 62 (1940): 2973–2980.

5. Gilbert N. Lewis, David Lipkin, and Theodore Magel, "Reversible Photochemical Processes in Rigid Media. A Study of the Phosphorescent State," *Journal of the American Chemical Society* 63 (1941): 3005–3018, on 3013.

6. Jacob Bigeleisen, "Gilbert N. Lewis and the Beginnings of Isotope Chemistry," *Journal of Chemical Education* 61 (1984): 205.

7. Michael Kasha, "The Triplet State: An Example of G. N. Lewis' Research Style," *Journal of Chemical Education* 61 (1984): 204–215.

8. James Franck, oral history interview by Thomas Kuhn and Maria Goeppert Mayer, 9–14 July 1962, *NBL*.

9. Göran Bergson, private communication, 2007.

10. Ludwig Ramberg, "1940 Special Report on Gilbert Lewis to the Nobel Chemistry Committee (Report 6)," *KVA*.

11. Arne Fredga, "1944 Special Report on Gilbert Lewis to the Nobel Chemistry Committee (Report 7)," *KVA*.

12. Ibid.

13. Nobel Chemistry Committee, "1944 Report of the Nobel Chemistry Committee to Royal Swedish Academy of Sciences," *KVA*.

14. István Hargittai, *The Road to Stockholm Nobel Prizes, Science, and Scientists* (Oxford: Oxford University Press, 2002), 228.

15. 1954 Nobel chemistry prize award to Linus Pauling, *KVA*.

16. Robert Marc Friedman, *The Politics of Excellence: Behind the Nobel Prizes in Science* (New York: Times Books, Henry Holt, 2001), 202–204.

17. Owen Richardson, Nominating letters for Gilbert Lewis, 27 Oct. 1941, 5 Oct. 1942, 19 Nov. 1943, *KVA*.

18. Author's interview with Jacob Bigeleisen, 2004.

19. Elisabeth Crawford, J.L. Heilbron, and Rebecca Ullrich, *The Nobel Population, 1901–1937* (Berkeley, Calif.: Office for History of Science and Technology, 1987).

20. Diana Kormos Barkan, *Walther Nernst and the Transition to Modern Physical Science* (Cambridge: Cambridge University Press, 1999), 234.

21. Crawford, Heilbron, and Ullrich, *The Nobel Population, 1901–1937*, 6.

22. 1949 Nobel chemistry prize award to William Giauque, *KVA*.

23. Arthur Lachman, *Borderland of the Unknown: The Life Story of Gilbert Newton Lewis, One of the World's Great Scientists* (New York: Pageant Press, 1955), 105.

24. William Jolly, *From Retorts to Lasers* (Berkeley: William Jolly, distributed by the College of Chemistry, University of California, Berkeley, 1987), 284.

25. Author's interview with Bigeleisen, 2004

26. Ibid.

27. Ibid.

28. Ibid.

29. Author's interview with Michael Kasha, 2004.

30. Author's interview with Bigeleisen, 2004.

31. Author's interview with Connick, 2004. I confirmed this account in an interview with Charles Auerbach, one of the two undergraduates, in 2006. When Jacob Bigeleisen reviewed this chapter, he said that he had witnessed the conversation.

32. Author's interview with Samuel Weissman, 2005.

33. Author's interview with Harold Johnston, 2004.

34. Author's interview with Bigeleisen, 2004.

35. Alameda County, California, coroner's inquest no. 10653, 1946.

36. Jolly, *From Retorts to Lasers*, 76.

37. Kasha, "The Triplet State," 204–215, on 213–214; interview with Kasha, 2004.

38. Author's interview with Frances Connick, 2004, and with Ted Geballe, 2006.

39. Robert Sproul, president of the University of California, to Langmuir, 13 Dec. 1945, *ILP.*

40. Author's interview with Kasha, 2004.

41. G.E. Coates and J.E. Coates, "Studies on Hydrogen Cyanide. Part XIII. The Dielectric Constant of Anhydrous Hydrogen Cyanide," *Journal of the Chemical Society* (1944): 77–81.

42. Joel Hildebrand, "The Humanism of Irving Langmuir," in *The Collected Works of Irving Langmuir*, Vol. 12, ed. C. Guy Suits and Harold Way (New York: Pergamon Press, 1962), 233–240, on 233.

43. Hildebrand, "The Humanism of Irving Langmuir," 233.

44. Antoine Lavoisier, *Elements of Chemistry in a New Systematic Order, Containing All the Modern Discoveries* (New York: Dover Publications, 1790), xx–xxi.

Epilogue

1. Kurt M. Mendelssohn, *The World of Walther Nernst: The Rise and Fall of German Science* (London: Macmillan, 1973), 38.

2. Daniel Charles, *Master Mind: The Rise and Fall of Fritz Haber, the Nobel Laureate Who Launched the Age of Chemical Warfare* (New York: Ecco, 2005), 252–253.

3. Ibid., 245–246.

4. James Arnold, Jacob Bigeleisen, and Clyde Hutchinson, "Harold Clayton Urey: April 24, 1893–January 5, 1981," in *National Academy of Sciences Biographical Memoirs* (Washington, D.C.: National Academy of Sciences, 1995), 363–411, on 393.

5. Marjorie Senechal, "A Prophet without Honor: Dorothy Wrinch, Scientist, 1894–1976," *Smith Alumnae Quarterly* 48 (1977): 18–23, on 23.

6. G. Lucente and A. Romeo, "Synthesis of Cyclols from Small Peptides via Amide-Amide Reaction," *Chemical Communications* (1971).

7. Wrinch to Pauling and Corey (marked "unsent"), 4 July 1951, *DWP.*

8. Undated handwritten letter from Scott to Wrinch; Scott to Pauling, 5 Jan. 1956; Pauling to Scott, 14 Jan. 1956; Scott to Wrinch, 29 Apr. 1956, *DWP.*

9. Harker to Wrinch, 19 Mar. 1954, *DWP.*

10. Pnina Abir-Am, "Disciplinary and Marital Strategies in the Career of Mathematical Biologist Dorothy Wrinch," in *Uneasy Careers and Intimate Lives, Women in Science 1789–1979*, ed. Pnina Abir-Am and D. Outram (New Brunswick, N.J.: Rutgers University Press, 1987), 239–280, on 272–276.

11. Author's interview with Marjorie Senechal, 2006.

12. Thomas Hager, *Force of Nature: The Life of Linus Pauling* (New York: Simon and Schuster, 1995), 565–627.

13. Langmuir to Elihu Lubkin, 17 Aug. 1945, *ILP.*

14. Author's interview with Clayton Heathock, 2005.
15. Author's interview with William Jolly, 2004.
16. Author's interview with Jacob Bigeleisen, 2004.
17. Author's interview with Michael Kasha, 2004.
18. Author's interview with Roger Hahn, 2005.
19. Author's interview with Edward Lewis, 2005.
20. Author's interview with Bigeleisen, 2004.

Index